AF581627

Fast-Growing Trees Species—Opportunities and Risks for Sustainable Agricultural and Forest Land Use Systems

Fast-Growing Trees Species—Opportunities and Risks for Sustainable Agricultural and Forest Land Use Systems

Editor

Dirk Landgraf

MDPI • Basel • Beijing • Wuhan • Barcelona • Belgrade • Manchester • Tokyo • Cluj • Tianjin

Editor
Dirk Landgraf
Faculty of Landscape
Architecture, Horticulture
and Forestry
Erfurt University of Applied
Sciences
Erfurt
Germany

Editorial Office
MDPI
St. Alban-Anlage 66
4052 Basel, Switzerland

This is a reprint of articles from the Special Issue published online in the open access journal *Forests* (ISSN 1999-4907) (available at: www.mdpi.com/journal/forests/special_issues/Fast-Growing_Trees).

For citation purposes, cite each article independently as indicated on the article page online and as indicated below:

LastName, A.A.; LastName, B.B.; LastName, C.C. Article Title. *Journal Name* **Year**, *Volume Number*, Page Range.

ISBN 978-3-0365-4458-8 (Hbk)
ISBN 978-3-0365-4457-1 (PDF)

Cover image courtesy of Dirk Landgraf

Contents

About the Editor

Dirk Landgraf

Dirk Landgraf is the holder of the professorship for renewable raw materials and wood market theory on the Faculty of Landscape Architecture, Horticulture and Forestry at the Erfurt University of Applied Sciences. He has been involved in the cultivation, management and use of fast-growing tree species in different land-use systems for many years.

Preface to "Fast-Growing Trees Species—Opportunities and Risks for Sustainable Agricultural and Forest Land Use Systems"

Cultivation systems with fast-growing tree species are often discussed, with much debate, in both science and practice. They are very diverse and pose opportunities as well as risks for land users and the surrounding ecosystems. Many different tree species and their varieties can be established on sites (usually) used for forestry; they can also be grown in short-rotation plantations (SRC) or in agroforestry systems (AFS) on agricultural soils. In most cases, the goals of the land users determine the choice of tree species and cultivation system, the rotation times, and their management. Thus, there is not only a large variety of management variants and technical and technological innovations, but also a large number of interactions with the surrounding ecosystems. In addition to the considerable effects on biodiversity in different landscapes, the tree–soil interactions also play an important role. For this reason, this Special Issue intends to present new findings from science and practice in this broad field of fast-growing tree species cultivation in all its forms.

Dirk Landgraf
Editor

MDPI

Editorial

Fast-Growing Trees Species—Opportunities and Risks for Sustainable Agricultural and Forest Land Use Systems

Dirk Landgraf

Faculty of Landscape Architecture, Horticulture and Forestry, University of Applied Sciences Erfurt, Leipziger Strasse 77, 99085 Erfurt, Germany; dirk.landgraf@fh-erfurt.de

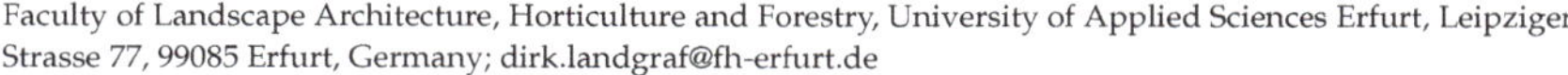

The cultivation of fast-growing tree species has noticeably increased worldwide in recent years. Fast-growing tree species are cultivated in different land use systems. In addition to being classically cultivated in forests, they are also cultivated in short-rotation coppices (SRC) with different rotation intervals on forest and agricultural land. Moreover, in recent years, people have become increasingly aware of cultivation in agroforestry systems.

Due to these intensive practical activities, many new questions concerning the management of fast-growing tree species are waiting to be answered using applied and basic research. Thus, a large number of publications by researchers worldwide can currently be found in a wide variety of journals.

In addition to numerous articles, Special Issues on fast-growing tree species have also published by several journals. I would particularly like to highlight the Special Issue in the *BioEnergy Research* journal on the subject of "Sustainable Biomass Value Chains Based on Poplar Plantations in European Rural Areas" [1] and the Special Issue that followed shortly thereafter in the *Forests* journal, entitled "Growth and Development of Short Rotation Woody Crops for Rural and Urban Applications" [2].

While these two Special Issues specifically focused on short-rotation woody crops, the focus of the current Special Issue is much broader. It contains a total of ten papers, which mainly come from Germany (7), but also from Poland (1), the Czech Republic (1) and China (1). It also needs to be emphasized that the authors of these articles often represent an international team.

The articles in this Special Issue cover a very wide range of topics. In addition to research on breeding [3] and on the influence of pruning practices on the height growth of paulownia [4], three articles deal with the influence of site characteristics and nutrient availability on the physiology and yield security of fast-growing tree species [5–7]. The article by Kalita et al. focuses on the modeling of soil carbon in *Salix* plantations [8], while the article by Boruszewski et al. reports on potentially suitable areas for the planting of fast-growing tree species in Poland [9]. Zitzmann and Rode [10] examine the impact of short-rotation plantation management on phytodiversity [10], while Helbig et al. deal with the influence of leaf feeding on the growth of poplars and willows [11]. Finally, Hernandez-Estrada et al. describe the dry matter loss of poplar wood chips during storage [12].

Funding: This research received no external funding.

Institutional Review Board Statement: Not applicable.

Informed Consent Statement: Not applicable.

Data Availability Statement: Not applicable.

Conflicts of Interest: The author declares no conflict of interest.

Citation: Landgraf, D. Fast-Growing Trees Species—Opportunities and Risks for Sustainable Agricultural and Forest Land Use Systems. *Forests* **2022**, *13*, 829. https://doi.org/10.3390/f13060829

Received: 4 May 2022
Accepted: 20 May 2022
Published: 26 May 2022

References

1. Meyer, M.; Wahren, F.P.; Weber, N.; Zalesny, R.S.; Weih, M. Sustainable Biomass Value Chains Based on Poplar Plantations in European Rural Areas. *BioEnergy Res.* **2021**, *14*, 355–356. [CrossRef]
2. Zalesny, R.S., Jr.; Pilipović, A. Growth and Development of Short Rotation Woody Crops for Rural and Urban Applications. *Forests* **2021**, *12*, 474.
3. Fladung, M. Targeted CRISPR/Cas9-Based Knock-Out of the Rice Orthologs TILLER ANGLE CONTROL 1 (TAC1) in Poplar Induces Erect Leaf Habit and Shoot Growth. *Forests* **2021**, *12*, 1615. [CrossRef]
4. Kadlec, J.; Novosadová, K.; Pokorný, R. Impact of Different Pruning Practices on Height Growth of Paulownia Clon In Vitro 112®. *Forests* **2022**, *13*, 317. [CrossRef]
5. Zhao, X.; Zhang, X.; Liu, Z.; Lv, Y.; Song, T.; Cui, J.; Chen, T.; Li, J.; Zeng, F.; Zhan, Y. Comparing the Effects of N and P Deficiency on Physiology and Growth for Fast- and Slow-Growing Provenances of *Fraxinus mandshurica*. *Forests* **2021**, *12*, 1760. [CrossRef]
6. Koczorski, P.; Furtado, B.; Hrynkiewicz, K.; Breezmann, M.; Weih, M.; Baum, C. Site-Effects Dominate the Plant Availability of Nutrients under Salix Species during the First Cutting Cycle. *Forests* **2021**, *12*, 1226. [CrossRef]
7. Lange, C.A.; Knoche, D.; Hanschke, R.; Löffler, S.; Schneck, V. Physiological Performance and Biomass Growth of Different Black Locust Origins Growing on a Post-Mining Reclamation Site in Eastern Germany. *Forests* **2022**, *13*, 315. [CrossRef]
8. Kalita, S.; Potter, H.K.; Weih, M.; Baum, C.; Nordberg, Å.; Hansson, P.-A. Soil Carbon Modelling in Salix Biomass Plantations: Variety Determines Carbon Sequestration and Climate Impacts. *Forests* **2021**, *12*, 1529. [CrossRef]
9. Boruszewski, P.; Laskowska, A.; Jankowska, A.; Klisz, M.; Mionskowski, M. Potential Areas in Poland for Forestry Plantation. *Forests* **2021**, *12*, 1360. [CrossRef]
10. Zitzmann, F.; Rode, M. Short-Rotation Coppice Managed According to Ecological Guidelines—What Are the Benefits for Phytodiversity? *Forests* **2021**, *12*, 646. [CrossRef]
11. Helbig, C.E.; Müller, M.G.; Landgraf, D. Effects of Leaf Loss by Artificial Defoliation on the Growth of Different Poplar and Willow Varieties. *Forests* **2021**, *12*, 1224. [CrossRef]
12. Hernandez-Estrada, A.; Pecenka, R.; Dumfort, S.; Ascher-Jenull, J.; Lenz, H.; Idler, C.; Hoffmann, T. Establishment of a Laboratory Scale Set-Up with Controlled Temperature and High Humidity to Investigate Dry Matter Losses of Wood Chips from Poplar during Storage. *Forests* **2022**, *13*, 459. [CrossRef]

 forests

MDPI

Article

Targeted CRISPR/Cas9-Based Knock-Out of the Rice Orthologs *TILLER ANGLE CONTROL 1* (*TAC1*) in Poplar Induces Erect Leaf Habit and Shoot Growth

Matthias Fladung 

Thünen Institute of Forest Genetics, Sieker Landstrasse 2, 22927 Grosshansdorf, Germany; matthias.fladung@thuenen.de; Tel.: +49-41-0269-6107

Abstract: Pyramidal-, erect- or upright-growing plant forms are characterized by narrow branch angles of shoots and leaves. The putative advantage of upright-leaf and shoot habit could be a more efficient penetration of light into lower canopy layers. Pyramidal genotypes have already been reported for various tree genotypes including peach. The paralogous rice ortholog *TILLER ANGLE CONTROL 1* (*TAC1*) has been proposed to be the responsible gene for upright growth. However, it has not really been demonstrated for any of the pyramidal tree genotypes that a knock-out mutation of the *TAC1* gene is causing pyramidal plant growth. By in silico analyses, we have identified a putative rice *TAC1* ortholog (Potri.014G102600, "TAC-14") and its paralog (Potri.002G175300, "TAC-2") in the genome of *P. trichocarpa*. Two putative *PcTAC1* orthologs in the *P.* × *canescens* clone INRA 717-1B4 were successfully knocked-out by applying a transgenic CRISPR/Cas9-approach. The mutants were molecularly analyzed and phenotyped over a period of three years in a glasshouse. Our results indicate that the homozygous knock-out of "TAC-14" is sufficient to induce pyramidal plant growth in *P.* × *canescens*. If up to twice as many pyramidal individuals were planted on short rotation coppices (SRCs), this could lead to higher wood yield, without any breeding, simply by increasing the number of trees on a default field size.

Keywords: CRISPR/Cas9; genome editing; *Populus*; INRA 717-1B4; pyramidal plant habitus; leaf petiole angle; branch angle

Citation: Fladung, M. Targeted CRISPR/Cas9-Based Knock-Out of the Rice Orthologs *TILLER ANGLE CONTROL 1* (*TAC1*) in Poplar Induces Erect Leaf Habit and Shoot Growth. *Forests* **2021**, *12*, 1615. https://doi.org/10.3390/f12121615

Academic Editor: Dirk Landgraf

Received: 27 October 2021
Accepted: 19 November 2021
Published: 23 November 2021

1. Introduction

Plants possess variable plant architectures and phenotypes which facilitate their optimal adaptation to specific environments, even if these do not always offer the best growing conditions. Plant architecture is species-specific and, therefore, has been used for decades as the basis for systematics and the taxonomic classification of plants [1]. However, variability of the plant phenotype and the degree to which the branching pattern is dependent on the growing habitat is well known and has already fascinated geneticists, developmental biologists, and plant breeders for more than 100 years. Plant architecture, in general, is characterized by the internodial and nodial organization of the main shoot, and location, orientation, and number of side-shoots, influenced by genetic components [1–6], and modulated by developmental and environmental factors like gravity, light, water, and touch [7,8].

An uncountable number of reports have been published in the past describing the spontaneous or induced alteration of the genetic constitution of an individuum, i.e., random or targeted mutagenesis mainly applied for the improvement of human food, e.g., barley [9] but also ornamental plants [10–14]. Here, artificial induction of mutations by X-ray irradiation or chemical treatments has been a key element of mutation breeding. Many of the mutant individuals obtained but also natural mutants showed alterations in plant habitus and architecture, e.g., nana, compacta, fastigiata, pendula, and pyramidalis (e.g. [14–17]).

Only very few mutations are dominant and become visible in a heterozygous allelic status. However, most mutations are of recessive nature and require allele homozygosity for phenotypic expression. Thus, a self-pollination cycle is needed for the generation of mutants. However, the timespan for vegetative growth until the generative phase is reached is, in particular in forest trees, usually very long, i.e., ranges from a few years to several decades. A solution to circumvent this problem is the application of modern biotechnological approaches, like genetic engineering or genome editing technology by transferring or modifying key genes. Both technologies offer a straight-forward way to alter plant architecture in a manageable timeframe in trees [18].

For example, the creation of dwarf and semi-dwarf mutants has been described following the genetic transfer of genes modifying hormone biosynthesis, e.g., gibberellic acid (GA), or conferring GA insensibility [19,20], or by hyperactivation of *the GA2-OXIDASE* (*GA2ox*) gene, the major GA catabolic enzyme in plants, in an activation tagging approach [21]. Dwarf poplar mutants have also been reported after transfer of the meristem identity gene *LEAFY* (*LFY*) [22,23], the *TaLEA* (*Tamarix androssowii LATE EMBRYOGENESIS ABUNDANT GENE*) [24], of two flowering time genes (*SUPPRESSOR OF CONSTANS* [*SOC1*] and *FRUITFUL* [*FUL*] [25], the *ROLC* gene from *Agrobacterium rhizogenes* [26–29], or by adopting an *Ac*/Ds-transposon activation tagging-based mutation system in poplar [30]. On the other hand, poplar with increased plant size have also been reported after e.g., transfer and overexpression of the *GA20-OXIDASE* gene [31–33] or knock-down of *PHENYLCOUMARAN BENZYLIC ETHER REDUCTASE1* (*PCBER1*), a gene of the neolignan biosynthesis pathway [34].

Pyramidal or erect plant growth has become of interest for commercial applications because of its possible advantage in plant yield, as reported for rice [35] and spring wheat [36]. Characteristic for the upright-growing branches is the narrow branch angle of shoots and leaves. An advantage of pyramidal growth could be that the erect-leaf habit allows more efficient penetration of light into lower canopy layers, and thus increases photosynthetic potential of the whole plant [35,37]; for review: [38]. Other papers, however, stated that factors like nitrogen and water availability more likely influence plant and grain yield [39]. But independently of possible physiological potential associated with erect leaf or shoot growth, plants with reduced need for individual space on a field would simply allow a higher number of plants to be planted on a default field size, as e.g., reported for *Prunus persica* (peach) [40] and *Malus domestica* (apple) [41].

In peach, mutant varieties with pronounced vertical growth of branches or upright stems have been investigated in detail [42,43]. This trait, originally named "broomy" (br) and later re-named "pillar", has been described as intermediate because heterozygous offspring revealed branch angles in between the two extremes [43]. It has been shown that the broomy or pillar phenotype is caused by a loss-of-function mutation in the ortholog of the rice *TILLER ANGLE CONTROL 1* (*TAC1*) gene [43,44]. On the other hand, overexpression of the *PpTAC1* gene in plum caused wider branch angles resulting in more horizontal oriented branches [44]. *TAC1* belongs to the IGT gene family which is characterised by a GφL (A/T)IGT motif [43]. This family contains other well-known members like *LAZY1* and *DEEPER ROOTING 1* (*DRO1*).

Other prominent examples showing pyramidal plant growth are e.g., *Cupressus sempervirens* 'Pyramidalis' (Italian Cypress [45]), *Taxus baccata* 'Pyramidalis' (pillar yew [46]), *Populus tremula* 'Erecta' (Swedish columnar aspen [47]), *P. alba* var. pyramidalis (*P. bolleana* [48]), and *P. nigra* cv. 'Italica' (Lombardy poplar [49]). Also, the commercial AC Sundancer™ hybrid poplar cultivar demonstrates vertical growth of branches [50]. For Lombardy poplar, ref. [43] has proposed that the pyramidal phenotype is possibly caused by a mutation in a poplar ortholog of *TAC1*. Interestingly, this pyramidal mutant poplar was detected in Central Asia in the 19th century [49], and since then it was propagated vegetatively on a very large scale and distributed worldwide as a single genotype [51,52]. However, for both, Lombardy poplar and for Swedish columnar aspen, or for any other pyramidal tree genotype, it has not really been demonstrated that a knock-out mutation of the poplar *TAC1*-orthologous gene is the cause of pyramidal plant growth.

In in silico studies, we have identified a putative rice *TAC1* ortholog in the *P. trichocarpa* genome and because of the quite recent complete genome duplication in poplar [53], its most homologous paralog. Based on these two *PtTAC1* sequences, we also determined the two putative *TAC1* homologous genes *PtrTAC1* and *PcTAC1* in the genomes of *P. tremula* and *P.* × *canescens*. To prove that one or the other, or both, *TAC1* ortholog(s) are responsible for pyramidal plant growth in poplar, we applied the CLUSTERED REGULARLY INTERSPACED SHORT PALINDROMIC REPEAT (CRISPR)/CRISPR-associated protein 9 (Cas9) editing system [54] to knock-out the two putative *PcTAC1* orthologs in the INRA 717-1B4 clone. We could show that the homozygous knock-out of only Potri.014G102600 ("TAC-14") is sufficient to induce pyramidal plant growth in *P.* × *canescens*.

2. Materials and Methods

2.1. Plant Material

The hybrid poplar clone INRA 717-1B4 or "P1" (*P.* × *canescens* = *P. tremula* × *P. alba* [55], grown in vitro on WPM medium without hormones at 23 °C and with 24 h light was used for *Agrobacterium*-mediated leaf-disc transformation (see below). Putative transgenic plants were regenerated after four to eight weeks culture on kanamycin-containing regeneration medium as described in [56]. Regenerated but not-transgenic plants (without any *PcTAC1* editing), as well as P1 from tissue culture, served as controls.

2.2. Identification of Putative PpeTAC1 Homologs in Poplar

Following blast of the coding sequence of the Prunus persica gene *PpeTAC1* (EMBL acc.no. KF218366; [43], in the full genome sequence of *P. trichocarpa* v3.1 (https://phytozome-next.jgi.doe.gov/info/Ptrichocarpa_v3_1, accessed on 22 November 2021), two significant hits were obtained: Potri.014G102600 (score: 183.4, E-value: 8.7×10^{-44}) and Potri.002G175300 (score: 172.6; E-value: 1.6×10^{-40}) (Supplementary Materials Table S1). The third hit was on chromosome 18 but with just a score of 46.4 and E-value of 1.6×10^{-2}. Therefore, we considered the two best hits as paralogous genes for the CRISPR/Cas-induced mutation approach.

The transcript sequences of Potri.014G102600 and Potri.002G175300 were blasted against the draft genome sequence of *P.* × *canescens* (clone INRA 717-1B4; P1;https://urgi.versailles.inra.fr/Species/Forest-trees/Populus/Clone-INRA-717-1B4, accessed on 22 November 2021) and *P. tremula* v1.1 PopGenIE (http://popgenie.org/blast, accessed on 22 November 2021), respectively, to find any putative homologous sequences for the used clone P1. The blast-results indicate that the Potri.014G102600 and Potri.002G175300 homologous sequences in *P.* × *canescens* are scaffold_72118, scaffold_37509, scaffold_1688, scaffold_170684 and scaffold_333464, scaffold_74020, scaffold_47155, respectively, and in *P. tremula* are Potra003475 and Potra001079, respectively (Supplementary Materials Table S1).

Obtained sequences were aligned to find identical positions in the first three exons of all *TAC1* homologs in *P. trichocarpa*, *P.* × *canescens* and *P. tremula* where target sequences in the gRNAs for simultaneous knock-out of both genes can be designed.

2.3. Guide RNA (gRNA) Design and Agrobacterium-Mediated Poplar Transformation

We employed the CRISPR/Cas9 system to generate *TAC1* mutations. Two guide RNAs (gRNAs) were designed manually. Secondary structures of the applied gRNAs were subsequently analysed with the RNAfold WebServer [57], accessible under http://rna.tbi.univie.ac.at/cgi-bin/RNAWebSuite/RNAfold.cgi (accessed on 30 October 2020), as described in [54]. This software relies on the RNA folding model of [58] and the minimum free energy (MFE) computing algorithm of [59].

Putative *PpeTAC1* (EMBL acc.no. KF218366; [43]) homologous genes in *P. trichocarpa*, *P.* × *canescens* and *P. tremula* were identified. For the CRISPR/Cas9-approach, two target sequences of the gRNAs were selected in exon 3 of the two paralogs Potri.014G102600 and

Potri.002G175300 (for each paralog 2R and 4F) (*P. trichocarpa* v3.1 (https://phytozome-next.jgi.doe.gov/info/Ptrichocarpa_v3_1, accessed on 22 November 2021). The reason to design two gRNAs was simply to double the chance of a successful editing. Both target regions are located in exon 3 of the two genes, only a few hundred base pairs apart.

Care was taken that both target sequences were mapped to identical positions in the two *TAC1* paralogs to ensure simultaneous knock-out of both genes. In addition, target sequences were checked for homology to the genomic sequence of used P1 in AspenDB (http://aspendb.uga.edu/index.php/databases/spta-717-genome, accessed on 22 November 2021) and (https://urgi.versailles.inra.fr/Species/Forest-trees/Populus/Clone-INRA-717-1B4, accessed on 22 November 2021 [60], and *P. tremula* (PopGenIE; http://popgenie.org/blast, accessed on 22 November 2021). The following two target sequences were chosen: (i) TAC1_4F_gRNA (target: CAC TGG CAG AAC TGT TCT TGG): in exon 3 of both Potri.002G175300 (13,448, 819 to 13,448,839) and Potri.014G102600 (8,029,047 to 8,029,067), (ii) TAC1_2R_gRNA (target: CCA TCA AGC ACA TCC ACC AGG): in exon 3 of both Potri.002G175300 (13,449,184 to 13,449,204) and Potri.014G102600 (8,029,382 to 8,029,402). The assembled multiplexing vector C672p9ioR-35s-CasWT-Nuc (14,257 bp; Supplementary Materials Figure S1) was transferred into *Agrobacterium tumefaciens* strain GV3101 (DNA Cloning Service, Hamburg, Germany) which was used for poplar leaf-disc transformation as described above. Details of the vector, enzymes, and antibiotics for *Agrobacterium*-growth are mentioned in the website of the company (https://dna-cloning.com/vectors/, accessed on 22 November 2021).

Plant transformation was carried out by applying the *Agrobacterium*-mediated leaf-disc co-cultivation method as described in [28,29,61]. Leaf-discs of in vitro grown P1-plants were harvested and soaked with *Agrobacterium tumefaciens* strain GV3101::pMP90RK (chromosomal background C58) carrying the CRISPR/Cas9 system (35S-driven Cas9 nuclease and gRNAs: C672p9ioR-35sCasWT-Nuc, Supplementary Materials Figure S1). The plasmid (provided by DNA Cloning Service, Hamburg, Germany) is also carrying the *NEOMYCIN PHOSPHOTRANSFERASE II* (*NPTII*) gene leading to kanamycin (kan) resistance. For transgenic plant selection, the regeneration media contained kanamycin (50 mg/L), and cefotaxime (500 mg/L) for bacterium removal [56].

2.4. Molecular Analyses of Transgenic Plants and Sequencing of CRISPR/Cas9-Induced Mutations

Genomic DNA was extracted from leaves of the 19 different putative Cas9, TAC1_4F_gRNA, and TAC1_2R_gRNA transgenic lines of the line N499 and the not-transgenic P1-control grown in vitro for PCR analyses according to [28,32]. PCR was performed as described in detail in [29] but with annealing temperatures of 58 °C for amplification of a partial *Cas9*-fragment and 60 °C for *NPTII*. The sequences of the primer pairs used in PCR reactions to amplify fragments of the *Cas9* and the *NPTII* genes are listed in Supplementary Materials Table S2.

CRISPR/Cas9-related mutations in the *P.* × *canescens* paralogs Potri.014G102600 and Potri.002G175300 were determined by Sanger sequencing of the two genomic target sites (for each paralog 2R and 4F). Primers used are summarized in Supplementary Materials Table S1. The mutations were homozygous and heterozygous single and double nucleotide insertions or deletions in one or both paralogous genes.

2.5. Morphological Investigations of Glasshouse Grown Plants

To experimentally test whether CRISPR/Cas9-mediated homozygous or heterozygous mutations in the *P.* × *canescens* paralogs Potri.014G102600 and Potri.002G175300 reveal a modified plant growth or habitus, plants from transgenic lines and controls, including two not-transgenic but in vitro regenerated lines (without any *TAC1*-mutation) and the wildtype clone P1, were multiplied in vitro by shoot cuttings.

Following in vitro culture, rooted shoots of about 5 to 10 cm in height were potted into soil in April 2019, transferred to growth chambers, fully covered with plastic lids, and cultivated in boxes on a 25 °C/16 °C and 16/8 h day/night cycle. For acclimatization to ambient air conditions (relative humidity of air: 70%), lids were opened daily for increasing time periods over one week. Following acclimatization, plants were watered daily and cultivated for 2 to 4 weeks in the growth chamber. At an age of about 4 to 6 weeks, plants were transferred into the glasshouse (min-mean-max air temperature, April to October: 15–19–24 °C, November to March: 8–17–21 °C). During the first and second growing season, the plants were cultivated in pots with sizes of 13 and 26 cm, respectively, watered daily, and supplemented with mineral fertilizer twice during the growing season.

Total height and diameter of the main shoot (1 cm above soil) were measured from five plants of each of the ten independent transgenic lines N499-1-2, -2-1, -3-1, -4-1, -8-3, -10-2, -11-1, 13-2, -14-1, -and 17-1, the control P1, and the regenerated but not-transgenic lines N499-6-1 and -9-1, cultivated in glasshouse after terminal bud formation (January 2020 and January 2021). Three to ten leaf petioles (July 2019) and side-shoots (August 2020 and April 2021) were photographed from five plants of each of the ten independent N499-transgenic lines mentioned before, the regenerated but not-transgenic lines and the control P1. Later, angles of leaf petioles and side-shoots were determined on a computer screen.

2.6. Data Analysis and Statistics

Data are presented in figures as mean + SD. Differences of the last measurements were tested for significance by ANOVA at $p < 0.05$ (Fisher's F-test). Values in the figures labelled with an asterisk differ significantly at $p < 0.05$ (Fisher's F-test) from the control.

3. Results

3.1. Regeneration and Molecular Analysis of Putative TAC1 Mutants

In total, 19 independent putatively transgenic lines were obtained after four to eight weeks on kanamycin-containing regeneration medium in the transformation experiment N499. Regenerated lines were termed as N499-X. All lines were investigated via PCR for presence of NPTII and Cas9 genes (Table 1). The results indicate that out of the 19 independent lines, ramets of eleven lines revealed both the NPTII and the Cas9 gene (Table 1).

Table 1. PCR analyses amplifying parts of the *NPTII* and *Cas9* genes of putative transgenic lines carrying the CRISPR/Cas9 system to modify the genomic sequences of the *P. trichocarpa* homologs Potri.014G102600 and Potri.002G175300. + = band appeared in PCR-experiments; - = no band.

Putative Transgenic Line	Ramet	*NPTII*	*Cas9*
N499-1	−2	+	+
N499-2	−1	+	+
N499-3	−1	+	+
N499-4	−1	+	+
N499-5	−1	-	-
N499-6	−1	-	-
N499-7	−1	-	-
N499-8	−3	+	+
N499-9	−1	-	-
N499-10	−2	+	+
N499-11	−1	+	+
N499-12	−1	-	-
N499-13	−2	+	+
N499-14	−2	+	+
N499-15	−1	-	-

Table 1. *Cont.*

Putative Transgenic Line	Ramet	*NPTII*	*Cas9*
N499-16	−1	-	-
N499-17	−1	+	+
N499-18	−1	-	-
N499-19	−1	+	+

The two target sites 2R and 4F in the paralog Potri.014G102600 were Sanger-sequenced in all obtained lines (Tables 2 and 3). The results indicate that all lines tested positive for *NPTII* and *Cas9*, PCR also reveal CRISPR/Cas9-caused mutations in the target 2R (Tables 2 and 3). Here, a preferential homozygous insertion of one single nucleotide was found, mostly T, but also A or G, or in lines N499-4-1, -8-3, -11-1 a different nucleotide in the two allelic strands, namely K (Table 3). In line N499-1-2, a deletion of two nucleotides was found. An undefinable heterozygous insertion or deletion was detected in line N499-13-2. On the other hand, not a single mutation in any line could be identified in the target 4F.

Table 2. Sanger sequencing of the two target regions of the two *P. trichocarpa* putative *PpeTAC1* paralogs Potri.014G102600 and Potri.002G175300 in *P.* × *canescens* (for each paralog targets 2R and 4F) to unravel putative CRISPR/Cas9−caused mutations. - = no modification (similar to P1), similar to wildtype control clone P1, del = deletion, ins = insertion, hom = homozygous, het = heterozygous, ? = not definable, n.d. = not determined. Grey shaded lines in the first column were transferred to the glasshouse for morphological inspections. Bold in Potri: lines with homozygous mutations in both genes. Underlined in Potri: lines with heterozygous mutations in both genes. Grey shaded, bold, and underlined in Potri: lines with homozygous mutation in Potri.014G102600 but with heterozygous mutation in Potri.002G175300.

P1/Putative Transgenic Line	Potri.014G102600 ("TAC-14")		Potri.002G175300 ("TAC-2")	
	2R	**4F**	**2R**	**4F**
P1 (Wildtype)	-	-	-	-
N499-1-2	**GT del hom**	-	**G ins hom**	-
N499-2-1	**T ins hom**	-	**R ins hom**	-
N499-3-1	**T ins hom**	-	**R ins hom**	-
N499-4-1	**K ins hom**	-	**? ins/del het**	-
N499-5-1	-	-	n.d.	-
N499-6-1	-	-	-	-
N499-7-1	-	-	n.d.	-
N499-8-3	K ins hom	-	R ins hom	-
N499-9-1	-	-	-	-
N499-10-2	T ins hom	-	A ins hom	-
N499-11-1	**K ins hom**	-	**? ins/del het**	-
N499-12-1	-	-	n.d.	-
N499-13-2	? ins/del het	-	? ins/del het	-
N499-14-2	**T ins hom**	-	**? ins/del het**	-
N499-15-1	-	-	n.d.	-
N499-16-1	-	-	-	-
N499-17-1	**T ins hom**	-	**? ins/del het**	-
N499-18-1	-	-	-	-
N499-19-1	T ins hom	-	K ins hom	-

Table 3. Types of mutations identified in the target region 2R of the two *P. trichocarpa* putative *PpeTAC1* paralogs Potri.014G102600 and Potri.002G175300 in *P.* × *canescens* following CRISPR/Cas9-editing. Heterozygous mutations were complex and were not resolvable in detail. Underlined: target sequence 2R (CCTGGTGGATGTGCTTGATGG), light-blue highlighted: PAM-sequence, light-green highlighted: homozygous GT deletion, underlined, bold-red letters: homozygous nucleotide insertions, hom = homozygous, het = heterozygous, ? = not definable.

Lines	Potri.014G102600 ("TAC-14"): Genomic Target Region 2R	Type of Mutation
P1, not-transgenic regenerated lines *	.....GAAACAAGTAGC CCT GGT-GGATGTGCTTGATGGTTG......	Wildtype
N499-1-2	...... GAAACAAGTAGCCCTG -- - GGATGTGCTTGATGGTTG......	GT-deletion, hom
N499-2-1, -3-1, -10-2, -14-2, -17-1	...... GAAACAAGTAGCCCTGGT**T**GGATGTGCTTGATGGTTG.......	T-insertion, hom
N499-3-3	...... GAAACAAGTAGCCCTGGT**G**GGATGTGCTTGATGGTTG.......	G-insertion, hom
N499-10-3	...... GAAACAAGTAGCCCTGGT**A**GGATGTGCTTGATGGTTG.......	A-insertion, hom
N499-4-1, -8-3, -11-1	...... GAAACAAGTAGCCCTGGT**K**GGATGTGCTTGATGGTTG.......	G/T-insertion, hom
N499-13-2	...... GAAACAAGTAGCCCTGGT-GGATGTGCTTGATGGTTG....... ...... ??????????????????	? bp insertion or deletion, het
	Potri.002G175300 ("TAC-2"): Genomic Target Region 2R	
P1, not-transgenic regenerated lines *	...... GAAACAAGTAGC CCT GGT-GGATGTGCTTGATGGTTG......	Wildtype
N499-1-2	...... GAAACAAGTAGCCCTGGT**G**GGATGTGCTTGATGGTTG.......	G-insertion, hom
N499-10-2	...... GAAACAAGTAGCCCTGGT**A**GGATGTGCTTGATGGTTG.......	A-insertion, hom
N499-2-1, -3-1, -8-3, -10-3	...... GAAACAAGTAGCCCTGGT**R**GGATGTGCTTGATGGTTG.......	A/G-insertion, hom
N499-4-1, -11-1, -13-2, -14-2, -17-1	...... GAAACAAGTAGCCCTGGT-GGATGTGCTTGATGGTTG....... ...... ??????????????????	? bp insertion or deletion, het

* = N499-6-1, -9-1, -16-1.

With exception of lines N499-5, -7, 12, and -15, the N499-lines were also analyzed for putative CRISPR/Cas9-caused mutations in the targets 2R and 4F of Potri.002G175300. Again, not a single mutation could be found for target 4F, but all transgenic lines carry mutations in target 2R (Tables 2 and 3). The preferential homozygous modifications in target 2R of Potri.002G175300 were a single insertion of either A or G (lines N499-1-2, -10-2; Tables 2 and 3) or of different nucleotides (mostly R and once K) in both allelic strands (lines N499-2-1, -3-1, and -8-3; Tables 2 and 3). In lines N499-4-1, -11-1, -13-2, -14-2, -17-1, heterozygous insertions or deletions were detected, which were unfortunately undefinable in most of these lines (Tables 2 and 3).

Taken together, the editing results in the target regions of the two *P. trichocarpa* putative *PpeTAC1* paralogs in *P.* × *canescens*, lines N499-1-2, -2-1, -3-1, -8-3, -10-2, and -19-1 reveal homozygous single or double nucleotide insertions/deletions putatively leading to knock-out mutations in both genes (bold in Potri in Table 2). On the other hand, lines N499-4-1, -11-1, -14-2, and -17-1 show homozygous mutations in Potri.014G102600 but heterozygous mutations in Potri.002G175300 (grey shaded, bold, and underlined in Potri in Table 2). And finally, line N499-13-2 reveals heterozygous mutations in both genes (underlined in Potri in Table 2).

3.2. Morphological Investigations of In Vitro and Glasshouse Grown Plants

During the growth of obtained putative transgenic N499-lines in vitro in cultivation boxes, a plant habitus with upright oriented leaves, indicating a smaller petiole angle, appeared in some lines (Figure 1a,b). However, this possible change of the plant phenotypes in vitro could not be quantified. Therefore, lines N499-1-2, -2-1, -3-1, -8-3, and -10-2 with

CRISPR/Cas9-mediated homozygous mutations in both *P. trichocarpa* paralogous genes Potri.014G102600 and Potri.002G175300 were transferred to soil and cultivated in the glasshouse for morphological inspections. In addition, lines N499-4-1, -11-1, -14-2, and -17-1 with a homozygous knock-out mutation in just Potri.014G102600 but a heterozygous one in Potri.002G175300 and line N499-13-2 with heterozygous mutations in both genes were also transferred to the glasshouse. Finally, the wildtype P1 and two regenerated but not-transgenic lines (without *TAC1*-mutation; N499-6-1, and -9-1) were included as controls in the experimentally tests in the glasshouse (lines transferred to the glasshouse are grey-shaded in the first column of Table 2).

Figure 1. Representative poplar plants grown in vitro. (**a**) Not-transgenic control plant of *P.* × *canescens* clone INRA 717-1B4. (**b**) Plant with upright oriented leaves carrying a homozygous knock-out editing in both of the poplar Potri.014G102600 ("TAC-14"; orthologous to the rice *TILLER ANGLE CONTROL 1*, (*TAC1*) and its paralog Potri.002G175300 ("TAC-2").

In the first year (2019), because of the lack of side-shoots, angles of representative leaf petioles were recorded from five plants of each of P1 wildtype and the two regenerated but not-transgenic controls, and the transgenic edited lines after two months' growth in the glasshouse (July 2019) (Figures 2a,b and 3a). Significantly smaller petiole angles (indicating more up-right growth) were found in all lines with homozygous knock-out in the two *P. trichocarpa* putative *PpeTAC1* paralogs in *P.* × *canescens* (N499-1-2, -2-1, -3-1, -4-1, -8-3, and -10-2) compared to the control clones (P1 and the regenerated but not-transgenic lines N499-6-1, and -9-1) (Figure 3a). In lines that were only homozygous in Potri.014G102600 but heterozygous in Potri.002G175300, the situation turned out to be inconsistent. Lines N499-4-1 and -17-1 reveal significantly smaller petiole angles, however, in lines N499-11-1 and -14-2, petiole angles were also reduced but not significant. In the line that is heterozygous in both genes (N499-13-2), petiole angles turned out to be slightly reduced or similar to the controls (Figure 3a).

Figure 2. Representative poplar plants grown in the glasshouse and in the field. (**a**) Transgenic plants of N499-3-1 and N499-10-3 (left two plants) with slightly upright oriented leaves in comparison with two INRA 717-1B4 control plants (right) after 1 month of cultivation in soil (2019). (**b**) Representative petioles of a control (upper image) and a homozygous "TAC-14"/"TAC-2" knock-out edited N499-transgenic plant (lower image) during growth in the first year in the glasshouse (2019). (**c**) Representative side-shoots of a control (upper image) and a homozygous *TAC-14*/*TAC-2* knock-out edited N499-transgenic plant (lower image) during growth in the second year in the glasshouse (2020). (**d**) Control (left) and homozygous "TAC-14"/"TAC-2" knock-out edited N499-transgenic plant in a glasshouse during the winter season after the second year (2020/21) of growth in a glasshouse. (**e**) Representative control (left) and homozygous "TAC-14"/"TAC-2" knock-out edited N499-transgenic plant during growth in the second year in the glasshouse (2020). (**f**) Erect *P. tremula* mutant grown in the field (left) compared to a homozygous "TAC-14"/"TAC-2" knock-out edited N499-transgenic plant during growth in the third year in the glasshouse (2021). Note: N499-transgenic plants with a homozygous editing event only in "TAC-14" revealed also a pronounced upright growth of side-shoots.

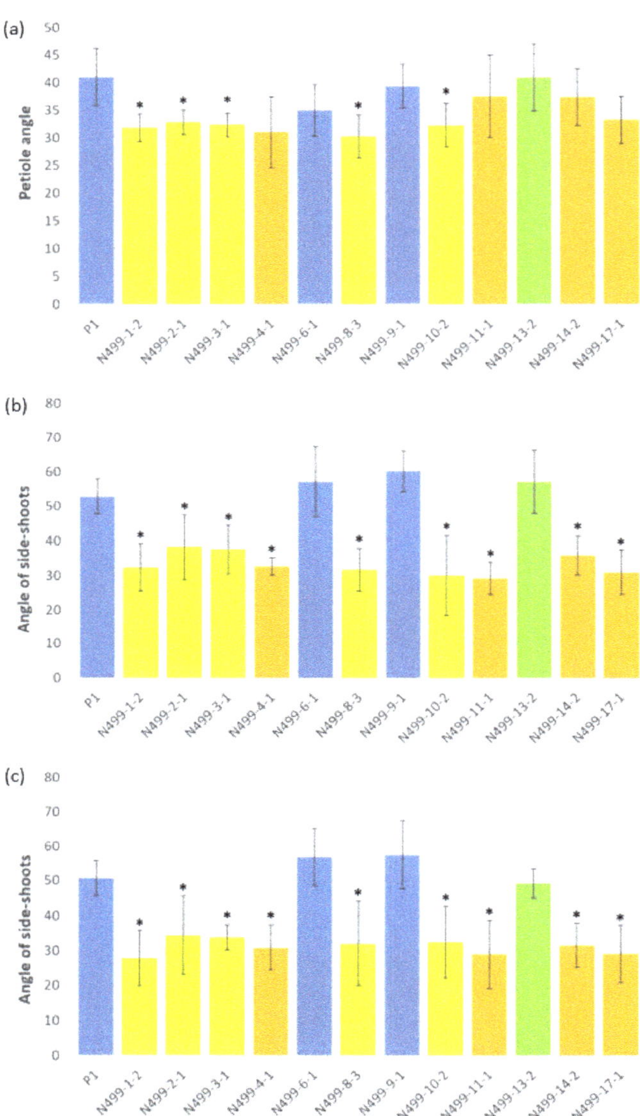

Figure 3. Measurements of the leaf petiole angles in the first year (2019 (**a**), and angles of side-shoots of plants of different independent N499 edited-transgenic lines and not-transgenic control plants grown in a glasshouse in the second (2020) (**b**) and third year (2021) (**c**) after transfer, respectively. Edited transgenic N499-lines reveal either homo- knock-out editing of both Potri.014G102600 ("TAC-14") and its paralog Potri.002G175300 ("TAC-2") (yellow-columns), homozygous in "TAC-14" but heterozygous in "TAC-2" (orange columns) or heterozygous in both *TAC*-genes (green columns). Controls: P1 and not-transgenic lines (blue columns). Angles of leaf petioles and side-shoots were determined from five plants of each line. Error bars represent the standard deviations of three biological replicates derived from five independent trees. * = Values differ significantly at $p < 0.05$ (Fisher's F-test) from the control P1.

Angles of representative side-shoots were determined in the second and third year (August 2020, end of April 2021) of glasshouse grown plants, again from five plants of the transgenic and of different control lines (Figure 2c,d and Figure 3b,c). Here, results again show that all lines with homozygous knock-out in the two *P. trichocarpa* putative *PpeTAC1* paralogs in *P.* × *canescens* but also all lines that are homozygous just in Potri.014G102600 but heterozygous in Potri.002G175300 show significantly smaller angles of the side-shoots than the P1 and the N499 control clones. Only the line N499-13-2, heterozygous in both genes, revealed side-shoot angles similar to the control lines. The phenotype of the homozygous Potri.014G102600 knock-out plants revealed an erect growth compared to the control plants (Figure 2e), and in particular, in the third year of glasshouse growth, the Potri.014G102600 knock-out plants are phenotypically quite similar to unrelated erect *P. tremula* individuals grown in the field (Figure 2f).

Total height and diameter of the main shoot (1 cm above soil) measured from all P1, N499-control and N499-transgenic plants after terminal bud formation in the first (January 2020) and the second year (February 2021) of cultivation in the glasshouse revealed no significant differences (data not shown).

4. Discussion and Conclusions

Upright or pyramidal growth of plants is a pronounced phenotype found in many plant species, e.g., *Prunus persica* (peach) [40], *Cupressus sempervirens* 'Pyramidalis' (Italian Cypress [45], *Taxus baccata* 'Pyramidalis' (pillar yew [46]), *Populus tremula* 'Erecta' (Swedish columnar aspen [47]), *P. alba* var. pyramidalis (*P. bolleana* [48]), and *P. nigra* 'Italica' (Lombardy poplar [49]), but also, a commercial-used hybrid poplar cultivar named Sundancer™ reveals vertical growth of branches [50]. Also, quite recently Fladung (unpublished) noticed a pyramidal growth in an aspen, *P. tremula*, male individual growing in the arboretum of the Thuenen Institute of Forest Genetics, Grosshansdorf, Germany. Control of branching, axillary meristem initiation, and the outgrowth of axillary buds have been intensively studied at the molecular level [62]. However, the molecular basis for different tree-growth habits and mechanisms determining the branch angle is still poorly understood, despite its possible importance for practical applications.

Among the poplar genotypes, without doubt, the most famous upright-growing tree mutant is *P. nigra* 'Italica' (Lombardy poplar), a cultivated clonal variety distributed worldwide for more than 300 years [49,52,63]. Even if the origin of the mutation is uncertain, the mutation was discovered in a single male *P. nigra* tree grown in central Asia [49]. In peach, a mutation in the orthologous gene of the rice *TILLER ANGLE CONTROL 1* (*TAC1*) was shown to be responsible for pyramidal plant growth [43]. However, the occurrence of the *TAC1* was, besides some other genes, originally thought to be restricted to monocots [64,65]. But *TAC1* homologs were soon identified as well in dicots [18,43], and overexpression of the *PdTAC1* gene in peach has led to more horizontal oriented branches [44]. Also [43] proposed that the pyramidal phenotype of the *P. nigra*, Lombardy poplar, is possibly caused by a mutation in a poplar ortholog of the *TAC1* gene. However, so far, not for any pyramidal tree genotype, has it experimentally been confirmed that a knock-out mutation of the rice *TAC1* orthologous gene(s) leads to pyramidal plant growth.

To experimentally test the role of putative *TAC1* ortholog(s) in poplar, we first determined in silico the most homologous gene and its paralogous one in *P. trichocarpa*, namely Potri.014G102600 ("TAC-14") and Potri.002G175300 ("TAC-2"), respectively. The orthologous genes were also determined in both *P. tremula* and *P.* × *canescens* by employing the two obtained *P. trichocarpa* sequences. Based on all sequences obtained from the three poplar species, two guide RNAs (gRNAs) were designed to generate *TAC1* mutations by use of the CRISPR/Cas9 system. We obtained a number of CRISPR/Cas9-mediated homozygous and heterozygous mutations in the *P.* × *canescens* clone INRA 717-1B4 in one or both of the *TAC1* genes. The *TAC1*-edited lines and (transgenic and not-transgenic) controls were analyzed molecularly and inspected morphologically for three years after transfer to the glasshouse. As an indicator for pyramidal growth, we focused on the prominent features, leaf and branch angle, which makes a major contribution to the overall tree architecture [66]. Our results indicate that the homozygous knock-out of just "TAC-14" is sufficient to induce pyramidal plant growth in *P.* × *canescens*.

During in vitro culture, a possible change of the plant phenotype towards pyramidal growth could be noticed but not quantified. Following transfer of the lines into the glasshouse, lines with CRISPR/Cas9-mediated homozygous mutations in both "TAC-14" and "TAC-2" genes formed leaves with a narrower angle and showed an upright growth of shoots in the second year of glasshouse cultivation. This finding is in agreement with the results in peach, where mutant varieties with pronounced vertical growth of branches or upright stems have been associated with a loss-of-function mutation in the *PdTAC1* gene [43]. In peach, but also in other woody plant species, obvious benefits of upright growth could be e.g., reduced need for pruning, easier application of chemicals (if needed), automation of harvest, and increased yield simply through high density planting [40,41]. Further [38] could show that erect-leaf habit allows more efficient penetration of light into lower canopy layers, thus increasing total plant net photosynthesis potentially leading to increased biomass formation. However, our data on plant height and stem diameter of poplar grown for three years in the glasshouse (of course not comparable to field growth conditions) revealed no significant difference between pyramidal and control plants.

Even though no increased biomass (just taking the two growth parameters height and stem diameter into account) could be noticed for pyramidal poplar, these individuals are, at least, similarly "productive" as "normal"-growing ones. This fact could be of high interest for short rotation coppices (SRCs), because due to erect leaf or shoot growth, pyramidal plants need reduced individual space, and thus, a higher number of plants could be planted on a default field size of SRCs, as e.g., reported for peach [40] and apple [41]. This could lead in consequence to increased yield per ha land area without any breeding activities. However, interestingly, with the possible exception of the hybrid poplar cultivar Sundancer™ [50], so far, all commercially used poplar clones revealed a non-pyramidal plant growth. However, theoretically at least, any commercially used poplar clone could serve as starting material for CRISPR-based "TAC-14"-knock-out to produce near-isogenic lines with pyramidal growth but unchanged yield. Unfortunately, in July 2018, the European Court of Justice (Case C-528/16) has judged that all plants derived by CRISPR-based modifications are to be regarded as genetically modified organisms (GMOs) within the meaning of Directive 2001/18 [67]. However, as a first step, already available erect poplar genotypes can be tested for growth and yield parameters because SRCs are replicable everywhere that the site conditions support poplar trees.

Supplementary Materials: The following are available online at https://www.mdpi.com/article/10.3390/f12121615/s1, Figure S1: C672p9ioR-35sCasWT-Nuc plasmid, Table S1: Putative *PpeTAC1* (EMBL acc.no. KF218366; [43]) homologous genes in *P. trichocarpa*., Table S2: Primers used in this study.

Funding: The project was funded with the internal budget of the Thünen Institute of Forest Genetics.

Acknowledgments: I deeply thank Katrin Groppe for tremendous technical assistance in the laboratory and the glasshouse staff (Wolfgang Graf, Sabine Benischek, Monika Spauszus) for plant cultivation and taking measurements.

Conflicts of Interest: The author of the research article has no conflict of interest to disclose.

References

1. Reinhardt, D.; Kuhlemeier, C. Plant architecture. *EMBO Rep.* **2002**, *3*, 846–851. [CrossRef]
2. Fladung, M.; Bossinger, G.; Roeb, G.W.; Salamini, F. Correlated alterations in leaf and flower morphology and rate of photosynthesis in a midribless (mbl) mutant of Panicum maximum (Jacq.). *Planta* **1991**, *184*, 356–361. [CrossRef]
3. Fladung, M. Genetic variants of Panicum maximum (Jacq.) in C4 photosynthetic traits. *J. Plant Physiol.* **1994**, *143*, 165–172. [CrossRef]
4. Godin, C. Representing and encoding plant architecture: A review. *Ann. For. Sci.* **2000**, *57*, 413–438. [CrossRef]
5. Wang, Y.; Li, J. Molecular basis of plant architecture. *Ann. Rev. Plant Biol.* **2008**, *59*, 253–279. [CrossRef]

6. Costes, E. Physiology and genetics of plant architecture. *Ann. Plant Rev. Online* **2019**, *2*, 1031–1068.
7. Barthélémy, D.; Caraglio, Y. Plant architecture: A dynamic, multilevel and comprehensive approach to plant form, structure and ontogeny. *Ann. Bot.* **2007**, *99*, 375–407. [CrossRef]
8. Chelakkot, R.; Mahadevan, L. On the growth and form of shoots. *J. R. Soc. Interface* **2017**, *14*, 20170001. [CrossRef]
9. Nybom, N. Mutation types in barley. *Acta Agricult. Scand.* **1954**, *4*, 430–456. [CrossRef]
10. Kharkwal, M.C.; Pandey, R.N.; Pawar, S.E. Mutation breeding for crop improvement. In *Plant Breeding*; Jain, H.K., Kharkwal, M.C., Eds.; Springer: Dordrecht, The Netherlands, 2004; pp. 601–645.
11. Pathirana, R. Plant mutation breeding in agriculture. *Plant Sci. Rev.* **2011**, *6*, 107–126. [CrossRef]
12. Oladosu, Y.; Rafii, M.Y.; Abdullah, N.; Hussin, G.; Ramli, A.; Rahim, H.A.; Miah, G.; Usman, M. Principle and application of plant mutagenesis in crop improvement: A review. *Biotechnol. Biotechnolog. Equip.* **2016**, *30*, 1–16. [CrossRef]
13. Holme, I.B.; Gregersen, P.L.; Brinch-Pedersen, H. Induced genetic variation in crop plants by random or targeted mutagenesis: Convergence and differences. *Front. Plant Sci.* **2019**, *10*, 1468. [CrossRef]
14. Lamo, K.; Bhat, D.J.; Kour, K.; Solanki, S.P.S. Mutation studies in fruit crops: A review. *Int. J. Curr. Microbiol. Appl. Sci.* **2017**, *6*, 3620–3633. [CrossRef]
15. Fladung, M.; Ziegenhagen, B. M13 DNA fingerprinting can be used in studies on phenotypical revisions of forest tree mutants. *Trees* **1998**, *12*, 310–314. [CrossRef]
16. Fladung, M.; Tusch, A.; Markussen, T.; Ziegenhagen, B. Analysis of morphological mutants in Picea. In Proceedings of the Applications of Biotechnology to Forest Genetics (Biofor 99), Vitoria-Gasteiz, Spain, 22–25 September 1999; Espinel, S., Ritter, E., Eds.; pp. 167–170.
17. Flachowsky, H.; Hanke, M.V.; Peil, A.; Strauss, S.H.; Fladung, M. A review on transgenic approaches to accelerate breeding of woody plants. *Plant Breed.* **2009**, *128*, 217–226. [CrossRef]
18. Teichmann, T.; Muhr, M. Shaping plant architecture. *Front. Plant Sci.* **2015**, *6*, 233. [CrossRef]
19. Zawaski, C.; Kadmiel, M.; Pickens, J.; Ma, C.; Strauss, S.H.; Busov, V. Repression of gibberellin biosynthesis or signaling produces striking alterations in poplar growth, morphology, and flowering. *Planta* **2011**, *234*, 1285–1298. [CrossRef]
20. Buhl, C.; Strauss, S.H.; Lindroth, R.L. Genetic down-regulation of gibberellin results in semi-dwarf poplar but few non-target effects on chemical resistance and tolerance to defoliation. *J. Plant Ecol.* **2019**, *12*, 124–136. [CrossRef]
21. Busov, V.B.; Meilan, R.; Pearce, D.W.; Ma, C.; Rood, S.B.; Strauss, S.H. Activation tagging of a dominant gibberellin catabolism gene (GA 2-oxidase) from poplar that regulates tree stature. *Plant Physiol.* **2003**, *132*, 1283–1291. [CrossRef]
22. Weigel, D.; Nilsson, O. A developmental switch sufficient for flower initiation in diverse plants. *Nature* **1995**, *377*, 495–500. [CrossRef]
23. Rottmann, W.H.; Meilan, R.; Sheppard, L.A.; Brunner, A.M.; Skinner, J.S.; Ma, C.; Cheng, S.; Jouanin, L.; Pilate, G.; Strauss, S.H. Diverse effects of overexpression of LEAFY and PTLF, a poplar (Populus) homolog of LEAFY/FLORICAULA, in transgenic poplar and Arabidopsis. *Plant J.* **2000**, *22*, 235–245. [CrossRef]
24. Yuan, H.M.; Chen, S.; Lin, L.; Wei, R.; Li, H.Y.; Liu, G.F.; Jiang, J. Genome-wide analysis of a TaLEA-introduced transgenic *Populus simonii* × *Populus nigra* dwarf mutant. *Int. J. Mol. Sci.* **2012**, *13*, 2744–2762. [CrossRef]
25. Bruegmann, T.; Fladung, M. Overexpression of both flowering time genes AtSOC1 and SaFUL revealed huge influence onto plant habitus in poplar. *Tree Genet. Genom.* **2019**, *15*, 1–13. [CrossRef]
26. Fladung, M. Transformation of diploid and tetraploid potato clones with the rolC gene of Agrobacterium rhizogenes and characterization of transgenic plants. *Plant Breed.* **1990**, *104*, 295–304. [CrossRef]
27. Schmülling, T.; Fladung, M.; Großmann, K.; Schell, J. Hormonal content and sensitivity of transgenic tobacco and potato plants expressing single rol genes of Agrobacterium rhizogenes T-DNA. *Plant J.* **1993**, *3*, 371–382. [CrossRef]
28. Fladung, M.; Muhs, H.J.; Ahuja, M.R. Morphological changes observed in transgenic Populus carrying the rolC gene from Agrobacterium rhizogenes. *Silv. Genet.* **1996**, *45*, 349–354.
29. Fladung, M.; Kumar, S.; Ahuja, M.R. Genetic transformation of Populus genotypes with different chimaeric gene constructs: Transformation efficiency and molecular analysis. *Transgen. Res.* **1997**, *6*, 111–121. [CrossRef]
30. Fladung, M.; Polak, O. Ac/Ds-transposon activation tagging in poplar: A powerful tool for gene discovery. *BMC Genom.* **2012**, *13*, 61. [CrossRef] [PubMed]
31. Eriksson, M.E.; Israelsson, M.; Olsson, O.; Moritz, T. Increased gibberellin biosynthesis in transgenic trees promotes growth, biomass production and xylem fiber length. *Nat. Biotechol.* **2000**, *18*, 784–788. [CrossRef]
32. Dünisch, O.; Funada, R.; Nakaba, S.; Fladung, M. Influence of overexpression of a gibberellin 20-oxidase gene on the kinetics of xylem cell development in hybrid poplar (*Populus tremula L* × *P. tremuloides Michx*). *Holzforschung* **2006**, *60*, 608–617. [CrossRef]
33. Jeon, H.W.; Cho, J.S.; Park, E.J.; Han, K.H.; Choi, Y.I.; Ko, J.H. Developing xylem-preferential expression of PdGA20ox1, a gibberellin 20-oxidase 1 from Pinus densiflora, improves woody biomass production in a hybrid poplar. *Plant Biotechnol. J.* **2016**, *14*, 1161–1170. [CrossRef] [PubMed]
34. Bruegmann, T.; Wetzel, H.; Hettrich, K.; Smeds, A.; Willför, S.; Kersten, B.; Fladung, M. Knockdown of PCBER1, a gene of neolignan biosynthesis, resulted in increased poplar growth. *Planta* **2019**, *249*, 515–525. [CrossRef]
35. Sakamoto, T.; Morinaka, Y.; Ohnishi, T.; Sunohara, H.; Fujioka, S.; Ueguchi-Tanaka, M.; Mizutani, M.; Sakata, K.; Takatsuto, S.; Yoshida, S.; et al. Erect leaves caused by brassinosteroid deficiency increase biomass production and grain yield in rice. *Nat. Biotechnol.* **2006**, *24*, 105–109. [CrossRef]

36. Richards, R.A.; Cavanagh, C.R.; Riffkin, P. Selection for erect canopy architecture can increase yield and biomass of spring wheat. *Field Crops Res.* **2019**, *244*, 107649. [CrossRef]
37. Burgess, A.J.; Retkute, R.; Herman, T.; Murchie, E.H. Exploring relationships between canopy architecture, light distribution, and photosynthesis in contrasting rice genotypes using 3D canopy reconstruction. *Front. Plant Sci.* **2017**, *8*, 734. [CrossRef]
38. Mantilla-Perez, M.B.; Salas Fernandez, M.G. Differential manipulation of leaf angle throughout the canopy: Current status and prospects. *J. Exp. Bot.* **2017**, *68*, 5699–5717. [CrossRef] [PubMed]
39. Sinclair, T.R.; Sheehy, J.E. Erect leaves and photosynthesis in rice. *Science* **1999**, *283*, 1456. [CrossRef]
40. Scorza, R.; Lightner, G.W.; Liverani, A. The pillar peach tree and growth habit analysis of compact x pillar progeny. *J. Amer. Soc. Hortic. Sci.* **1989**, *114*, 991–995.
41. Kelsey, D.F.; Brown, S.K. 'McIntosh Wijcik': A columnar mutation of 'McIntosh' apple proving useful in physiology and breeding research. *Fruit Var. J.* **1992**, *46*, 83–87.
42. Tworkoski, T.; Scorza, R. Root and shoot characteristics of peach trees with different growth habits. *J. Amer. Soc. Hortic. Sci.* **2001**, *126*, 785–790. [CrossRef]
43. Dardick, C.; Callahan, A.; Horn, R.; Ruiz, K.B.; Zhebentyayeva, T.; Hollender, C.; Whitaker, M.; Abbott, A.; Scorza, R. *PpeTAC1* promotes the horizontal growth of branches in peach trees and is a member of a functionally conserved gene family found in diverse plants species. *Plant J.* **2013**, *75*, 618–630. [CrossRef] [PubMed]
44. Hollender, C.A.; Waite, J.M.; Tabb, A.; Raines, D.; Chinnithambi, S.; Dardick, C. Alteration of *TAC1* expression in Prunus species leads to pleiotropic shoot phenotypes. *Horticult. Res.* **2018**, *5*, 1–9. [CrossRef]
45. Papageorgiou, A.C.; Panetsos, K.P.; Hattemer, H.H. Genetic differentiation of natural Mediterranean cypress (Cupressus sempervirens L.) populations in Greece. *For. Genet.* **1994**, *1*, 1–12.
46. Hoffman, M.H.A. Cultivar classification of *Taxus* L. (Taxaceae). *Acta Horticult.* **2004**, *634*, 91–96. [CrossRef]
47. Kwantlen Polytechnic University. School of Horticulture Plant Database—*Populus tremula* 'Erecta'. Available online: https://plantdatabase.kpu.ca/plant/plantDetail/284 (accessed on 22 November 2021).
48. Xu, X.; Tong, L.; Li, F.; Kang, S.; Qu, Y. Sap flow of irrigated Populus alba var. pyramidalis and its relationship with environmental factors and leaf area index in an arid region of Northwest China. *J. For. Res.* **2011**, *16*, 144–152. [CrossRef]
49. Zsuffa, L. The genetics of *Populus nigra* L. *Ann. For.* **1974**, *6*, 29–53.
50. Schroeder, W.; Soolanayakanahally, R.; Lindquist, C. AC Sundancer™ Poplar. *Can. J. Plant Sci.* **2013**, *93*, 1285–1287. [CrossRef]
51. Wood, C.D. "A Most Dangerous Tree": The Lombardy Poplar in landscape gardening. *Arnoldia* **1994**, *54*, 24–30.
52. Vanden Broeck, A.; Cox, K.; Brys, R.; Castiglione, S.; Cicatelli, A.; Guarino, F.; Heinze, B.; Steenackers, M.; Vander Mijnsbrugge, K. Variability in DNA methylation and generational plasticity in the Lombardy Poplar, a single genotype worldwide distributed since the Eighteenth century. *Front. Plant Sci.* **2018**, *9*, 1635. [CrossRef]
53. Tuskan, G.A.; DiFazio, S.; Jansson, S.; Bohlmann, J.; Grigoriev, I.; Hellsten, U.; Putnam, N.; Ralph, S.; Rombauts, S.; Salamov, A.; et al. The genome of black cottonwood, *Populus trichocarpa* (Torr. & Gray). *Science* **2006**, *313*, 1596–1604.
54. Bruegmann, T.; Deecke, K.; Fladung, M. Evaluating the efficiency of gRNAs in CRISPR/Cas9 mediated genome editing in poplars. *Int. J. Mol. Sci.* **2019**, *20*, 3623. [CrossRef]
55. Leple, J.C.; Brasileiro, A.C.M.; Michel, M.F.; Delmotte, F.; Jouanin, L. Transgenic poplars: Expression of chimeric genes using four different constructs. *Plant Cell Rep.* **1992**, *11*, 137–141. [CrossRef]
56. Briones, M.V.; Hoenicka, H.; Cañas, L.A.; Beltrán, J.P.; Hanelt, D.; Sharry, S.; Fladung, M. Efficient evaluation of a gene containment system for poplar through early flowering induction. *Plant Cell Rep.* **2020**, *39*, 577–587. [CrossRef] [PubMed]
57. Hofacker, I.L. Vienna RNA secondary structure server. *Nucl. Acids Res.* **2003**, *31*, 3429–3431. [CrossRef] [PubMed]
58. Andronescu, M.; Condon, A.; Hoos, H.H.; Mathews, D.H.; Murphy, K.P. Efficient parameter estimation for RNA secondary structure prediction. *Bioinformatics* **2007**, *23*, i19–i28. [CrossRef] [PubMed]
59. Zuker, M.; Stiegler, P. Optimal computer folding of large RNA sequences using thermodynamics and auxiliary information. *Nucl. Acids Res.* **1981**, *9*, 133–148. [CrossRef]
60. Mader, M.; Le Paslier, M.C.; Bounon, R.; Bérard, A.; Faivre Rampant, P.; Fladung, M.; Leplé, J.C.; Kersten, B. Whole-genome draft assembly of Populus tremula × Populus alba clone INRA 717-1B4. *Silv. Genet.* **2016**, *65*, 74–79. [CrossRef]
61. Bruegmann, T.; Polak, O.; Deecke, K.; Nietsch, J.; Fladung, M. Poplar Transformation. In *Transgenic Plants*; Kumar, S., Barone, P., Smith, M., Eds.; Methods in Molecular Biology; Humana Press: New York, NY, USA, 2019; pp. 165–177.
62. Bennett, T.; Leyser, O. Something on the side: Axillary meristems and plant development. *Plant Mol. Biol.* **2006**, *60*, 843–854. [CrossRef]
63. Henry, A. The Black poplars. *Gard. Chron.* **1914**, *56*, 46–47.
64. Doebley, J.; Wang, R.L. Genetics and the evolution of plant form: An example from maize. Cold Spring Harbor Symp. *Quant. Biol.* **1997**, *62*, 361–367.
65. Yu, B.; Lin, Z.; Li, H.; Li, X.; Li, J.; Wang, Y.; Zhang, X.; Zhu, Z.; Zhai, W.; Wang, X.; et al. *TAC1*, a major quantitative trait locus controlling tiller angle in rice. *Plant J.* **2007**, *52*, 891–898. [CrossRef] [PubMed]
66. Hallé, F.; Oldeman, R.A.; Tomlinson, P.B. Opportunistic tree architecture. In *Tropical Trees and Forests*; Hallé, F., Oldeman, R.A., Tomlinson, P.B., Eds.; Springer: Berlin/Heidelberg, Germany, 1978; pp. 269–331.
67. Carreño, I.; Dolle, T. The Court of Justice of the European Union's judgment on mutagenesis and international trade: A case of GMO, mutagenesis and international trade. *Glob. Trade Cust. J.* **2019**, *14*, 91–101.

MDPI

Article

Impact of Different Pruning Practices on Height Growth of Paulownia Clon in Vitro 112®

Jiří Kadlec *, Kateřina Novosadová and Radek Pokorný

Department of Silviculture, Faculty of Forestry and Wood Technology, Mendel University in Brno, Zemědělská 3, 613 00 Brno, Czech Republic; katerina.novosadova@mendelu.cz (K.N.); radek.pokorny@mendelu.cz (R.P.)
* Correspondence: jiri.kadlec.uzpl@mendelu.cz; Tel.: +420-545134553

Abstract: We focused on the ability of one-year-old and two-year-old plants of Paulownia Clon in vitro 112® to sprout and grow branches, and on their pruning for their best possible growth on a plantation in Střelice u Brna. Furthermore, we carried out pruning on selected parts that comprised: spring pruning; reduction in the angle between the stem axis and one new growing sprout; and year-long pruning. The sprouting capacity of Paulownia was high—up to 56% (one-year-old plants) and 50% (two-year-old plants). Branches grew on 34% of all one-year-old plants and on 57% of all two-year-old plants. The best possible spring pruning method seems to be the keeping of one stem sprout or one stump sprout for one-year-old plants and one stem sprout for two-year-old plants. The newly growing stem sprout should be formed by bandaging it to the stem, and, as a result of this, the angle between the stem and the sprout can be reduced to 20° in contrast to 50° when the sprout is not bandaged. Our results suggest that it is best to take off the lower $^{1}/_{3}$ of the branches and leaves, which leads to faster height growth of the plant.

Keywords: spring pruning; year-long pruning; branching; angle diversion of sprout

Citation: Kadlec, J.; Novosadová, K.; Pokorný, R. Impact of Different Pruning Practices on Height Growth of Paulownia Clon in Vitro 112®. *Forests* **2022**, *13*, 317. https://doi.org/10.3390/f13020317

Academic Editor: Dirk Landgraf

Received: 22 January 2022
Accepted: 14 February 2022
Published: 15 February 2022

1. Introduction

Paulownia spp. Siebold & Zucc. is a fast-growing deciduous tree; it can grow up to a height of 12 to 30 m [1] and can reach a stem diameter of around 1 m at breast height [2]. The area of origin of *Paulownia* spp. is China [3], Laos and Vietnam [4]. It was first used as a decorative tree in other areas such as Japan and the Korean peninsula [2]. Today, it grows on all continents, except for Antarctica. The first time that *Paulownia tomentosa* was planted in the region of today's Czech Republic (CR) was in 1844 [2].

New types of hybrids were made from original Paulownias, especially to increase timber production. This hybrid clone cultivation widened the ecological valence and spread its growth beyond its original habitat. These new types of hybrids were planted in the CR in the past ca. 10 years [5]. Paulownia Clon in vitro 112® has high resistance to extremely low temperatures down to −25 °C [6] and shows high adaptability to local climatic conditions [7]; therefore, it seems to be the best choice for the CR. Additionally, Paulownia Clon in vitro 112® grows faster compared to other Paulownia hybrids, and its volume is around 0.3 m^3 after three years from planting [6]. Its wood is easy to work (with) and therefore suitable for carving [8], and it can be used to produce wood fiber products or for cellulose and biofuel production [9], for building construction [10] and in the energy and sawmill industries [11].

Based on information from Zhao-Hua et al. [12], it forms thick branches already on two-year-old plants after the first winter period. Furthermore, it has a tendency to form many widely spread branches, if grown in an open space [6]. Trees of Paulownia are always planted in open spacing in plantations (for example, according to UCLM [13], 5 × 5 m for sawmill logs or 3.3 × 1 m for biomass). Another factor influencing the growth habit is damage to the buds (i.e., the apical bud, similar to the second to fourth pairs of axillary buds) caused by frost, which is typical for Paulownia [12]. The frost can be early, late [14] or hard frost [15]. All these types of frost can cause damage, which leads to hormonal imbalance, thanks to which the surviving lateral buds accelerate their growth, and the sprouts originating from them take over the function of the dead apical sprout [12,16]. As a rule, only one of the newly formed lateral sprouts is significantly dominant and assumes the function of a terminal sprout, and the other sprouts turn into branches. However, in the event of unclear dominance, multiple terminal sprouts occur [17]. Moreover, Paulownia is a plant with a very strong sprouting ability [12] to create root, stem [18] and stump sprouts [6]. With only a few exceptions, the root and stem sprouts appear on the trees every year, and frost damage to the trees in their first years after planting stimulates the formation of both types of sprouts. Due to a combination of these factors, the stem can become crooked, and we could say that it manifests bushy (broom-like) growth, which was described by Narovcová et al. [17].

This is why pruning is applied to attain an upright stem without sprouts and branches in the lower part of the tree [19,20]. There can be two different types of pruning. The first type is applied for the formation of the tree habit in the way that undesirable newly created sprouts are removed (i.e., "spring pruning"). These sprouts come in three types: root sprouts, which grow from adventitious buds on the roots [21]; stump sprouts, which grow in multiple places around the edge of stump sprouts [22]; and stem sprouts, which grow from adventitious buds on the stem [23]. When all the sprouts are left to grow, a tree with multiple stems growing from the roots, stem or stump is formed, and such a tree grows more slowly and creates dense bushy growth. The occurrence of multiple stems can be successfully reduced via spring pruning [24], and the newly formed tree consists of one strong terminal sprout [25]. The second type is the elimination of the leaves and branches at the start of or during the vegetation period (i.e., "year-long pruning"). Open-space-planted Paulownia, such as *Eucalyptus nitens* (H.Deane & Maiden) Maiden, generates and retains large branches and, due to this, does not produce sawlog, veneer-quality timber [26] or knot-free timber known as clearwood [27]. Year-long pruning is carried out for the purpose of increasing wood quality [26,27] and, also, accelerating the height increment [28]. However, severe year-long pruning can reduce height growth and have a negative impact on the quality of the wood [29,30].

The reason why we researched the ability of young Paulownia trees to generate sprouts and branches and compared the effects of spring and year-long pruning was to grow trees with upright stems and crowns set high, which is lucrative, especially for plantation growing.

We investigated the following:

- Whether one-year-old and two-year-old plants produce branches and root, stem and stump shoots, and, if yes, how many.
- Whether bandaging of the stem sprout to the stem influences the straightness of the stem while pruning—we tried to reduce the angle between the axis of the original stem and that of the new sprout.
- Whether it is possible to achieve a greater height growth of the tree in the event that a different type of sprout (stem, stump, root) is supported after apical bud frost injury (i.e., spring pruning).
- Whether the plant grows faster with year-long pruning (i.e., the removal of different volumes of branches and leaves).

2. Materials and Methods

The research plot was situated on a private, fenced, agricultural area in Střelice u Brna (49°15′460.56″ N; 16°47′179.14″ E) with an altitude of around 300 a.s.l. and a slope from 0 to 10°. The soil is Cambisol with a loam soil texture; it contains about 1.6% organic matter, has a pH/KCl of around 6.3 and pH_{H_2O} of about 7.1 and goes down to a depth of 110 cm. In 2018, the average yearly air temperature was 8.3 °C, the minimum yearly air temperature was −10.2 °C and the annual rainfall was 448 mm [31]. The plantation was established with ramets of Paulownia Clon in vitro 112®, which were bought from the company Oxytree Solution s.r.o., which is registered in the CR [32].

Paulownia Clon in vitro 112® is a hybrid that originated as a natural cross between *Paulownia elongata* S.Y.Hu and *P. fortunei* (Seem.) Hemsl., and the newly created plants, which had the required properties, were reproduced using the in vitro method in order for their cultivated properties to be maintained [33]. This hybrid was created in the laboratory of the company IN VITRO S.L. (Sant Feliu de Llobregat, Spain), and the World Intellectual Property Organization has given it the trademark application number of 1,181,727, expiration date 25 September 2023 [34]. The adaptability of this hybrid to various soils and climates was tested by the University of Castilla-La Mancha [35].

A total of 686 in vitro container-grown ramets (height: 20 ± 2 cm; root neck thickness: 5 ± 1 mm; mean $\pm$ SD) were planted in 2016 and 737 in 2017, with a spacing of 4×4 m. After the planting of the ramets, we referred to them as plants. Each plant was assigned a unique code according to the row and its position in each for improved identification during the measurements.

We performed three experiments in the plantation to better document the natural development of one- and two-year-old trees without any silvicultural treatments and the influence of pruning on the heights of the plants.

Gardening shears were used for cutting off the above-ground part of the plant. Spring pruning and year-long pruning were carried out with the use of a sharp knife (or pruning shears), in accordance with Arborist Standards [36]. Cut wounds were not treated and were left to heal naturally. The height of the plant and the length of the sprout were measured with a folding rule, and the angles between the rest of the stem and the sprouts were measured using a protractor.

2.1. Spring Pruning

The height was measured on all plants in February 2018. Those that were chosen in each of the two years (i.e., 125 one-year-old plants and 125 two-year-old plants) were all within a specified range of heights and thicknesses, close to the average (Table 1). The ground around each of these plants was marked with a color spray so that the plant was easy to find for each successive measurement.

Before the start of the vegetation period (the 70th day of the year (DOY)), we performed spring pruning in five variants on the plants with frost-damaged stems planted in 2016 and in five variants on the plants planted in 2017, where each variant was conducted on 25 plants:

- Stump pruning, where the plant was cut on the 70th DOY (leaving a maximum 5 cm stump) and, on the 100th DOY, the thickest stump sprout was left, and all the other sprouts were removed (Figure 1).
- Root pruning, where the plant was cut on the 70th DOY (leaving a maximum 5 cm stump) and, on the 100th DOY, the thickest root sprout was left, and all the other sprouts were removed (Figure 1).
- Stem pruning, where, on the 100th DOY, the thickest stem sprout was left, and the rest of the plant (approximately 10 cm above that sprout) was removed, together with all the other sprouts. Stem pruning was carried out in two different ways:
    - Stem pruning without bandaging, where the sprout that was saved was not bandaged to the remaining part of the stem (Figure 1).
    - Stem pruning with bandaging, where the sprout that was saved was bandaged to the remaining part of the stem to ensure straight growth (Figure 1).
- Reference plant, where the plant with sprouts was left to grow naturally (Figure 1).

Table 1. Average values of the heights of the trees and the thicknesses of the stems measured 10 cm above the ground for experimental trees, and the statistical significance of the differences in these parameters between the variants.

Experiment	Variant	Average Height (±SD) [cm]	SS	Average Thickness (±SD) [mm]	SS
		Plants planted in 2016			
Spring pruning	Stump	72.5 (±24.9)	Ns	17.6 (±5.6)	ns
	Root	72.6 (±28.2)	Ns	17.4 (±6.1)	ns
	Stem with bandaging	75.2 (±19.1)	Ns	16.8 (±5.4)	ns
	Stem without bandaging	74.6 (±18.6)	Ns	17.9 (±4.8)	ns
	Reference plant	76.0 (±17.9)	Ns	16.0 (±4.2)	ns
Year-long pruning	Pruning 1/3	77.5 (±22.7)	Ns	18.9 (±5.0)	ns
	Pruning 2/3	74.9 (±20.3)	Ns	17.1 (±5.1)	ns
	Regular pruning	73.4 (±19.7)	Ns	15.9 (±5.2)	ns
	Reference plant	72.8 (±21.2)	Ns	17.8 (±4.5)	ns
		Plants planted in 2017			
Spring pruning	Stump	61.6 (±14.4)	Ns	12.2 (±4.6)	ns
	Root	62.3 (±18.4)	Ns	12.4 (±6.3)	ns
	Stem with bandaging	64.6 (±15.7)	Ns	11.3 (±4.2)	ns
	Stem without bandaging	63.6 (±16.6)	Ns	12.4 (±4.1)	ns
	Reference plant	62.1 (±15.1)	Ns	12.3 (±4.9)	ns
Year-long pruning	Pruning 1/3	62.4 (±13.6)	Ns	12.4 (±4.1)	ns
	Pruning 2/3	64.1 (±11.5)	Ns	13.4 (±4.7)	ns
	Regular pruning	62.3 (±12.8)	ns	13.1 (±4.5)	ns
	Reference plant	62.1 (±14.3)	ns	12.9 (±4.5)	ns

SD—standard deviation; SS—statistical significance; ns—no statistical significance (p = 0.95).

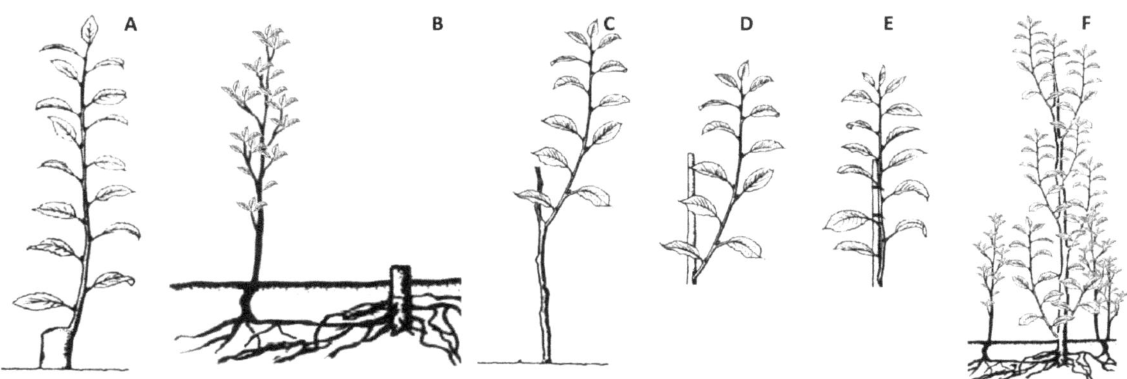

Figure 1. Line drawings of the variants of spring pruning (**A**—stump pruning; **B**—root pruning; **C**—stem pruning; **D**—stem pruning without bandaging; **E**—stem pruning with bandaging; **F**—reference plant).

The height of the entire plant and the length of the sprout that was left to grow were measured on the 100th DOY. The angle diversion between the rest of the stem (i.e., the stem axis) and the axis of the sprout that was left to grow was also measured—the smaller the angle between the stem axis and the sprout axis, the smaller the sprout diversion from the rest of the stem. The stem sprout (in this variant) was bandaged to the rest of the stem with a jute cord, and a paper bag was inserted between the sprout and the jute cord to protect the sprout. The height of the plant was measured repeatedly at about one-month intervals during the entire vegetation period. The sprout bandage was removed for the time of the measurement, the diversion angle of both stem variants was measured (with and without bandaging) and the sprout was re-bandaged after measurement. The last measurement of the height of the plant was carried out on the 301st DOY, the bandage was permanently removed and the angle between the rest of the stem and the sprout that was left to grow was also measured.

We used the length of the new sprouts for a better comparison of the spring pruning results, instead of the total height of each plant. We deducted the height of the stump and/or stem from the growth from the previous year (or two years) from the total plant height.

2.2. Year-Long Pruning

The height was measured on all plants in February 2018. Those that were chosen in each of the two years (i.e., 100 one-year-old plants and 100 two-year-old plants) were all within a specified range of heights and thicknesses, close to the average (Table 1). The ground around each of these plants was marked with a color spray so that the plant was easy to find for each successive measurement.

Year-long pruning proceeded on the 70th DOY: the above-ground part of the chosen plant was removed (leaving stumps with heights not exceeding five centimeters). After 30 days from the date of removal, the strongest stump sprout was left, and all other new stump sprouts or new root sprouts were removed. Year-long pruning was carried out in four variants on the plants planted in 2016 and on the same number of those planted in 2017 (Table 1). Each variant was conducted on 25 plants. These variants were as follows:

- Pruning $^{1}/_{3}$, where the lowest $^{1}/_{3}$ of the leaves and branches was removed from the plant on the 151st DOY.
- Pruning 2/3 , where the lowest 2/3 of the leaves and branches were removed from the plant on the 151st DOY.
- Regular pruning, where the lowest leaves were removed from the plant (on the 151st DOY), so only the last four rows of the leaves growing opposite each other remained, and each began regular removal of leaves approximately 14 days from this date.
- Reference plant, where the plant with sprouts was left to grow naturally.

The height of the entire plant (including the stump) and the length of the new growing sprout were measured on the 100th DOY. At approximately one-month intervals from this date, the heights of all of the plants were measured. On the 151st DOY, the first removal of leaves and branches was carried out on the plants with the pruning $^{1}/_{3}$, pruning 2/3 and regular pruning variants. When measuring the heights of the plants, in the cases where there were more than the last four rows of pairs of leaves growing opposite each other, the bottom rows of leaves were removed using the regular pruning variant. The measurements were carried out until the 301st DOY, when the last measurement of the plant height was conducted on the plants with year-long pruning.

For a better comparison of the results of the year-long pruning, we used the length of the new sprouts instead of the total height of each plant, so we deducted the height of the stump from the total plant height.

2.3. Sprouting Capacity and Branching

The plantation owner removed the above-ground parts of all one-year-old and two-year-old trees (which were not used for spring and year-long pruning) on the 70th DOY. The thickest stump sprout was left, and all other sprouts were removed on the 100th DOY. From this date, the owner did not perform any silvicultural treatment, and the plants were left to grow naturally. We recorded the numbers of newly emerged stem, stump and root sprouts and branches (i.e., "sprouting capacity and branching") on 270 of the plants planted in 2016 and on 423 of those planted in 2017 (i.e., on a total of 693 plants). The stump sprouts, stem sprouts, root sprouts and branches on each of the plants were counted on the 301st DOY. Consequently, the plants were sorted into groups based on the numbers of root, stem and stump sprouts, and also on the numbers of branches that were found on each plant (0—no sprout/branch; 1—one sprout/branch; n—n sprouts/branches). Sprouting capacity and branching were investigated, also according to the age of the plant, because those planted in 2016 were two years old at the time, and those planted in 2017 were one year old.

Statistical data were evaluated in TIBCO Statistica™, with a reliability interval of 95%. The Shapiro–Wilk test was employed to find out the data dispersion normality and homogeneity. Analysis of variance (ANOVA) was used to evaluate the suitability of the annual and biennial plants selected for spring pruning and year-long pruning. The nature of the results of the experiment did not make it possible for parametric testing to be applied for the evaluation of the differences; therefore, the Kruskal–Wallis non-parametric test was employed next, after which the Dunn test was applied to identify differences among the main effects and interactions.

3. Results

3.1. Spring Pruning

On the 186th DOY, stump sprouts (56 ± 6 cm; average length ± SD) and stem sprouts (with bandaging 50 ± 2 cm; without bandaging 49 ± 4 cm) grew best on the one-year-old plants (Figure 2). These differences between the first group (of stem sprouts with/without bandaging and the stump sprouts) and the second group (of root sprouts and the reference plants) on the one-year-old plants grew with each new measurement during the season. During the last measurement (the 301st DOY), the average length of the stem sprouts without bandaging was 119 ± 7 cm, that of the stem sprouts with bandaging was 115 ± 6 cm and that of the stump sprouts was 114 ± 8 cm. These lengths were statistically different from those of the root sprouts and reference plants. Percentage differences among the individual variants are presented in Table 2.

Table 2. Percentage differences and statistically significant values among the spring pruning variants of one- and two-year-old plants. The percentages indicate how much shorter the sprout named in the row is, compared to that named in the column, e.g., the reference plant was 46% shorter than the stem sprout with the bandaging.

		One-year-old plants			
DOY	Variant	Stem sprout with bandaging		Stem sprout without bandaging	Stump sprout
186th	Reference plant	46% (p = 0.0001)		44% (p = 0.0018)	52% (p = 0.0001)
	Root sprout	30% (p = 0.0062)		28% (p = 0.0058)	38% (p = 0.0086)
301st	Reference plant	46% (p = 0.0001)		48% (p = 0.0001)	45% (p = 0.0001)
	Root sprout	49% (p = 0.0009)		51% (p = 0.0001)	48% (p = 0.0001)
		Two-year-old plants			
DOY	Variant	Stem sprout with bandaging	Stump sprout	Stem sprout without bandaging	Root sprout
186th	Stem sprout without bandaging	24% (p = 0.0058)			
	Stump sprout	22% (p = 0.0068)			
	Root sprout	42% (p = 0.0032)	26% (p = 0.0075)	24% (p = 0.0089)	
	Reference plant	61% (p = 0.0004)	50% (p = 0.0032)	48% (p = 0.0002)	32% (p = 0.0045)
301st	Stem sprout without bandaging	14% (p = 0.0174)			
	Root sprout	19% (p = 0.0168)			
	Stump sprout	18% (p = 0.0423)			
	Reference plant	54% (p = 0.0021)	44% (p = 0.0023)	46% (p = 0.0025)	43% (p = 0.003)

DOY—day of year.

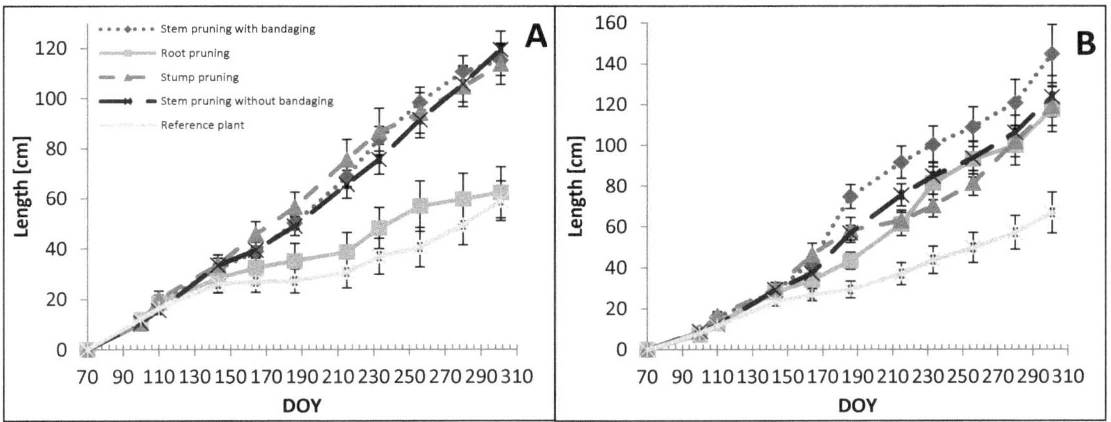

Figure 2. Average sprout length. **A**—one-year-old plants; **B**—two-year-old plants. Whiskers denote SD.

On the 186th DOY, we divided the two-year-old-plant results into four groups, based on statistical significance (Figure 2). The best growth was found on the stem sprouts with bandaging (75 ± 6 cm). The stump sprouts and the stem sprouts without bandaging made up the second group of the longest sprouts. The root sprouts made up the third group, and the reference plants created the last group. The order of the groups changed during the measurement period. On the 301st DOY, it was possible to divide the average sprout lengths into three groups, based on their statistical differences. The longest sprouts were stem sprouts with bandaging (145 ± 14 cm), and they made up the first group. The stump sprouts and the stem sprouts without bandaging created the second group. The reference plants made up the last group. Percentage differences among variants are presented in Table 2.

We conducted an experiment on the stem sprouts to change the angle diversion from the straight-stem axis (Figure 3). The closer the top of the sprout to the stem axis, the straighter the alternative apical sprout. On the 100th DOY, the angle diversion was similar: one-year-old plants with bandaging 59 ± 5°, without bandaging 61 ± 4°; two-year-old plants with bandaging 58 ± 4°, without bandaging 59 ± 4°.

The difference in the angle diversion on the one-year-old plants became visible on the 143rd DOY. The stem sprout with bandaging had an average angle of 47 ± 4°, which was about 16% smaller than that of the stem sprout without bandaging (p = 0.0047). On the 301st DOY, the average angle of the stem sprout with bandaging was 21 ± 4°, and it was about 51% smaller than the other (p = 0.0003).

The measured results on the two-year-old plants were similar to those on the one-year-old plants. The difference in the angle diversion became visible from the 143rd DOY. The stem sprout with bandaging had an angle of 49 ± 4°, and it was about 16% smaller than the average angle on the stem sprout without bandaging (p = 0.0326). On the 301st DOY, the difference between both variants was 37% (p = 0.0031), and the sprout which was straighter was the one with bandaging (29 ± 5°).

In addition, we compared what influence the age of the plant had on the angle diversion. At the start of the measurement, there were no statistical differences in the angles between the ages of the plants or the variants of bandaging. On the 301st DOY, the difference in angle diversion of the sprout without bandaging was 13% (p = 0.0498), where the average angle of the one-year-old plants was smaller than that of the two-year-old plants. Additionally, the average angle diversion of the sprout with bandaging was different. On the 301st DOY, the difference in the angle between the one-year-old and two-year-old plants was 29% (p = 0.0254), where the average angle of the one-year-old plants was smaller (21 ± 4°) than that of the two-year-old plants.

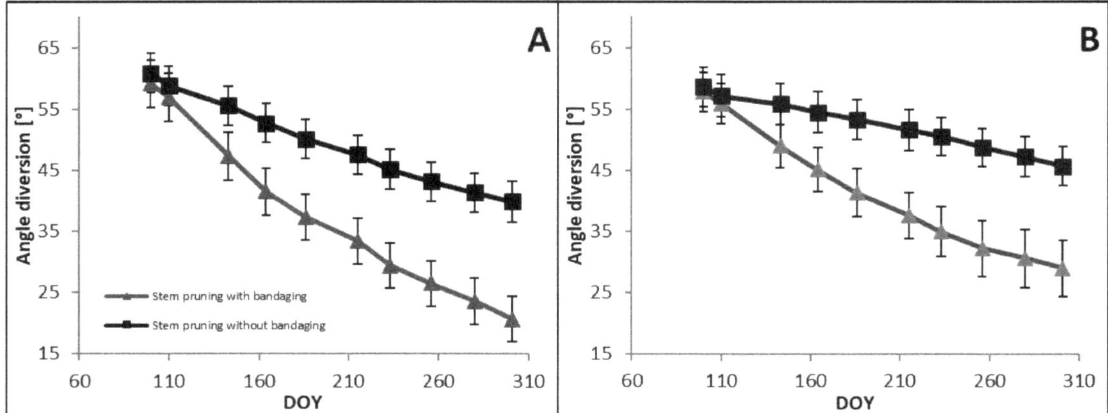

Figure 3. Average angle diversion between substitute stem sprout and stem axis. **A**—one-year-old plants; **B**—two-year-old plants. Whiskers denote SD.

3.2. Year-Long Pruning

The start of growth was the same for all variants of the one-year-old plants (Figure 4). We found three statistical groups on the 186th DOY. The plants with the pruning $^1/_3$ variant (84 ± 6 cm) had the longest average sprout. The plants with the pruning 2/3 variant and the reference plants had the second longest sprout and created the second statistical group. The plants with regular pruning made up the third statistical group. The statistical groups were the same on the 301st DOY as on the 186th DOY. The plants with the pruning $^1/_3$ variant (157 ± 7 cm) had the longest average sprout, and those with the pruning 2/3 variant and the reference plants had the second longest. The plants with regular pruning made up the third statistical group.

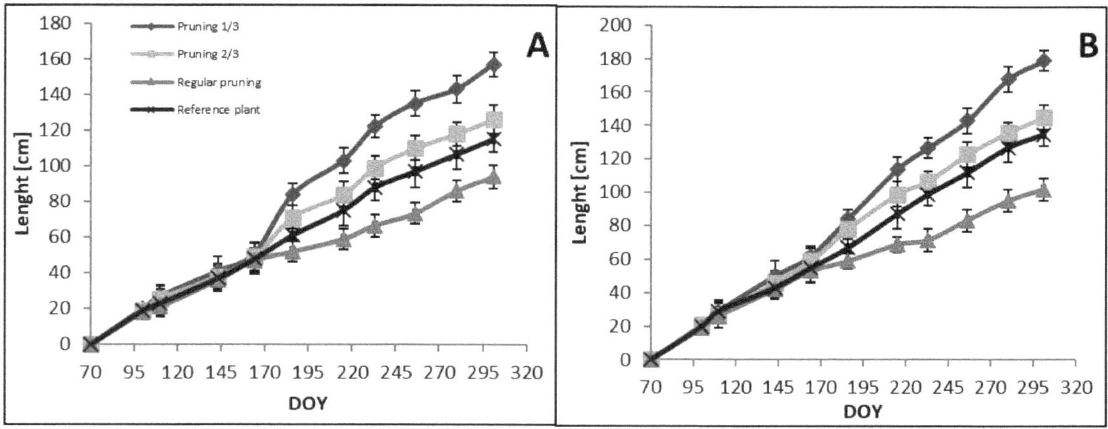

Figure 4. Average length of sprouts supplying stems during the vegetation period according to the variant of year-long pruning ($^1/_3$ pruning, 2/3 pruning, regular pruning and reference plants). **A**—one-year-old plants; **B**—two-year-old plants. Whiskers denote SD.

The start of growth was the same for all variants of the two-year-old plants (Figure 4). The first measurement that revealed visible differences was carried out on the 186th DOY and showed statistically significant differences among the variants. The plants with the pruning 1/3 variant (84 ± 6 cm) and those with the pruning 2/3 variant (78 ± 6 cm) created the first statistically homogenous group, and the plants with regular pruning and the reference plants created the second statistically homogenous group. The last measurement (on the 301st DOY) showed that the plants with the pruning 1/3 variant (179 ± 6 cm) had the longest average length. The plants with the pruning 2/3 variant and the reference plants made up the second statistical group. The plants with regular pruning created the third statistical group. Percentage differences among the variants are presented in Table 3.

Table 3. Percentage differences and statistically significant values among the year-long pruning variants of one- and two-year-old plants. The percentages indicate how much shorter the sprout named in the row is, compared to that named in the column, e.g., the sprout with 2/3 pruning was 15% shorter than that with 1/3 pruning.

		One-year-old plants		
Day of year	Variant	1/3 pruning	2/3 pruning	Reference plant
186th	2/3 pruning	15% (p = 0.0365)		
	Reference plant	27% (p = 0.0097)		
	Regular pruning	38% (p = 0.0071)	27% (p = 0.0104)	14% (p = 0.0345)
301st	2/3 pruning	20% (p = 0.0145)		
	Reference plant	27% (p = 0.0066)		
	Regular pruning	40% (p = 0.0024)	25% (p = 0.0128)	18% (p = 0.0172)
		Two-year-old plants		
DOY	Variant	1/3 pruning	2/3 pruning	Reference plant
186th	Reference plant	30% (p = 0.0249)	24% (p = 0.0217)	
	Regular pruning	20% (p = 0.0347)	14% (p = 0.0422)	
301st	2/3 pruning	23% (p = 0.0004)		
	Reference plant	25% (p = 0.0002)		
	Regular pruning	43% (p = 0.0001)	30% (p = 0.0001)	24% (p = 0.0003)

3.3. *Sprouting Capacity and Branching*

We can conclude that Paulownia has a high sprouting capacity (Table 4). The total sprouting capacity of the one-year-old plants was 80.6%, where the most frequent number of sprouts per plant was three (on 16.7% of the plants) and the maximum number of sprouts per plant was nine (on 1.2% of the plants), regardless of their origin. Depending on the origin of the sprout, the one-year-old plants had the lowest sprouting capacity from the stump (48%) and the highest sprouting capacity from the roots (56%). The most frequent number of root, stem and stump sprouts per plant was one (on 11.4%, 13.5% and 15.2% of the plants, respectively). The maximum number of sprouts per plant was seven (root sprouts on 6.1% of the plants; stem sprouts on 5.9% of the plants) and six (stump sprouts on 3.9% of the plants).

The total sprouting capacity of the two-year-old plants was 69.8%, where the most frequent number of sprouts per plant was three (on 16.1% of the plants) and the maximum number of sprouts per plant was eight (on 0.9% of the plants), regardless of their origin. Depending on the origin of the sprout, the two-year-old plants had the lowest sprouting capacity from the stump (34%) and the highest sprouting capacity from the roots (50%). The most frequent number of root, stem and stump sprouts per plant was one (on 10.1%, 12.5% and 14.7% of the plants, respectively). The maximum number of sprouts per plant was seven root sprouts (on 1.0% of the plants), six stem sprouts (on 5.1% of the plants) and five stump sprouts (on 2.5% of the plants).

Branches grew on 34% of all one-year-old plants. The most frequent was the occurrence of one pair of mutually opposite branches (on 12% of all plants), and the maximum number of mutually opposite branches was five (on 0.2% of all plants). Branches grew on 57% of all two-year-old plants. The most frequent was the occurrence of one pair of mutually opposite branches (on 23% of all plants), and the maximum number of mutually opposite branches was five (on 0.1% of all plants).

Table 4. Percentage shares of the numbers of each type of sprout (root, stem and stump) and of branches per plant of the plants planted in 2017 and 2016—one year and two years after planting.

	Number of Sprouts/Branches (Groups)										
	0	1	2	3	4	5	6	7	8	9	Sum
	One-year-old plants (planted in 2017)										
All sprouts	19.4	6.2	9.4	16.7	12.2	10.4	10.4	8.4	5.7	1.2	100.0
Root sprout	44.4	11.4	8.9	8.4	7.5	6.5	6.8	6.1	0.0	0.0	100.0
Stem sprout	49.2	13.5	8.4	6.1	6.0	5.5	5.4	5.9	0.0	0.0	100.0
Stump sprout	52.1	15.2	9.5	7.5	6.0	5.8	3.9	0.0	0.0	0.0	100.0
Branches	66.5	11.9	10.9	7.1	3.4	0.2	0.0	0.0	0.0	0.0	100.0
	Two-year-old plants (planted in 2016)										
All sprouts	30.2	8.4	9.6	15.1	12.0	9.9	8.7	5.2	0.9	0.0	100.0
Root sprout	50.0	10.9	9.4	9.1	8.1	6.2	5.3	1.0	0.0	0.0	100.0
Stem sprout	59.6	12.5	6.5	5.5	5.4	5.4	5.1	0.0	0.0	0.0	100.0
Stump sprout	65.9	14.7	7.6	5.2	4.1	2.5	0.0	0.0	0.0	0.0	100.0
Branches	43.4	23.0	18.2	11.2	4.1	0.1	0.0	0.0	0.0	0.0	100.0

4. Discussion

The presented results can be marked as unique and applicable to the growing of Paulownias—especially in areas where Paulownia Clon in vitro 112® does not reach the common annual growth, as it does in its homeland or in areas with optimal soil and climatic conditions.

Paulownia has a strong sprouting capacity [6,12,18]. Sprouts can grow from the roots [12,18], the stem [18] and/or the stump [6,12]. Our results confirm the frequent occurrence of sprouts, even in the climatic and common soil conditions of the CR. Paulownia has sprouts of all three of the above-mentioned types. At least one type of sprout grew on almost every second plant, be it a one-year-old or a two-year-old plant. Unfortunately, the above-mentioned authors do not state if some type of sprout appears more often than others. The results indicate that the most common are root sprouts, and, on the other hand, the least common seem to be stump sprouts.

Our results show that Paulownia in the CR grows branches in one third of all one-year-old plants. In most cases, on plants which had branches, there was only one row of branches growing opposite each other. There were very few plants that had as many as five rows of branches growing opposite each other. There were branches growing on more than half of the two-year-old plants. The number of branches was from one to five rows of branches growing opposite each other on the plants which had branches. Our results confirm the high branchiness described above. The results of measurements conducted on the one-year-old plants, in contrast to those conducted on the two-year-old plants, were not confirmed by the study of [12], who described that thick branches do not appear until the second year after planting. We assume that, in our plantation, as well as in other open spacing plantations (where the recommended spacing is, for example, 3.3 × 1 m or 2 × 1.5 m [6], 3 × 2 m or 5 × 5 m for timber [13] or 550 pcs/ha [37]), such a low density of trees can lead to a very low set crown and a larger number of branches at a young age.

In view of the fact that Paulownia has the ability to create root, stem and stump sprouts, and that the climate in the CR is the cause of apical sprout frost damage, we have come up with an experiment that would show the growth curve based on the selected type of apical sprout. This method is named spring pruning [19], where undesirable sprouts are removed. Spring pruning was carried out at the beginning of the vegetation period, and several different types of sprouts were left to grow. The best growing proved to be stem sprouts and stump sprouts on one-year-old plants and stem sprouts on two-year-old plants. It is assumed that the reason for the best stem sprout growth is hormonal imbalance caused by death of the apical sprout [17]. This imbalance could be the reason for the situation where the buds that were closest to the apical bud were supported by chemical processes, so they could replace the apical sprout as soon as possible, to keep the plant growing. Even Zhao-Hua et al. [12] found that the first buds under the apical bud took over its function once the terminal bud was damaged by frost. There is no mention of the length and increment of a stem sprout originating from an axial bud in the available literature. On the other hand, these authors described the length and increment of a stump sprout. The reason why they mention this, in general, is that the above-ground part of the plant is cut after the first year [6,12,13]. This is carried out to support the root system or to support the growth of the sprouts after the tree has been felled upon reaching the desired dimensions [12], which can be performed in three to five cycles [36–38]. Icka et al. [6] indicated that we can expect a four-meter stump sprout of Paulownia Clon in vitro 112®, even in the first year after planting. Zhao-Hua et al. [12] (without specifying the species) stated that an average stump sprout grows to a length of five to six meters, but individually, they can also grow to a length of up to ten meters. The average length of the sprouts on our plantation was 50–160 cm, which was much less than that described by the authors above. It is the different climatic and pedological conditions that could be responsible for this (when we compare them with those in the homeland of Paulownia). Above all, this can be caused by different temperature and rain progress according to what Icka et al. [6] or Zhao-Hua et al. [12] describe.

The experiment with the bandaging of the stem sprout to the stem was conducted together with the spring pruning. So-called bayonet growth will follow when the closest live bud takes over the apical sprout function on the stem that was damaged by frost [39]. Diversion can (sporadically) reach up to 80° from the stem axis (according to our measurements), but the average diversion reaches about 45° on one-year-old plants and 50° on two-year-old plants at the end of the vegetation period. On the other hand, when the sprout was bandaged to the rest of the stem that had been damaged by frost from the very beginning, and it was straightened up in this way, it showed an angle of diversion of about 10° on the one-year-old plants and about 30° on the two-year-old plants. This technique (i.e., process) helped the stems to be more upright.

Year-long pruning was another experiment designed to affect the height growth speed and stem straightness. This approach is applied when there is a small number of plants per hectare, together with the greater branching ability of these plants [20]. As a result, the increment is lower because the plant sends the necessary nutrients into its branches and sprouts [40]. On the other hand, when too many branches and leaves are taken away, height growth slows down because the assimilation part is eliminated. The taller the plant, the more it needs twigs and branches, as those create supportive tissue for tree stability, and the leaves support the transport of water from the ground up to the leaves via a sufficient negative water potential [41–43]. The reason behind our experiment was to find the most effective year-long pruning method that would support the growth of the plant, even if we reduce the assimilation part. Our results suggest that the best approach to year-long pruning is to remove all leaves and branches in the lowest third of the stem. This variant took away the unnecessary ineffective leaves and branches because the investment of nutrition and water into these ineffective leaves and branches reduces the utilization of nutrients and water, thanks to the higher-up leaves and branches that encourage accelerated growth of the entire plant. The limited growth, as a result of the preservation of the ineffective leaves and branches, is visible on the reference plants. We can say that the number of leaves that were taken away was not reduced more than necessary. Photosynthetic production, nutrition and energy income were not as low as those in the regular pruning variant with a greatly reduced assimilation part. We were able to see it even on the 2/3 pruning variant where the branches up to 2/3 of the plant height above the ground were taken away.

5. Conclusions

This work was aimed at influencing the growth of Paulownia Clon in vitro 112®, which was grown in the conditions of the CR, outside of its homeland. This study can be considered unique; however, it covers only one plot in the CR, and therefore the presented results are only preliminary.

Based on the height growth measured in Spain, Albania, Hungary and Romania, it can be assumed that straight stems without so-called elbow bends can grow thanks to high one-year-old sprout growth, even when the apical bud is replaced by a lateral bud. Paulownias growing in Střelice u Brna are smaller and their average height is far below one meter. As a result of this, we can expect a decrease in the quality of the assortment because there will be no long straight stems, and the length of the potential saw log will be less than 3 m, which is the minimum length of a saw log according to the rules recommended in the CR [44].

Spring pruning seems to influence the speed of growth of the plant. The reaction to year-long pruning was not similar for the one-year-old and two-year-old plants.

In terms of the one-year-old plants:

- The plants with stem or stump pruning grew higher than those with root pruning and the reference plants.
- The reference plants became bushy and had slow growth. We do not recommend leaving the plants without pruning.
- When we used root pruning, the new plants grew slowly, and they were smaller than those with the other variants of pruning (including the reference plants) at the end of the vegetation period. We do not recommend root pruning.

In terms of the two-year-old plants:

- We found out that the best pruning (according to the growth of the plants) was stem pruning.
- Next, with an almost insignificant difference, were the plants with root pruning and stump pruning.
- It is important to carry out some pruning and not leave the plant to grow naturally. We do not recommend leaving the plants without pruning.

However, when we use stem pruning, we should use bandaging. This treatment straightens the stem and reduces the angle between the elbow bend and the rest of the stem.

Additionally, year-long pruning appears to influence the speed of growth of the plant. The reaction to year-long pruning was similar for the one-year-old and two-year-old plants.

- The best year-long pruning (according to the growth of the plants) seems to be pruning $^1/_3$.
- The reference plants became bushy and had slow growth. We do not recommend leaving the plants without pruning.
- Plants with the pruning 2/3 variant had similar growth to the reference plants. We do not recommend pruning 2/3 .
- The plants with regular pruning were smaller, even compared to the reference plants. We definitely do not recommend using regular pruning.

Our current results indicate that, in the following years, it would be advisable to leave the stem sprout and direct its growth by bandaging the apical part to the stem that is damaged by frost. This approach would be applied until the apical bud reaches a height beyond the reach of frost (i.e., approximately 4 m). It is also advisable that the newly growing branches, together with the leaves that grow up to the height of the lowest third of the stem, be regularly removed.

These silvicultural measures support the growth of plants that grow beyond the frost layer faster, thereby reducing the extent of damage by frost and the shape unevenness (elbow bends) of the stem. This could result in an increased economic valuation of wood cultivated in a plantation.

Author Contributions: J.K.—main idea, methodology, work coordination, measurement, software, validation, formal analysis, writing—original draft preparation, writing—editing, project administration, funding acquisition. K.N.—methodology, measurement, writing—original draft preparation, writing—editing. R.P.—modification of methodology, supervision, funding acquisition. All authors have read and agreed to the published version of the manuscript.

Funding: This research was funded by the Internal Grant Agency (IGA) of the Faculty of Forestry and Wood Technology, Mendel University in Brno, grant number LDF_VP_2018018 ("The influence of different variants of wintering and pruning felling on the growth and quality of Paulownia wood"), number LDF_VP_2019011 ("Determination of optimum irrigation for the growth of Paulownia") and number LDF_VP_2020034 ("Optimalization of production of planting material *Paulownia* spp. propagation by seed (including presowing treatment) and vegetative propagation (cutting of belowground and aboveground parts").

Informed Consent Statement: Not applicable.

Data Availability Statement: The data is available upon request from jiri.kadlec.uzpl@mendelu.cz.

Acknowledgments: The authors thank Dalibor Veverka for the possibility to carry out this research experiment on his plantation, Petra Skočdopolová and Jan Hobl for the revision of the English language and Oldřich Mauer for the line drawing.

Conflicts of Interest: The authors and the owner of the plantation give their consent for this article to be published, agreeing that there is no competitive interest or conflict of interest.

References

1. Smiley, C.J. A Record of Paulownia in the Tertiary of North America. *Am. J. Bot.* **1961**, *48*, 175–179. [CrossRef]
2. Úradníček, L. Paulovnie plstnatá [*Paulownia tomentosa*]. *Lesnická Práce* **2013**, *5*, 36–41. (In Czech)
3. Bergmann, B.A.; Rubin, A.R.; Campbell, C.R. Potential of *Paulownia elongata* trees for swine waste utilization. *Am. Soc. Agric. Eng.* **1997**, *40*, 1733–1738. [CrossRef]
4. Hieke, K. *Praktická Dendrologie 2 [Practical dendrology Vol 2]*; SZN: Praha, Czech Republic, 1978; p. 590. (In Czech)
5. Görner, T. Pěstování paulovnie pro energetické účely [Planting of Paulownia for energy purposes]. *Ochrana Přírody* **2017**, *3*, 23–24. (In Czech)
6. Icka, P.; Damo, R.; Icka, E. *Paulownia tomentosa*, a Fast Growing Timber. 2016. Available online: https://www.degruyter.com/downloadpdf/j/agr.2016.10.issue-1/agr-2016-0003/agr-2016-0003.pdf (accessed on 20 January 2022).
7. Bikfalvi, M. Paulownia—The Intelligent Tree. 2014. Available online: http://www.paulowniagreene.ro/wp-content/uploads/2012/11/Paulownia-Clon-in-Vitro-and-Paulownia-Cotevisa-2-Presentation-EN.pdf (accessed on 20 January 2022).
8. Ates, S.; Ni, Y.; Akgul, M.; Tozluoglu, A. Characterization and evaluation of *Paulownia elongata* as a raw material for paper production. *Afr. J. Biotechnol.* **2008**, *7*, 4153–4158.
9. Yadav, N.K.; Vaidya, B.N.; Henderson, K.; Frost Lee, J.; Stewart, W.M.; Dhekney, S.A.; Joshee, N. A Review of Paulownia Biotechnology: A Short Rotation, Fast Growing Multipurpose Bioenergy Tree. *Am. J. Plant Sci.* **2013**, *4*, 2070–2082. [CrossRef]
10. García-Morote, F.A.; López-Serrano, F.R.; Martínez-García, E.; Andrés-Abellán, M.; Dadi, T.; Candel, D.; Rubio, E.; Lucas-Borja, M.E. Stem Biomass Production of *Paulownia elongata* × *P. fortunei* under Low Irrigation in a Semi-Arid Environment. *Forests* **2014**, *5*, 2505–2520. [CrossRef]
11. Akyildiz, M.H.; Hamiyet, S.K. Some technological properties and uses of paulownia (*Paulownia tomentosa* Steud.) wood. *J. Environ. Biol. Triveni Enterp.* **2010**, *31*, 351–355. (In Indian)
12. Zhu, Z.-H.; Chao, C.-J.; Lu, X.-Y.; Xiong, Y.G. Paulownia in China: Cultivation and Utilization. 1986. Available online: https://paulowniamp.files.wordpress.com/2010/05/paulownia-in-china.pdf (accessed on 20 January 2022).
13. UCLM. Badanie wartości gospodarczej odmian (VCU). Testowanie clon in vitro 112. Hibrid *Paulownia elongace* × *Paulownia fortunei*. Sprawozdanie techniczne (Czerwiec 2013). [The Economic Value of Varieties (VCU) Survey. Testing clon in vitro 112. Hybrid *Paulownia elongace* × *Paulownia fortunei*. Technical Report (June 2013).]. 2013. Available online: http://x-plus.pl/wp-content/uploads/oxytree-vcu_report-PL.pdf (accessed on 20 January 2022). (In Polish)
14. Středa, T.; Středová, H.; Rožnovský, J. Podmínky pro přezimování polních plodin v kontextu vývoje klimatu [Conditions for overwintering field crops in the context of climate development]. In *Úroda. Sborník Referátů*; Salaš, P., Ed.; Lednice 20; 2011; pp. 582–589. Available online: www.cbks.cz/rostliny2011/prispevky/StredaStredovaRoznovsky.pdf (accessed on 20 January 2022). (In Czech)

15. Mauer, O. *Zakládání Lesů I. Učební Text [Forest Establishment I. Learning Text]*; Mendelova Univerzita: Brno, Czech Republic, 2009; p. 172. (In Czech)
16. Kincl, M.; Krpeš, V. *Základy Fyziologie Rostlin [Basics of Plant Physiology]*; Montanex: Ostrava, Czech Republic, 2000; p. 221. (In Czech)
17. Nárovcová, J.; Nárovec, V.; Čermák, M. Netvárnost borovice lesní v nejmladších kulturách. *Lesnická Práce* **2004**, *8*, 8–9. (In Czech)
18. Rahman, M.A.; Rahman, F.; Rahmatulla, M. In vitro regeneration of *Paulownia tomentosa* Steud. plants through the induction of adventitious sprouts in explants derived from selected mature trees, by studying the effect of different plant growth regulators. *Am.-Eurasian J. Sustain. Agric.* **2013**, *7*, 280–289.
19. Polanský, B. *Pěstění Lesů II [Silviculture Vol. II]*; SZN: Praha, Czech Republic, 1955; p. 427. (In Czech)
20. Vyskot, M. *Pěstění Lesů [Silviculture]*; SZN: Praha, Czech Republic, 1978; p. 432. (In Czech)
21. Hrynkiewicz-Sudnik, J.; Sękowski, B.; Wilczkiewicz, M. *Rozmnażanie Drzew i Krzewów Liściastych 1–3 [Reproduction of Deciduous Trees and Bushes 1–3]*; Państwowe Wydawnictwo Naukowe: Warsaw, Poland, 1990; p. 855. (In Poland)
22. Leonardsson, J.; Götmark, F. Differential survival and growth of stumps in 14 woody species after conservation thinning in mixed oak-rich temperate forests. *Eur. J. For. Res.* **2014**, *134*, 199–209. [CrossRef]
23. Tesař, V.; Chroust, Z.; Kantor, P.; Peňáz, J.; Vacek, S.; Henžlík, V.; Kouba, J.; Materna, V.; Poleno, Z.; Worel, J.; et al. *Pěstování v Heslech. Studijní Příručka [Silviculture in Keyword. Study Guide]*; MZLU: Brno, Czech Republic, 1996; p. 95. (In Czech)
24. Souček, J. Podrost v dubovém porostu s rozdílnou výchovou [Undergrowth in oak stands with different thinning]. *Zprávy Lesnického Výzkumu* **2009**, *54*, 17–22. (In Czech)
25. Burda, P.; Nárovcová, J.J.; Nárovec, V.; Kuneš, I.; Baláš, M.; Machovič, I. *Technologie Pěstování Listnatých Poloodrostků a Odrostků nové Generace v Lesních Školkách. Certifikovaná Metodika [Technology of Planting Deciduous Semi-Sapling and Sapling of New Generation in forest nurseries. Certified Methodology]*; Výzkumý ústav lesního hospodářství a myslivosti, v.v.i.; Jíloviště-Strnady: Jíloviště-Strnady, Czech Republic, 2015; p. 56. (In Czech)
26. Neilsen, W.A.; Pinkard, E.A. Developing silvicultural regimes for sawlog and veneer production from temperate eucalypt plantations in Tasmania. In Proceedings of the Conference on XII SILVOTECNA. Eucalypts in Chile, Present and Future Conception, Valdivia, Chile; 1999; p. 27.
27. Waugh, G.; Yang, J.L. Opportunities for sawn products from Tasmanian plantation eucalypts. In Proceedings of the Conference on Faces of Farm Forestry, Australian Forest Growers, Launceston, Tasmania, Australia, 25th April–1st May 1994; pp. 215–220.
28. Funk, D.T. Stem form response to repeated pruning of young black walnut trees. *Can. J. For. Res.* **1979**, *9*, 114–116. [CrossRef]
29. Lückhoff, H.A. Pruning of *Eucalyptus grandis*. *For. S. Afr.* **1967**, *8*, 75–85.
30. Bredenkamp, B.V.; Malan, F.S.; Conradie, W.E. Some effects of pruning on growth and timber quality of *Eucalyptus grandis* in Zululand. *S. Afr. For. J.* **1980**, *11*, 29–34. [CrossRef]
31. Historical Data—Weather—Historical Data. 2020. Available online: http://portal.chmi.cz/historicka-data/pocasi/zakladni-informace?l=en (accessed on 20 January 2022).
32. OXYTREE SOLUTIONS s.r.o., Brno IČO 05211468 Obchodní Rejstřík Firem [OXYTREE SOLUTIONS Ltd., Brno CIN 05211468 European Business Register]. 2021. Available online: https://rejstrik-firem.kurzy.cz/05211468/oxytree-solutions-sro/ (accessed on 20 January 2022).
33. Paulownia Clon In Vitro 112®. 2021. Available online: https://www.paulownia112.com/#paulownia (accessed on 20 January 2022).
34. Paulownia Clon In Vitro 112. Paulownia Clon In Vitro 112 International Trademark (WIPO) Information. 2013. Available online: https://trademark.trademarkia.com/wipo/trademark-1181727.htm?fbclid=IwAR1dQ61B8u38SRXBZAUJhgFl8nSCpihv7dSPguuIh2Ipbp01IHoc77Eo9OM (accessed on 20 January 2022).
35. Cerro-Barja Del, A. Informe del Proyecto de Investigación: "Forestación de Zonas Semiáridas de Castilla-La Mancha noc *Paulownia* spp.". [Information from the Investigation Project: "Forest of Semi-Arid Areas of Castilla-La Mancha noc *Paulownia* spp."]. 2009. Available online: https://www.paulownia112.com/wp-content/uploads/2015/10/INFORME-FINAL-PAULOW.pdf (accessed on 20 January 2022). (In Spain)
36. SPPK A02. *Nature and Landscape Management Standards. Pruning of Trees*; MENDELU: Brno, Czech Republic, 2015; p. 32.
37. Jabłoński, D. Surowiec Drzewny. Oxytree: Drewno do Przerobu w Tartaku w 6 lat od Posadzenia Drzewa [Wood Raw Material. Oxytree: Wood for Processing at a Sawmill 6 Years after Planting the Tree]. 2016. Available online: http://www.drewno.pl/artykuly/10535,oxytree-drewno-do-przerobu-w-tartaku-w-6-lat-od-posadzenia-drzewa.html (accessed on 20 January 2022). (In Polish)
38. Oxytree. Investice do Budoucnosti! [Investing in the Future!]. 2016. Available online: http://oxytree.cz/co-je-oxytree/ (accessed on 20 January 2022). (In Czech)
39. Josten, E.; Reiche, T.; Wittechen, B. *Dřevo a Jeho Obrábění [Wood and Its Machining]*; Grada: Praha, Czech Republic, 2010; p. 336. (In Czech)
40. Vyskot, M. *Základy Růstu a Produkce Lesů [Basics of Forest Growth and Production]*; SZN: Praha, Czech Republic, 1971; p. 440. (In Czech)
41. Givnish, T.J. Biochemical constraints on crown geometry in forest herbs. In *On the Economy of Plant Form and Function. Proceedings of the Sixth Maria Moors Cabot Symposium, Evolutionary Constraints on Primary Productivity: Adaptive Patterns of Energy Capture in Plants*; Givnish, T.J., Ed.; Cambridge University Press: Cambridge, UK, 1986; pp. 525–583.

42. Friend, A.D. The prediction and physiological significance of tree height. In *Vegetation Dynamics and Global Change*; Solomon, M., Shugart, H.H., Eds.; Springer: Boston, MA, USA, 1993; pp. 101–115.
43. Niinemets, U. Distribution patterns of carbon and nitrogen as affected by tree dimensions and relative light conditions in the canopy of *Picea abies*. *Trees* **1997**, *11*, 144–154. [CrossRef]
44. UEWPI. *Doporučená Pravidla pro Měření a Třídění Dříví v ČR 2008 [Recommended Rules for Measuring and Sorting Wood in the Czech Republic 2008]*; Lesnická Práce: Praha, Czech Republic, 2007; p. 147. (In Czech)

Article

Comparing the Effects of N and P Deficiency on Physiology and Growth for Fast- and Slow-Growing Provenances of *Fraxinus mandshurica*

Xingtang Zhao [1,2], Xu Zhang [1,2], Zhang Liu [1,2], Yipin Lv [1,2], Tingting Song [1,2], Jinghong Cui [1,2], Tianchi Chen [1,2], Jianxia Li [1,2], Fansuo Zeng [1,2,3] and Yaguang Zhan [1,2,3,*]

1 National Key Laboratory of Tree Genetics and Breeding, Northeast Forestry University, Harbin 150040, China; longhuxingsheng@163.com (X.Z.); zhangxunefu@163.com (X.Z.); Lz2021@nefu.edu.cn (Z.L.); lyp314216@163.com (Y.L.); S1806592067@126.com (T.S.); cjh19970122@163.com (J.C.); ybace007@163.com (T.C.); LiJX15032319387@126.com (J.L.); zengfansuo@126.com (F.Z.)
2 College of Life Sciences, Northeast Forestry University, Harbin 150040, China
3 Heilongjiang Touyan Innovation Team Program, Tree Genetics and Breeding Innovation, Northeast Forestry University, Harbin 150040, China
* Correspondence: zhanyaguang2014@126.com

Abstract: With the continuous increase in atmospheric carbon dioxide emissions, nitrogen (N) and phosphorus (P) as mineral elements increasingly restrict plant growth. To explore the effect of deficiency of P and N on growth and physiology, *Fraxinus mandshurica* (hereafter "*F. mandshurica*") Rupr. annual seedlings of Wuchang (WC) provenance with fast growth and Dailing (DL) provenance with slow growth were treated with complete nutrition or starvation of N (N-), P (P-) or both elements (NP-). Although P- and N- increased the use efficiency of P (PUE) and N (NUE), respectively, they reduced the leaf area, chlorophyll content and activities of N assimilation enzymes (NR, GS, GOGAT), which decreased the dry weight and P or N amount. The free amino acid content and activities of Phosphoenolpyruvate carboxylase (PEPC) and acid phosphatase enzymes were reduced by N-. The transcript levels of NRT2.1, NRT2.4, NRT2.5, NRT2.7, AVT1, AAP3, NIA2, PHT1-3, PHT1-4 and PHT2-1 in roots were increased, but those of NRT2.1, NRT2.4, NRT2.5, PHT1-3, PHT1-4, PHT2-1 and AAP3 in leaves were reduced by P-. WC was significantly greater than DL under P- in dry weight, C amount, N amount, leaf area, PUE, NUE, which related to greater chlorophyll content, PEPC enzyme activity, N assimilation enzyme activities, and transcript levels of N and P transporter genes in roots and foliage, indicating a greater ability of WC to absorb, transport and utilize N and P under P-. WC was also greater than DL under N- in terms of the above indicators except the transcript levels of N and P assimilation genes, but most of the indicators did not reach a significant level, indicating that WC might be more tolerant to N- than DL, which requires further verification. In summary, WC was identified as a P-efficient provenance, as the growth rate was greater for the genetic type with high than low tolerance to P-.

Keywords: nitrogen; phosphorus; carbon; physiology; *F. mandshurica*

Citation: Zhao, X.; Zhang, X.; Liu, Z.; Lv, Y.; Song, T.; Cui, J.; Chen, T.; Li, J.; Zeng, F.; Zhan, Y. Comparing the Effects of N and P Deficiency on Physiology and Growth for Fast- and Slow-Growing Provenances of *Fraxinus mandshurica*. *Forests* **2021**, *12*, 1760. https://doi.org/10.3390/f12121760

Academic Editor: Dirk Landgraf

Received: 9 November 2021
Accepted: 8 December 2021
Published: 13 December 2021

1. Introduction

With the continuous increase in atmospheric carbon dioxide emissions, among the three major chemical elements—carbon (C), nitrogen (N) and phosphorus (P)—required for plant growth, N and P [1–3] as mineral elements increasingly restrict plant growth.

N is the mineral element with the greatest demand for plant growth and development, and it is the basic component of macromolecules such as protein, nucleic acid and chlorophyll. In forests with short rotation period and rapid growth, the demand for N is large, and N deficiency is the main nutrient element that restricts plant growth [4]. There are two main forms of nitrogen absorption by plants: nitrate N and ammonium N. Nitrate N is transported from the soil to the cells by root nitrate transporters (NRTs) and ammonium is carried out using ammonium transporters (AMTs) through active transport, respectively. Different NRTs respond differently to N deficiency. In Arabidopsis, AtNRT1 and AtNRT2 are low-affinity and high-affinity transporters, respectively [5]. The absorption of nitrate ions by roots is mainly performed by NRT2.1 [6], and additionally, AtNRT2.4, which is reduced by N starvation, will also absorb some nitrate ions [7]. The transcript levels of AtNRT2.7 are smaller in roots than in shoots [8]. High transcript levels of AtNRT2.7 have been detected in Arabidopsis seeds [9,10]. The AtNRT2.5 transcript is induced by N starvation [11,12]. However, OsNRT2.4 is a dual-affinity nitrate transporter [13].

After being absorbed into the cell, a small amount of nitrate N is reduced in roots, and a large amount is transported to leaves for reduction in poplar [14]. Nitrate N is first reduced to ammonium N by nitrate reductase (NR) and nitrite reductase, and then further synthesized into amino acids through the glutamate synthase (GOGAT) and glutamine synthase (GS) cycles [15]. N deficiency reduces the activities of N assimilation enzymes such as NR and GOGAT in poplar [16]. Free amino acids are the main transport form of N in plants, which can reflect the supply and demand of N. In Arabidopsis, AtAAP3 is a plasma membrane amino acid transporter and is expressed in root phloem [17]. OsAAP3 can transport nine kinds of amino acids, and the high transcription level of OsAAP3 in rice increases the amino acid content and rice yield [18].

P is the second largest mineral element required for plant growth and development [3]. In forestland, the main sources of P are organic P and inorganic phosphorus (Pi). Current research shows that plants can only directly absorb Pi. However, due to absorbance by soil cations and low solubility in the rhizosphere, most Pi in the soil is unavailable to plants, which often causes forest trees to suffer from P deficiency [19,20]. Therefore, the selection of afforestation materials with poor P tolerance will increase the growth of forest.

The absorption and transport of Pi are regulated by different phosphate transporters (PHTs) in plants. The transcript level of HvPHT1;3 is higher in barley P-efficient strains [21]. AtPHT1;1 to AtPHT1;4 is responsible for the absorption of phosphate from the rhizosphere [22]. Different P transporters respond differently to P deficiency. The transcript level of PHO1 (PHOSPHATE 1) homologs H1 (PHO1;H1) is induced by P starvation but inhibited by sufficient P [23], while transcripts of AtPHO1;H9 are only found in pollen and flowers [24]. Excluding AtPHT1;6, the transcript level of all AtPHT1 genes increases under phosphate starvation [25]. PHT2;1 regulates the P starvation response [26].

Plants can increase the availability of soil Pi by secreting organic acids [27,28] and phosphatase such as acid phosphatases (APs) from the roots to the soil under P deficiency [16,29]. The phosphoenolpyruvate carboxylase (PEPC, EC 4.1.1.31) enzyme in the tricarboxylic acid cycle plays an important role in the synthesis of organic acids under P deficiency [30]. Knowledge about the mechanism of tolerance to deficiency of N and P in plants is mostly from studies on herbs and model plants. However, research on woody non-model plants is rarely reported.

To obtain sufficient C skeleton and energy, N and P assimilation requires consumption of carbohydrates [31,32]. Therefore, the effects of deficiency of N and P on plant growth can be measured not only by directly measuring dry weight but also by indirectly measuring the soluble sugar content. For example, in oilseed flax, N addition reduces the soluble sugar content of leaves, and P addition increases the soluble sugar content of leaves, stems and seeds [33]. The deficiency of N and P causes the leaves to accumulate more carbohydrates, and the transport of carbohydrates from the stem to the root increases [34].

F. mandshurica Rupr. is one of the three hard broad-leaved tree species in Northeast China and is regarded as one of the main tree species used in afforestation, representing 10% of planted hardwoods [35]. There have been many reports on the selection of the superior provenance of *F. mandshurica* based on growth rate. However, there is little information on the physiological and biochemical regulatory mechanisms related to the rapid growth of superior provenances. The C amount of 20 provenances of *F. mandshurica* in three provinces of northeastern China was determined, and Wuchang (WC) with the largest amount of C sequestration as well as Dailing (DL) with a low level were selected. To explore the effects of deficiency of P and N on growth and physiology of woody non-model plants, *F. mandshurica* annual seedlings of WC with fast growth and DL with slow growth were treated with N starvation (N-), P starvation (P-) and both N and P starvation (NP-). We hypothesized that (i) N- and P- would suppress the absorption and metabolism of N and P for both of *F. mandshurica* provenances and (ii) the tolerance to deficiency of N and P of fast-growing *F. mandshurica* provenance would be greater.

2. Materials and Methods

2.1. Plant Materials and Nutritional Treatments

This research was conducted in the greenhouse of Northeast Forestry University in Harbin, Heilongjiang, China. In the experiment, pots with 180 mm upper diameter, 130 mm lower diameter and 150 mm height were used uniformly. On 11 May 2500 seeds of Dailing (DL) and Wuchang (WC) were planted directly in pots filled with sand. To ensure the moisture in the sand, 200 mL water was poured every two days. On 28 May, there were 703 seedlings (305 for DL and 398 for WC) with two true leaves, and 480 seedlings (240 for DL and 240 for WC) with similar heights were selected and transplanted into pots with sand, with two seedlings in each nutrient pot. There were 60 seedlings at each level of nutrient treatment for each provenance. The study contained four nutrition treatment groups, namely, nitrogen starvation (N-), phosphorus starvation (P-), nitrogen and phosphorus starvation (NP-) and complete nutrition (Control). With reference to the composition of Hoagland nutrient solution and Jarkko and Toini nutrient solution, the composition of the nutrient solution in this study was as follows: 8 mM NH_4NO_3, 1 mM KH_2PO_4, 1 mM KCl, 1 mM $CaCl_2 \cdot 6H_2O$, 0.6 mM $MgSO_4 \cdot 7H_2O$, 0.02 mM $FeCl_3 \cdot 6H_2O$, 6 μM $MnCl_2 \cdot 4H_2O$, 166 μM H_3BO_3, 0.36 μM $ZnCl_2$, 0.36 μM $CuCl_2 \cdot 2H_2O$, 0.36 μM $NaMoO_4 \cdot 2H_2O$. P- was achieved by removing KH_2PO_4, and then KCl was used to adjust the potassium ion balance. The other components were the same as described for complete nutrition. N- was achieved by removing NH_4NO_3, and the other components were the same as described for complete nutrition. NP- was achieved by removing KH_2PO_4 and NH_4NO_3, and then KCl was used to adjust the potassium ion balance. The other components were the same as described for complete nutrition. To ensure the health of the tested seedlings, from 28 May to 29 June, we poured 200 mL complete nutrient solution into each pot every 3 days. To ensure the effectiveness of nutrient deficiency on the test seedlings, from 1 July to 5 August, we poured 200 mL treatment nutrient solution into the corresponding pots every 3 days. As the test duration extended through the hottest part of July, the water in the sand culture evaporated very quickly, and the nutrient pot was supplemented with enough water at 5 p.m. every day.

2.2. Sample Harvesting and Determination of Dry Weight and Leaf Area

On 5 August, the total foliage was cut off with branch shears, and the total foliar fresh weight (FFW) was weighed. The stems were then cut at the rhizome boundary, and the total stem fresh weight (SFW) was weighed. Finally, the sand on the roots was washed off with water, the water on the roots was absorbed by filter paper and the total root fresh weight (RFW) was determined. The subsamples of roots, stems and foliage were sampled, and the fresh weight of the subsamples was determined. All the subsamples were placed in an oven at 70 °C to dry until the dry weight of the subsamples no longer changed, and the dry weight was recorded. The total dry weight of root (RDW), stem (SDW) and foliage (FDW) was calculated by multiplying RFW, SFW and FFW by the ratio of the dry weight to the fresh weight of subsamples, respectively.

Simultaneously, other subsamples of roots and leaves were collected for physiological and enzyme activity determination and RNA extraction. These subsamples were precooled with liquid nitrogen and stored in a refrigerator at −80 °C.

The leaf area was calculated by the ratio of leaf mass to area. First, 20 small disks with the same size were punched out with a puncher, and the fresh weight of 20 small discs was measured and recorded as W0. The area of the 20 small discs was determined according to the circle area calculation formula and recorded as A0. The total leaf area was calculated by multiplying FFW by the ratio of A0 to W0.

2.3. Determination of the Content and Amount of C, N and P

The total N concentration was determined using the sulfuric acid–hydrogen peroxide digestion Kjeldahl method [36]. The total P concentration was determined using the sulfuric acid–hydrogen peroxide digestion and vanadium molybdenum yellow colorimetric methods [37]. The total organic matter concentration was determined using the potassium dichromate oxidation external heating method [38].

The amounts of N, P and C were calculated by multiplying the dry weight using the concentration of N, P and C. The total amounts of N, P and C of whole seedlings were obtained by adding the amounts of the elements from the roots, stems and leaves. The concentrations of C, N and P per plant were obtained by dividing the amounts of C, N and P by the dry weight, respectively. N utilization efficiency (NUE) and P utilization efficiency (PUE) were obtained by dividing the dry weight of individual seedlings by their amounts of N and P.

The root-to-shoot ratio of dry weight (DRS), N amount (NRS) and P amount (PRS) were calculated by dividing the dry weight, N amount and P amount of roots by that of aboveground (foliar and stem), respectively.

2.4. Determination of Physiological Traits

Chlorophyll content was determined using the alcohol extraction method [39]. The soluble sugar content was determined using the anthrone colorimetric method [40]. The nitrate ion content was determined by the nitrosalicylic acid colorimetric method [41]. The soluble protein content was determined with Coomassie Brilliant Blue G-250 method [42]. The free amino acids content was determined as described by Rosen with modifications [43]: 0.5 g fresh leaves or roots were boiled in 20 mL deionized water for 20 min. To remove the solid residue, the extraction solution was filtered with filter paper. To completely extract the remaining amino acids, the boiled solids were rinsed with distilled water 3 times, and the rinse solution was also filtered into the extraction solution. Finally, the volume of the extraction solution was adjusted to 100 mL with distilled water. Then, 0.5 mL extraction solution was added to 1 mL of the measurement solution containing 10 mM sodium acetate and 1.2% (m/v) ninhydrin. After boiling in water for 12 min, 5 mL 95% ethanol was added to the cooled reaction solution. To determine the free amino acid content, the absorbance of the mixed solution was determined at 570 nm.

2.5. Determination of Enzyme Activities

The determination of enzyme activities was divided into the extraction of crude enzyme solutions and determination of enzyme activity. The difference in enzyme molecular size and physiological activity led to specific extraction and activity determination methods for each enzyme.

The same extraction of crude enzyme solution from frozen tissues following the protocol of Turnao et al. [44] and Brugière et al. [45] with slight modifications was used for determination of enzyme activity of both glutamine synthetase (GS) and NADH-GOGAT (EC 1.4.1.14) enzyme. Approximately 0.5 g tissue was ground to a fine powder with a mortar and pestle in liquid nitrogen. The powder was transferred to an ice-cold mortar containing 2 mL extraction buffer (100 mM Tris-HCl, pH 7.6, 1 mM EDTA, 1.0 mM $MgCl_2$, 10 mM β-mercaptoethanol). The samples were incubated on ice for 15–30 min. Debris was removed from the sample by centrifugation at 13,000× g for 10 min. The supernatant was assayed for GS activity, as described by Guiz et al. [46]. The total volume of the enzyme reaction solution was 2.5 mL (100 mM Tris-HCl, pH 7.8, 50 mM sodium glutamate, 5 mM hydroxylamine hydrochloride, 50 mM magnesium sulfate and 20 mM adenosine triphosphate). Next, 0.3 mL crude enzyme solution was added to start the reaction. After 20 min, 1.5 mL ferric chloride reaction solution (0.67 M ferric chloride, 0.37 M hydrochloric acid and 20% (*w*/*v* trichloroacetic acid) was added to stop the reaction. The absorbance of the supernatant at 540 nm was measured after centrifugation (10,000× *g* for 5 min).

The determination liquor of NADH-GOGAT enzyme activity consisted of 5 components: 20 mM L-glutamine (A), 100 mM α-ketoglutarate (B), 10 mM KCl (C), 25 mM Tris-HCl buffer pH 7.6 (D) and 3 mM NADH (now equipped and used) (E). First, 0.05 mL B, 0.1 mL C and 1.95 mL D were added to the 4 mL centrifuge tube in 30 °C water for 5 min. To determine NADH-GOGAT activity, 0.2 mL E, 0.3 mL enzyme solution and 0.4 mL A were added to the reaction solution. The absorbance of the supernatant at 340 nm was measured and recorded, and when appropriate, it was poured back into the test tube in 30 °C water. The absorbance of the supernatant at 340 nm was measured and record at 340 nm again after 3 min, and the difference between the two absorbances was calculated.

Nitrate reductase (NR) crude enzyme solution was obtained by extracting 0.5 g fresh powder of roots or leaves in 4 mL phosphoric acid buffer (pH 8.7) containing 0.01 mM cysteine and 1 mM EDTA. After centrifugation (4000 rpm, 15 min at 4 °C), to determine the NR enzyme activity, 0.4 mL supernatant was added to 1.6 mL reaction solution (0.05 M KNO_3 phosphate buffer and 1 mg/mL NADH) in 25 °C water for 30 min. To terminate the reaction, 1 mL 1% (*w*/*v*) yellow amine solution and then 1 mL 0.02% (*w*/*v*) naphthyl vinylamine solution were added to the reaction solution sequentially, and the color reaction was developed for 15 min. The absorbance of the supernatant at 540 nm was measured after centrifugation (4000 rpm, 5 min).

The acid phosphatase (AP) crude enzyme solution was obtained by extracting 1.2 g fresh root powder in 8 mL 0.2 M sodium acetate buffer (pH 5.8) on ice for 5 min. After centrifugation (12,000 rpm, 15 min), to determine the AP enzyme activity, 1 mL supernatant was added to 2 mL 0.05 M p-nitrophenol disodium phosphate reaction solution in 37 °C water for 30 min. To terminate the reaction, 2 mL 0.5 M $CaCl_2$ and 2 mL 2 M NaOH were added to the reaction solution. The absorbance of the supernatant at 410 nm was measured after the mixed solution was centrifuged at 2500 rpm and 4000 rpm for 5 min, respectively.

Phosphoenolpyruvate carboxylase (PEPC) activity was determined with the kit (G0606W) supplied by Suzhou Grace Technology Co., Ltd. in accordance with kit instructions.

2.6. Determination of Gene Transcripts Involved in N and P Assimilation

Determination of the transcript level of the key genes (*NRT2.1*, *NRT2.4*, *NRT2.5*, *NRT2.7*, *NIA2*, *AAP3*, *PHT1-3*, *PHT1-4*, *PHO1;H1*, *PHO1;H9*, *PHT2-1*) in N and P assimilation first required the extraction of mRNA. In this study, the reverse transcription kit (AT341) of TransGen Biotech was used for RNA reverse transcription. According to the manufacturer's instructions, 0.5 µg of total RNA was reverse transcribed into single-stranded cDNA, and then the genomic DNA was digested with gDNA Remover. To verify whether the cDNA synthesis was successful, PCR of the tubulin gene was performed using single-stranded cDNA as the template. Supplementary Table S1 shows the gene-specific primers used for qPCR. Quantitative real-time PCR was performed using the ABI7500 system with TransStart Tip Green qPCR Super Mix (+Dye I/+Dye II) (TransGen Biotech, AQ142). According to the manufacturer's instructions, the reaction system was 20 µL, including 10 µL 2 × TransStart Tip Green qPCR SuperMix, 0.4 µL 10 µM forward primer and reverse primer, 1 µL cDNA and 8.2 µL nuclease-free water. The following program was used for qPCR amplification: initial denaturation at 95 °C for 30 s, followed by 40 cycles of 95 °C for 5 s and 60 °C for 30 s. The tubulin gene was used as endogenous reference gene to estimate the relative transcript levels of specific genes in three biological replications.

2.7. Statistical Analysis and Graphic Production

Analysis of variance, mean, mean error and multiple comparisons were determined using SPSS19.0. Unless otherwise specified, all significance levels in this study were at the $p < 0.05$ level. To estimate provenance, nitrogen, phosphorus and their interaction effects, the variance was decomposed according to the following model:

$$Y_{ijkm} = \mu + G_i + N_j + P_k + G_i \times N_j + G_i \times P_k + N_j \times P_k + G_i \times N_j \times P_k + \varepsilon_{ijkm} \quad (1)$$

where Y_{ijkm} represents the trait of i_{th} provenance, the j_{th} nitrogen level, the k_{th} phosphorus level and the m_{th} repeat (i = DL and WC; j = N starvation and complete nutrition; k = P starvation and complete nutrition; m = all of the positive integer within 60; μ represents the overall mean; G_i represents the i_{th} provenance effect; N_j represents the j_{th} nitrogen level effect; P_k represents the k_{th} phosphorus level effect; $G_i \times N_j$ represents the interaction effect between the i_{th} provenance and the j_{th} nitrogen level; $G_i \times P_k$ represents the interaction effect between the i_{th} provenance and the k_{th} phosphorus level; $N_j \times P_k$ represents the interaction effect between the j_{th} nitrogen level and the k_{th} phosphorus level; $G_i \times N_j \times P_k$ represents the interaction effect of the i_{th} provenance, the j_{th} nitrogen level and the k_{th} phosphorus level; ε_{ijkm} represents random error of the i_{th} provenance, the j_{th} nitrogen level, the k_{th} phosphorus level and the m_{th} repeat.

The transcript level of gene was calculated by analyzing the qPCR cycle threshold (CT). To obtain the delta CT value (ΔCT) of the target gene, the CT value of the internal reference gene was subtracted from the CT value of the target gene. The value calculated when the exponential function of 2 takes $-\Delta$CT as the exponent represents the multiple of the target gene to the internal reference gene. To ensure homogeneity of variance, we took the logarithm to base 2 of the multiple of the transcript of target gene to the internal reference gene.

The column charts in this study were drawn using Origin 8.0 and the line charts were drawn using Excel.

3. Results

3.1. Dry Weight, DRS, Foliar Area and Chlorophyll Content

Annual seedlings of two *F. mandshurica* provenances, Dailing (DL) and Wuchang (WC), were tested for N starvation (N-), P starvation (P-), N and P starvation (NP-) and complete nutrition (control) (Supplementary Figure S6). The results showed that the dry weight of WC was 2.45 times that of DL under P-, although there was no significant difference in the dry weight of annual seedlings between DL and WC under complete nutrition (Figure 1a). Compared with the control, N-, P- and NP- reduced the dry weight, and the reduction in DL was greater than that of WC. For example, the dry weight of DL and WC under P- was reduced to 25.6% and 69.2% of that under complete nutrition, respectively. The leaf area of WC was 1.95 times that of DL under P- (Figure 1c), although there was no significant difference in leaf area between DL and WC under complete nutrition. Compared with the control, N-, P- and NP- significantly reduced the leaf area. The leaf area of DL and WC under P- was reduced to 28.8% and 64.9% compared with that under complete nutrition. For chlorophyll content, DL was significantly greater than that of WC under complete nutrition (Figure 1d). Compared with the control, N-, P- and NP- reduced the chlorophyll content. The chlorophyll contents of DL and WC under N- and P- were reduced to 67.4%, 77.5% and 80.9%, 95% of that under complete nutrition, respectively. Moreover, WC was significantly greater than DL in terms of the dry-weight root-to-shoot ratio (DRS) under N- (Figure 1b). Compared with complete nutrition, P-, N- and NP- reduced DRS of DL and increased DRS of WC.

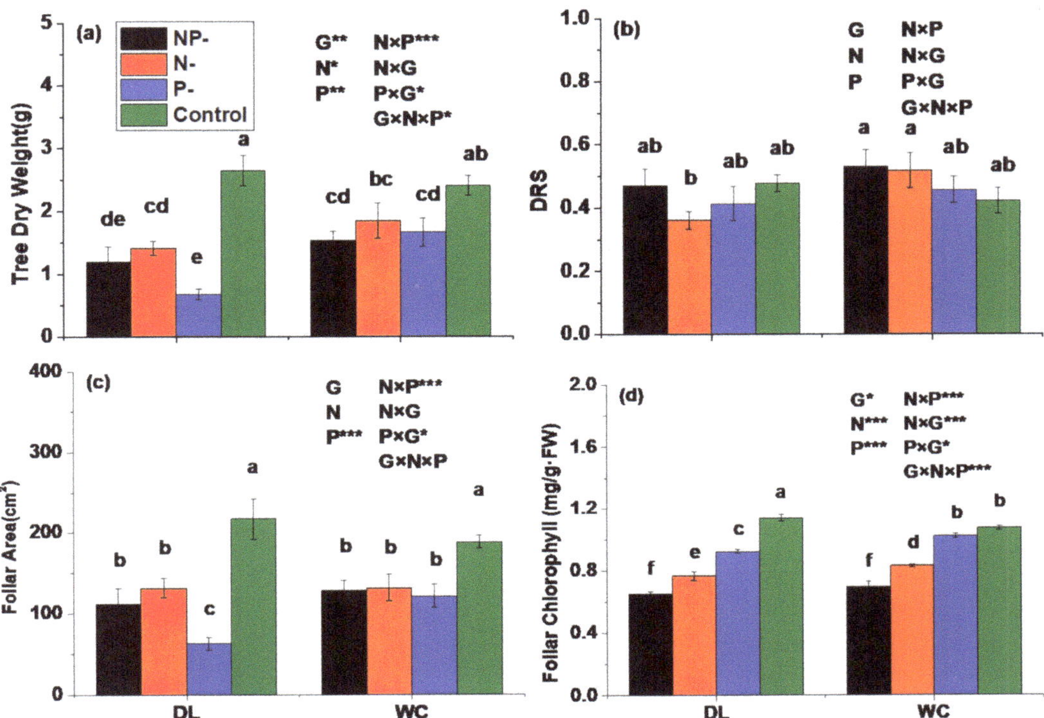

Figure 1. Dry weight (**a**) and leaf area (**c**) in whole trees, chlorophyll content (**d**) in leaves and ratio of root to shoot in dry weight (DRS (**b**)) for Dailing (DL) and Wuchang (WC) provenances of *F. mandshurica* under complete nutrients (control) and starvation of N (N-), P (P-) or both elements (NP-). Bars indicate means ± SE ($n = 13$). Different letters on the bars indicate significant differences. *p*-values obtained from the ANOVAs for provenances (G), nitrogen (N), phosphorus (P) and their interactions (N × P, N × G, P × G, N × P × G). * $p < 0.05$; ** $p < 0.01$; *** $p < 0.001$; no * means insignificant.

3.2. Nitrogen Status

N- and NP- significantly reduced the free amino acid content and nitrate ion content in root (Figure 2b,d), indicating that the N deficiency in sand culture was transformed into N deficiency in seedlings. However, in leaf, N- and NP- increased the nitrate ion content (Figure 2a) but decreased the free amino acid content (Figure 2c), indicating that the conversion of nitrate ions to amino acids was inhibited, consistent with the decrease in activity of N assimilation enzymes (Figure 3). Moreover, the NUE of WC was significantly greater than that of DL under N- (Figure 2g). For example, the N amount per seedling of WC was slightly greater than that of DL (Figure 2f), although the nitrate ion content in root and N concentration per seedling (Figure 2e) of DL under N- were significantly greater than that of WC. N- and NP- significantly increased the NUE of DL and WC. The increase in NUE of WC under P- was greater than that of DL. For example, the NUE of DL and WC under P- was 1.06 times and 1.69 times that of complete nutrition.

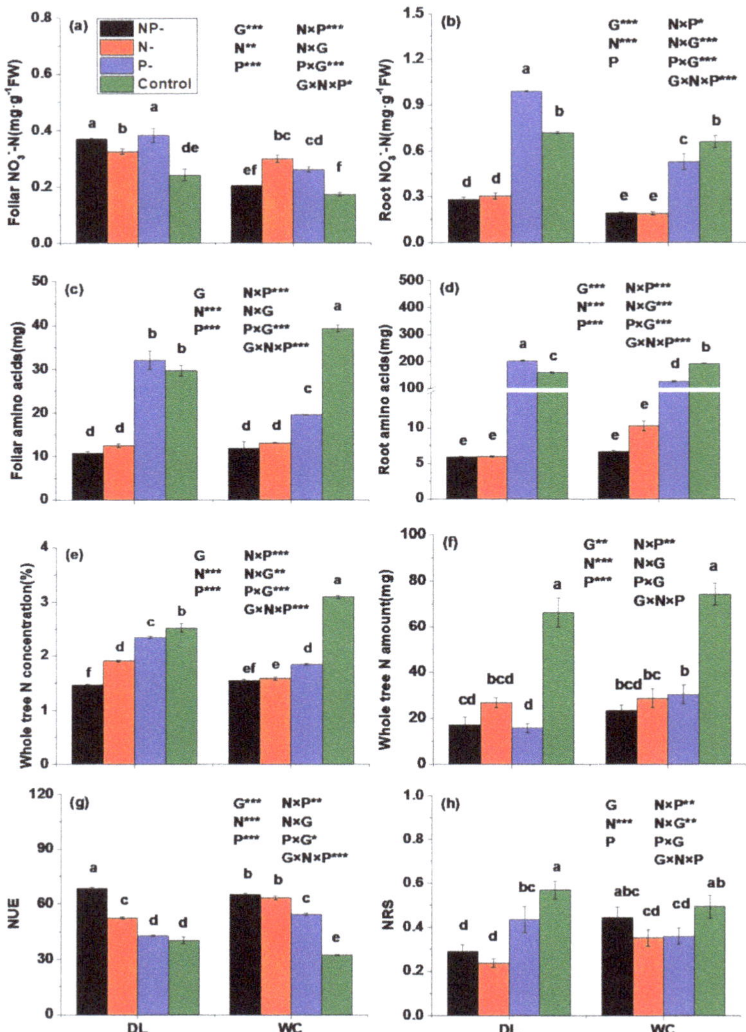

Figure 2. NO_3^- content and total free amino acids content in leaves (**a**,**c**) and roots (**b**,**d**), nitrogen concentration (**e**) and N amount (**f**) in whole trees, NUE (**g**) and ratio of root to shoot in N amount (NRS (**h**)) for Dailing (DL) and Wuchang (WC) provenances of *F. mandshurica* under complete nutrients (control) and starvation of N (N-), P (P-) or both elements (NP-). Bars indicate means ± SE (n = 13). Different letters on the bars indicate significant differences. p-values obtained from the ANOVAs for provenances (G), nitrogen (N), phosphorus (P) and their interactions (N × P, N × G, P × G, N × P × G). * $p < 0.05$; ** $p < 0.01$; *** $p < 0.001$; no * means insignificant.

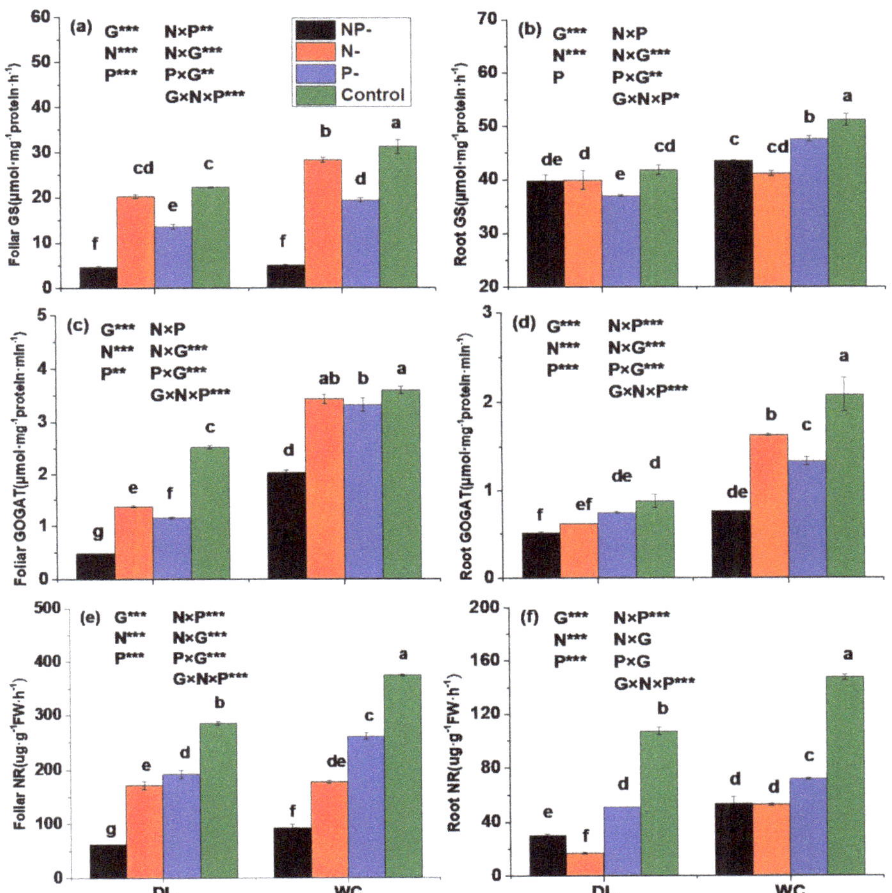

Figure 3. Activities of GS, GOGAT and NR in leaves (**a**,**c**,**e**) and roots (**b**,**d**,**f**) for Dailing (DL) and Wuchang (WC) provenances of *F. mandshurica* under complete nutrients (control) and starvation of N (N-), P (P-) or both elements (NP-). Bars indicate means ± SE (n = 13). Different letters on the bars indicate significant differences. p-values obtained from the ANOVAs for provenances (G), nitrogen (N), phosphorus (P) and their interactions (N × P, N × G, P × G, N × P × G). * $p < 0.05$; ** $p < 0.01$; *** $p < 0.001$; no * means insignificant.

The effects of P- on nitrate ion content and free amino acid content in DL and WC roots were different (Figure 2b,d). P- significantly increased the nitrate ion and amino acid contents of DL, but significantly reduced those of WC. The nitrate ion content and amino acid content in leaf and root and N concentration per plant of DL under P- were significantly greater than those of WC. However, due to the greater dry weight, the N amount of WC was significantly greater than that of DL, indicating that the NUE of WC was greater than that of DL under P-.

There was no significant difference in the N-amount root-to-shoot ratio (NRS) between DL and WC provenances under complete nutrition (Figure 2h). WC was significantly larger than DL in NRS and 1.53 times DL under NP-. Compared with the control, N-, P- and NP- reduced NRS, and the reduction in NRS of DL was greater than that of WC. NRS of DL, WC under N- and NP- was 41.8%, 71.5% and 51.2%, respectively, and 90.1% for the control.

3.3. Activities of Enzymes Involved in N Assimilation

Overall, the activities of N assimilation enzymes (NR, GS, GOGAT) in leaves and roots of seedlings of WC were significantly greater than those of DL under N-, P-, NP- and complete nutrition (Figure 3), indicating that the N assimilation capacity of WC was greater than that of DL. Compared with the control, N-, P- and NP- reduced the activity of N assimilation enzymes (NR, GS, GOGAT) in leaves and roots. However, P- decreased the GOGAT enzyme activity in leaves of DL more than WC. For example, the GOGAT enzyme activity of DL and WC under P- was reduced to 45.7% and 92.4% of the control, respectively (Figure 3c).

3.4. Phosphorus Status

Significant differences in P use efficiency (PUE) were observed between provenances N and P (Figure 4c). For PUE, WC was significantly greater than DL under N- and P-, which was consistent with the significantly greater P concentration of DL than WC under N- and P-, respectively (Figure 4a). However, there was no significant difference in the amount of P in annual seedlings between DL and WC under N- and P- (Figure 4b), which was attributed to the greater dry weight of WC than DL. Compared with complete nutrition, N-, P- and NP- significantly increased PUE, and the increase in WC was greater than that of DL. However, it significantly reduced the P concentration, and the reduction in WC was greater than that of DL and significantly reduced the P amount, and the reduction in DL was greater than that of WC. For example, PUE of DL and WC under P- was 1.69 and 2.69 times that of complete nutrition; P of DL and WC under P- was reduced to 15.1% and 25.8% of that under complete nutrition; and the P concentration was reduced to 58.9% and 37.2%, respectively. For the P-amount root-to-shoot ratio (PRS), there was no significant difference between DL and WC under complete nutrition (Figure 4d). PRS of WC was significantly greater than DL, which was 1.78 and 1.58 times that of DL under N- and NP-. Compared with the control, N- and NP- increased the PRS of WC by 24% and 12% and reduced PRS of DL by 25.8% and 26.9%, respectively.

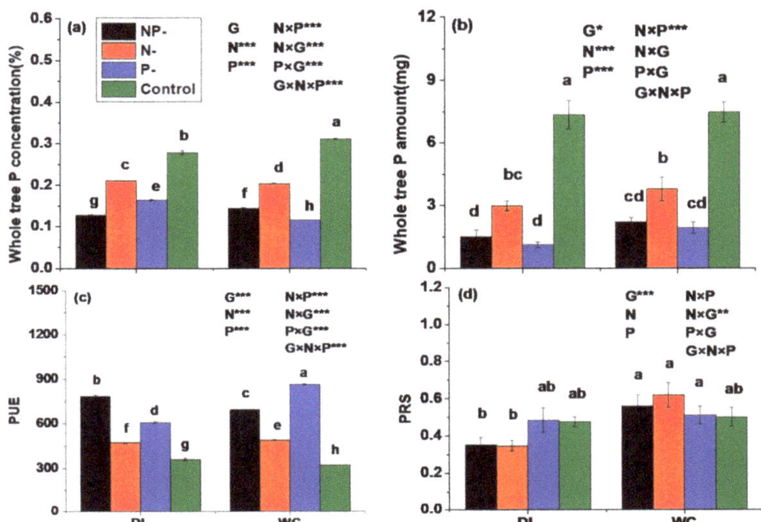

Figure 4. Phosphorus concentration (**a**) and P amount (**b**) in whole trees, PUE (**c**) and ratio of root to shoot in P amount (PRS (**d**)) for Dailing (DL) and Wuchang (WC) provenances of *F. mandshurica* under complete nutrients (control) and starvation of N (N-), P (P-) or both elements (NP-). Bars indicate means ± SE (n = 13). Different letters on the bars indicate significant differences. p-values obtained from the ANOVAs for provenances (G), nitrogen (N), phosphorus (P) and their interactions (N × P, N × G, P × G, N × P × G). * $p < 0.05$; ** $p < 0.01$; *** $p < 0.001$; no * means insignificant.

3.5. Activities of Enzymes Involved in P Assimilation

PEPC enzyme activity of WC was significantly greater than that of DL under P- and N-, but there was no significant difference between DL and WC under complete nutrition (Figure 5a). Compared with the control, P- significantly increased the PEPC activity. N- significantly reduced the PEPC enzyme activity. PEPC enzyme activity of DL and WC under N- (NP-) was 68.6% (60.1%) and 74.7% (61.5%) of the control, respectively. For the root AP enzyme activity, WC was significantly greater than that of DL under complete nutrition (Figure 5b). Compared with the control, P- increased the AP enzyme activity of DL and WC, and N- significantly reduced the AP enzyme activity. The AP enzyme activity of DL and WC under N- (NP-) was reduced to 21.1% (35.1%) and 29% (40.1%) of the control.

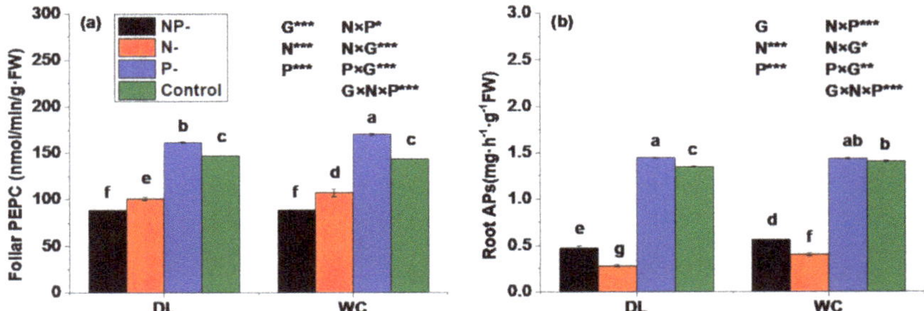

Figure 5. Activities of PEPC (**a**) in leaves and activities of APs (**b**) in roots for Dailing (DL) and Wuchang (WC) provenances of *F. mandshurica* under complete nutrients (control) and starvation of N (N-), P (P-) or both elements (NP-). Bars indicate means ± SE (n = 13). Different letters on the bars indicate significant differences. p-values obtained from the ANOVAs for provenances (G), nitrogen (N), phosphorus (P) and their interactions (N × P, N × G, P × G, N × P × G). * $p < 0.05$; ** $p < 0.01$; *** $p < 0.001$; no * means insignificant.

3.6. Carbon Status

The C amount was greater in of WC than DL under P- and NP- (Figure 6d), although the C concentration of WC was significantly reduced in WC than DL (Figure 6c), which was mainly due to the greater dry weight of WC compared with DL. N-, P- and NP- reduced the amount of C but increased the soluble sugar content in roots (Figure 6b), indicating that nutrition starvation inhibited C assimilation. Moreover, the inhibition of C assimilation of DL by nutrient starvation was greater than that of WC. The increase in soluble sugar content of DL was greater than that of WC. For example, the soluble sugar content in the roots of DL, WC under N- and P- was 1.88 times, 1.21 times and 1.53 times, respectively, and 1.31 times that of the control. The decrease in the C amount was greater in the DL than WC. For example, the C amount in the roots of DL, WC under N- and P- was 46.8%, 95.8% and 23.5%, respectively, compared with 49% for the control. Leaves are the organs that absorb and sequestrate C. N- significantly reduced the soluble sugar content in the leaves. However, the soluble sugar content in leaves of DL was insignificantly different from that of WC under complete nutrition, N-, P- and NP-.

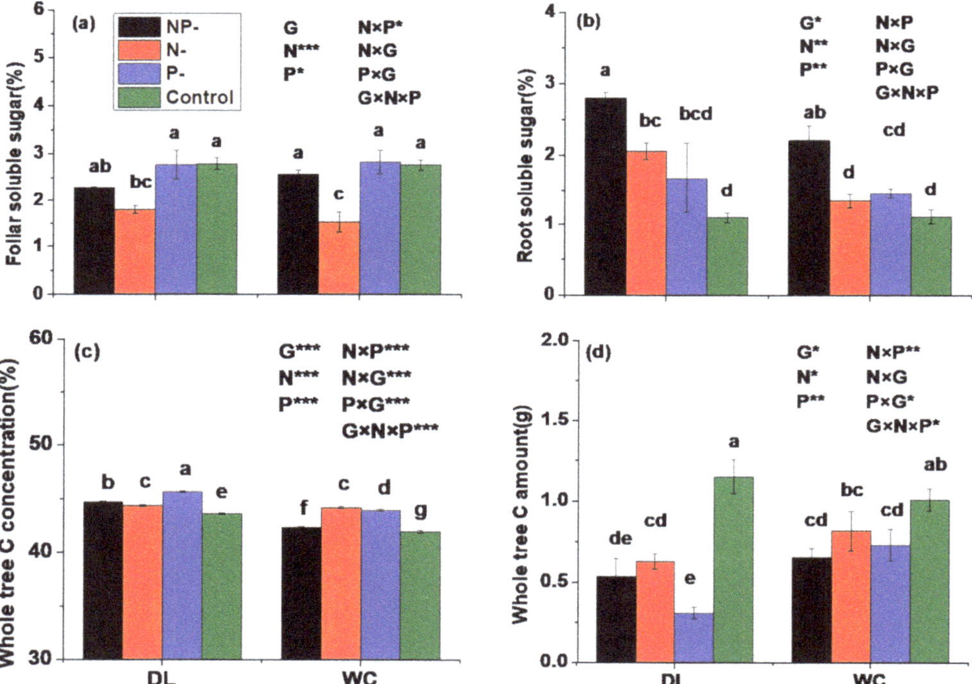

Figure 6. Soluble sugar content in leaves (**a**) and roots (**b**), carbon concentration (**c**) and C amount (**d**) in whole trees for Dailing (DL) and Wuchang (WC) provenances of *F. mandshurica* under complete nutrients (control) and starvation of N (N-), P (P-) or both elements (NP-). Bars indicate means ± SE (n = 13). Different letters on the bars indicate significant differences. p-values obtained from the ANOVAs for provenances (G), nitrogen (N), phosphorus (P) and their interactions (N × P, N × G, P × G, N × P × G). * $p < 0.05$; ** $p < 0.01$; *** $p < 0.001$; no * means insignificant.

3.7. Changes in Transcript Levels of Key Genes Involved in P/N Assimilation

Nitrate transporters (NRTs) and phosphate transporters (PHTs and PHOs) play a key role in the absorption and transport of N and P. The transcript levels of five phosphate transporters including PHT1-3, PHT1-4, PHT2-1, PHO1;H1 and PHO1;H9, four NRTs including NRT2.1, NRT2.4, NRT2.5, NRT2.7, two amino acid transporters including AVT1, AAP3 and one nitrate reductase [NADH] 2 (NIA2) were determined in leaves and roots (Figures 7 and 8).

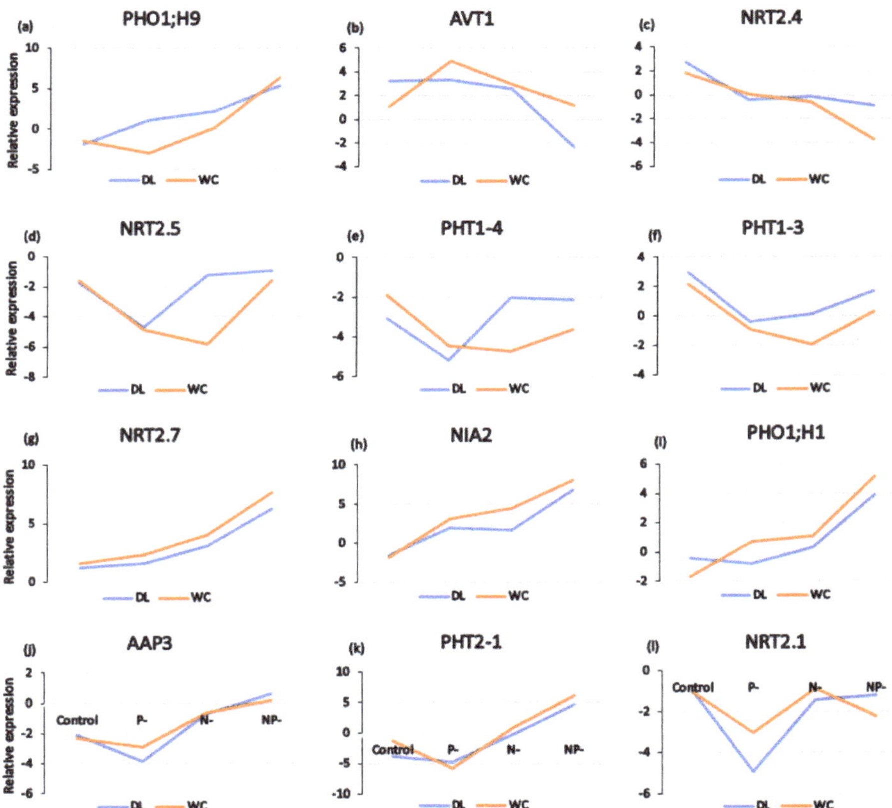

Figure 7. The transcriptional fold changes in genes involved in the uptake and assimilation of P and N in leaves for Dailing (DL) and Wuchang (WC) provenances of *F. mandshurica* under complete nutrients (control) and starvation of N (N-), P (P-) or both elements (NP-). (**a**) PHO1;H9, phosphate transporter PHO1 homolog 9; (**b**) AVT1, vacuolar amino acid transporter 1; (**c**) NRT2.4, high affinity nitrate transporter 2.4; (**d**) NRT2.5, high affinity nitrate transporter 2.5; (**e**) PHT1-4, inorganic phosphate transporter 1-4; (**f**) PHT1-3, inorganic phosphate transporter 1-3; (**g**) NRT2.7, high affinity nitrate transporter 2.7; (**h**) NIA2, nitrate reductase [NADH] 2; (**i**) PHO1;H1, phosphate transporter PHO1 homolog 1; (**j**) AAP3, amino acid transporter; (**k**) PHT2-1, inorganic phosphate transporter 2-1; (**l**), NRT2.1, high affinity nitrate transporter 2.1.

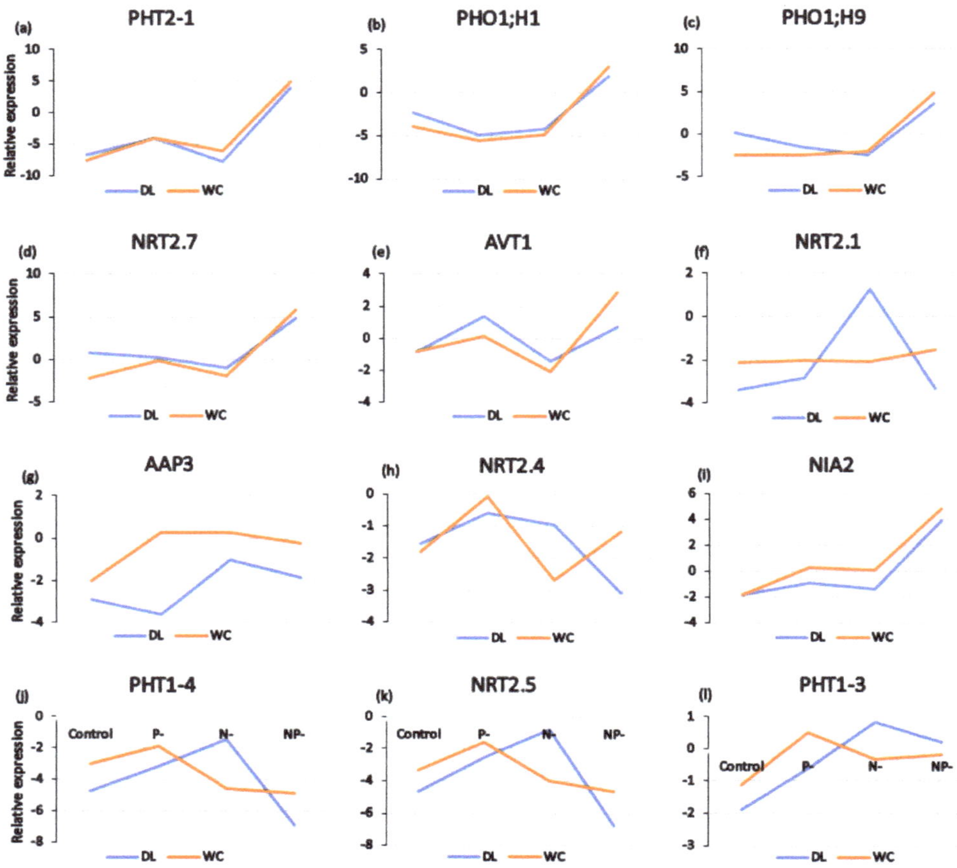

Figure 8. The transcriptional fold changes in genes involved in the uptake and assimilation of P and N in roots for Dailing (DL) and Wuchang (WC) provenances of *F. mandshurica* under complete nutrients (control) and starvation of N (N-), P (P-) or both elements (NP-). (**a**) PHT2-1, inorganic phosphate transporter 2-1; (**b**) PHO1;H1, phosphate transporter PHO1 homolog 1; (**c**) PHO1;H9, phosphate transporter PHO1 homolog 9; (**d**) NRT2.7, high affinity nitrate transporter 2.7; (**e**) AVT1, vacuolar amino acid transporter 1; (**f**) NRT2.1, high affinity nitrate transporter 2.1; (**g**) AAP3, amino acid transporter; (**h**) NRT2.4, high affinity nitrate transporter 2.4; (**i**) NIA2, nitrate reductase [NADH] 2; (**j**) PHT1-4, inorganic phosphate transporter 1-4; (**k**) NRT2.5, high affinity nitrate transporter 2.5; (**l**) PHT1-3, inorganic phosphate transporter 1-3.

The transcript levels of NRT2.1, NRT2.5, PHT1-3, PHT1-4 and AAP3 were greater in roots of WC than DL under complete nutrition (Figure 8f,g,j–l), indicating a greater absorption capacity of N and P in roots of WC than DL. P starvation indicates a relative increase in N, which increases the transcript levels of N assimilation genes in roots (NRT2.1, NRT2.4, NRT2.5, NRT2.7, AVT1, AAP3 and NIA2) (Figure 8d–k). P- also increases the transcript levels of PHT1-3, PHT1-4 and PHT2-1 (Figure 8a,j,l). However, the mRNA levels of PHO1;H1 and PHO1;H9 decreased under P- (Figure 8b,c).

Response patterns of transporters of N and P to deficiency of N and P in roots differed between DL and WC. Compared with DL, the transcript levels of PHT1-3, PHT1-4 and NRT2.5 were greater in roots of WC under P- but reduced under N-, indicating that WC and DL were relatively strong in response to P- and N-, respectively. Compared with DL, the transcript levels of AAP3, NRT2.1, NRT2.4, NIA2 and AAP3 were greater in roots of WC under P-, indicating that WC was relatively strong in response to P- in roots, but the mRNA levels of NRT2.1 and NRT2.4 were lower than in WC, indicating that DL was relatively strong in response to N- in roots.

Most of NRT and PHT genes of DL and WC had a similar transcript level in leaves under complete nutrition (Figure 7). Compared with complete nutrition in leaves, N- increased the transcript levels of AAP3, PHT2-1, PHO1; H1, PHO1; H9, NRT2.7 and NIA2 (Figure 7a,g–k) and decreased that of NRT2.4 (Figure 7c), while P- increased the transcript levels of NRT2.7 and NIA2 (Figure 7g,h) and reduced those of NRT2.1, NRT2.4, NRT2.5, PHT1-3, PHT1-4, PHT2-1 and AAP3 (Figure 7c–f,j–l), indicating that transporter genes were induced by N- and inhibited by P- in leaves, respectively.

Compared with DL, excluding PHO1;H9 and PHT1-3, the transcript levels of PHT1-4, NRT2.1, NRT2.7, NIA2, AAP3, AVT1 and PHO1;H1 were greater in leaves of WC under P-, indicating that WC was relatively strong in response to P- in leaves. However, compared with DL, the mRNA levels of NRT2.5, PHT1-3, PHT1-4 and PHO1;H9 were reduced in WC, but NRT2.1, NRT2.7, NIA2 and PHO1;H1 were elevated under N-, indicating that WC and DL responded differently to N- in leaves.

4. Discussion

N and P are the first and second largest mineral elements required for plant growth [47]. Deficiency of N or P is often the limiting factor for plant growth [48]. In this research, the effects of N and P deficiency on the growth and physiology of *F. mandshurica* was studied. Simultaneously, the relationship between growth rate and tolerance to deficiency of N and P was analyzed.

4.1. *Effects of N- and P- on the Growth, Physiology and Transcript Levels of N and P Assimilation Genes of F. mandshurica*

In this study, the free amino acid and nitrate ion contents in root were significantly reduced by N- and NP-, indicating that N deficiency in sand culture had been transformed into N deficiency in seedlings. The N concentration and N amount of the whole tree was significantly reduced. The leaves of seedlings of *F. mandshurica* turned smaller and yellow under N- (Supplementary Figures S4 and S5), which resulted from the decrease in chlorophyll content and caused the reduction in dry weight and leaf area of the whole tree. In maize, the chlorophyll content was significantly decreased by low-N stress [36]. However, in leaves, N- and NP- increased the nitrate ion content but decreased the free amino acid content, indicating that the conversion of nitrate ions to amino acids was inhibited, which was consistent with the decrease in activity of N assimilation enzymes (NR, GS and GOGAT). In poplar, N acquisition and assimilation was reduced by limiting the N supply [49]. Studies have shown that an increase in the activity of N-metabolizing enzymes is induced by short-term N deficiency but inhibited by long-term N deficiency [50].

To enhance the adaptation to N deficiency, externally, plants increase the absorption of N by increasing the distribution of underground biomass; internally, plants increase NUE by transporting N from senescent to vigorous growth parts [51,52]. Many members of the nitrate transporter family have been identified as N deficiency responsive genes, such as *AtNRT2.1* [6], *CmNRT2.1* [53], *AtNRT2.4* [7], *OsNRT2.4* [13] and *NRT2.5* [12], which play an important role in the absorption and transport of N. In this study, three genes (*AAP3*, *NRT2.7*, *NIA2*) in leaves and three genes (*NRT2.1*, *AAP3*, *NIA2*) in roots showed induced transcript levels by N- in the two provenances.

Seedlings of *F. mandshurica* were more affected by P- than N-. The sizes of the leaves became smaller under P- (Supplementary Figure S4), and the chlorophyll content was reduced by P-, which resulted in a significant decrease in dry weight. The decrease in plant growth rate and resistance caused by P deficiency was attributed to the decrease in content of P-containing compounds [54]. The P concentration and P amount of DL and WC under P- were reduced to 58.9% and 37.2% and to 15.1% and 25.8% compared with the values under complete nutrition.

P is often adsorbed by the cations of iron, aluminum and calcium salts in the soil, and thus, its direct use by plants is difficult [55]. Plants can increase the availability of P by secreting organic acids and phosphatase to decompose organic P into inorganic P and increase the solubility of inorganic P, respectively [56]. P starvation increases the activities of AP [29] and PEPC enzymes in Chinese fir [57]. In the present study, the activities of PEPC and AP enzymes were significantly increased by P-.

To enhance the adaptation to P deficiency, externally, plants increase the absorption of P by increasing the distribution of underground biomass, and internally, plants increase PUE by transporting P from senescent to vigorous growth parts [51,52]. P starvation increases the PUE in Chinese fir [57]. PUE of DL and WC under P- is 1.69 and 2.69 times that of complete nutrition. Phosphate transporters play an important role in the absorption and transport of P. The transcript levels of *PHT1.4* and *PHO1* genes in Chinese fir are increased by P starvation. In Arabidopsis, P deficiency increases the expression of PHO1;H1 [24,58] and PHT2;1 [26]. AtPht1;1 and AtPht1;4 play significant roles in Pi acquisition at both low- and high-Pi environments [22]. In the present study, P- increased the transcript levels of PHT1-3, PHT1-4 and PHT2-1 in roots and leaves. However, the mRNA levels of PHO1;H1 and PHO1;H9 decreased under P- in roots.

The coupled relationships between P and N are tightly linked under nutrient-limited conditions [59]. Plant growth depends not only on the absolute amount of N and P supply but also on the balance of N and P [60]. For example, certain forests are subjected to P deficiency driven by N deposition in Southwest Sweden [61]. N deposition could result in greater P deficiency in legume plantations compared with non-legume plantations [62]. In the present study, the transcript levels of N and P assimilation genes were not only regulated by the absolute amount of N and P supply but also the relative amount. P starvation indicates a relative increase in N, which increases the transcript levels of N assimilation genes in roots (*NRT2.1*, *NRT2.4*, *NRT2.5*, *NRT2.7*, *AVT1*, *AAP3* and *NIA2*) and leaves (*NRT2.7* and *NIA2*) but decreases the other genes in leaves (*NRT2.1*, *NRT2.4*, *NRT2.5* and *AAP3*).

C sequestration is the essence of plant growth. N induces new net primary production and C sequestration in global forests [63]. Therefore, the effect of deficiency of N and P on growth can be studied by measuring C sequestration. Deficiencies of N and P result in the accumulation of carbohydrate in leaves [64,65]. The content of leaf-soluble sugar is decreased by N but increased by P addition in oilseed flax, respectively [33]. In this study, N-, P- and NP- reduced the C amount but increased the soluble sugar content in roots, indicating that nutrition starvation increased the translocation of C to roots.

4.2. Relationship between the Growth Rate and Tolerance to N- and P- for F. mandshurica

In larch, a rapidly growing genotype is less tolerant to unbalanced nutritional conditions than a slow-growing genotype [66]. However, our research results showed that the tolerance of WC to nutrient deficiency was greater than DL, which resulted in a greater growth rate of WC than DL. The C amount was greater in WC with fast growth than DL with slow growth under P- and NP-, although the C concentration was significantly more reduced in WC than DL, which was mainly due to the greater dry weight of WC compared with DL. P- and NP- reduced the DRS of DL and increased the DRS of WC, indicating that a larger proportion of biomass was translocated below-ground by WC to obtain sufficient phosphorus. The availability of P for WC was greater than that of DL, as PEPC enzyme activity of WC was significantly greater than that of DL under P-. The high P- tolerance of WC was also related to the greater transcript levels of N and P transporters [16]. The transcript levels of PHT1-3 and PHT1-4 in roots and PHT1-4, NRT2.1, NRT2.7, NIA2, AAP3, AVT1 and PHO1;H1 were greater in leaves of WC than DL under P-, which strengthened WC compared with DL in N and P absorption and transport capacity.

Not only phosphorus but also nitrogen metabolism was affected by P deficiency. In poplar, the NUE, activities of NR and GOGAT enzymes and responsiveness of the key genes encoding N transporters is greater in the fast-growing genotype than the slow-growing genotype under P starvation [16]. Nitrate ion and amino acid contents in leaf and root and N concentration are significantly greater in DL on a per-plant basis under P- than WC.

However, due to the greater dry weight, the N amount was significantly greater in WC than DL, which was consistent with the greater NUE and assimilation enzymes (NR, GS, GOGAT) in leaves and roots of WC seedlings than DL under P-. Moreover, the transcript levels of N transport and assimilation genes (*NRT2.1*, *NRT2.4*, *NRT2.5*, *NIA2* and *AAP3*) were greater in roots of WC than DL under P-.

Compared with the control, WC performed better than DL under P-. Compared with complete nutrition, P- decreased the dry weight, C amount, P amount, leaf area, chlorophyll content and GOGAT enzyme activity, but the reduction in WC traits was smaller than DL. P- significantly increased the soluble sugar content, PUE and NUE, but the increase in WC was greater than in DL.

Compared with DL, the tolerance to N- was greater in WC, which was attributed to its greater NUE. The C amount was elevated in WC compared with DL under N-, although this difference was not significant, mainly due to the greater dry weight of WC compared with DL. Chlorophyll was the main factor in photosynthesis, and its content determined the growth rate. Compared with DL, the chlorophyll content was greater in WC. NUE was mainly related to the activities of N metabolizing enzymes and N transporters. The activities of N assimilation enzymes (NR, GS, GOGAT) were significantly greater in WC than DL under N-. However, the transcript levels of the N transporters could not fully support the greater tolerance of WC than DL to N-. Compared with DL, the transcript levels of NRT2.1, NRT2.7 and NIA2 were greater in leaves of WC, but NRT2.1, NRT2.4 and NRT2.5 levels were reduced in roots of WC under N-. Therefore, whether tolerance to nitrogen deficiency was greater in WC than DL requires further confirmation considering the transcript levels of N transporters. Moreover, the reduction in chlorophyll content and C amount caused by N- was greater in WC than in DL.

5. Conclusions

In summary, the growth rates of WC with fast growth and DL with slow growth under complete nutrition conditions was insignificantly different; under N and P deficiency, especially P deficiency, the dry weight, C amount, N amount and P amount were greater in WC than DL, which was related to the greater leaf area, chlorophyll content, DRS, NRS, PRS, NUE, PUE, N assimilation enzyme activity (NR, GS GOGAT), P absorption enzyme activity (PEPC and APs) and transcript levels of N and P transporter genes in WC compared with DL. Therefore, the growth rate of genetic types with high tolerance to N- and P- was greater than that with low tolerance.

Supplementary Materials: The following are available online at https://www.mdpi.com/article/10.3390/f12121760/s1, Figure S1: Soluble protein content in foliar (a) and roots (b) for Dailing (DL) and Wuchang (WC) provenances of *F. mandshurica* under complete nutrients and starvation of N, P or both elements. Bars indicate means ± SE (n = 13). Different letters on the bars indicate significant differences. P-values obtained from the ANOVAs for Provenances (G), nitrogen (N), Phosphorus (P) and their interactions (N × P, N × G, P × G, N × P × G). * $p < 0.05$; ** $p < 0.01$; *** $p < 0.001$; no * means insignificant, Figure S2: The concentration of carbon, nitrogen and phosphorus for root, shoot, foliar and whole tree of seedlings for Dailing (DL) and Wuchang (WC) provenances of *F. mandshurica* under complete nutrients and starvation of N, P or both elements. Bars indicate means ± SE (n = 13). Different letters on the bars indicate significant differences, Figure S3: Dry weight and amount of carbon, nitrogen and phosphorus for root, shoot, foliar and whole tree of seedlings for Dailing (DL) and Wuchang (WC) provenances of *F. mandshurica* under complete nutrients and starvation of N, P or both elements. Bars indicate means ± SE (n = 13). Different letters on the bars indicate significant differences. P-values obtained from the ANOVAs for Provenances (G), nitrogen (N), Phosphorus (P) and their interactions (N × P, N × G, P × G, N × P × G). * $p < 0.05$; ** $p < 0.01$; *** $p < 0.001$; no * means insignificant, Figure S4: Variation in leaf size and color of seedlings for Dailing (DL) and Wuchang (WC) provenances of *F. mandshurica* under complete nutrients and starvation of N or P, Figure S5: Variation in height of seedlings for Dailing (DL) and Wuchang (WC) provenances of *F. mandshurica* under complete nutrients and starvation of N or P, Figure S6: Seedlings for Dailing(DL) and Wuchang(WC) provenances of *F. mandshurica* under complete nutrients and starvation of N, P or both elements, Table S1: Primers used for qRT-PCR.

Author Contributions: X.Z. (Xingtang Zhao) selected samples, measured data, analyzed data and wrote the paper; Y.Z. conducted the overall design and approved the paper; F.Z. carried out the design and collected seeds; X.Z. (Xu Zhang), Z.L., Y.L., T.S., J.C., T.C. and J.L. participated in data determination. All authors have read and agreed to the published version of the manuscript.

Funding: This research was jointly supported the Fundamental Research Funds for the Central Universities of China (No. 2572017AA15) and Heilongjiang Province Applied Technology Research and Development Program Key Project (GA19B201).

Conflicts of Interest: The authors declare no conflict of interest.

References

1. Fleischer, K.; Rammig, A.; De Kauwe, M.G.; Walker, A.P.; Domingues, T.F.; Fuchslueger, L.; Garcia, S.; Goll, D.S.; Grandis, A.; Jiang, M.; et al. Amazon forest response to CO_2 fertilization dependent on plant phosphorus acquisition. *Nat. Geosci.* **2019**, *12*, 736–741. [CrossRef]
2. He, M.; Dijkstra, F.A. Phosphorus addition enhances loss of nitrogen in a phosphorus-poor soil. *Soil Biol. Biochem.* **2015**, *82*, 99–106. [CrossRef]
3. Schachtman, D.P.; Reid, R.J.; Ayling, S.M. Phosphorus uptake by plants: From soil to cell. *Plant Physiol.* **1998**, *116*, 447–453. [CrossRef] [PubMed]
4. Li, H.; Li, M.; Luo, J.; Cao, X.; Qu, L.; Gai, Y.; Jiang, X.; Liu, T.; Bai, H.; Janz, D.; et al. N fertilization has different effects on the growth, carbon and nitrogen physiology, and wood properties of slow- and fast-growing *Populus* species. *J. Exp. Bot.* **2012**, *63*, 6173–6185. [CrossRef]
5. Okamoto, M.; Vidmar, J.J.; Glass, A.D.M. Regulation of NRT1 and NRT2 gene families of Arabidopsis thaliana responses to nitrate provision. *Plant Cell Physiol.* **2003**, *44*, 304–317. [CrossRef]
6. Li, W.; Wang, Y.; Okamoto, M.; Crawford, N.M.; Siddiqi, M.Y.; Glass, A.D. Dissection of the AtNRT2.1:AtNRT2.2 inducible high-affinity nitrate transporter gene cluster. *Plant Physiol.* **2007**, *143*, 425–433. [CrossRef]
7. Kiba, T.; Feria-Bourrellier, A.B.; Lafouge, F.; Lezhneva, L.; Boutet-Mercey, S.; Orsel, M.; Brehaut, V.; Miller, A.; Daniel-Vedele, F.; Sakakibara, H.; et al. The *Arabidopsis* nitrate transporter NRT2.4 plays a double role in roots and shoots of nitrogen-starved plants. *Plant Cell* **2012**, *24*, 245–258. [CrossRef]
8. Wang, R.; Okamoto, M.; Xing, X.; Crawford, N.M. Microarray analysis of the nitrate response in *Arabidopsis* roots and shoots reveals over 1000 rapidly responding genes and new linkages to glucose, trehalose-6-phosphate, iron, and sulfate metabolism. *Plant Physiol.* **2003**, *132*, 556–567. [CrossRef]
9. David, L.C.; Dechorgnat, J.; Berquin, P.; Routaboul, J.M.; Debeaujon, I.; Daniel-Vedele, F.; Ferrario-Méry, S. Proanthocyanidin oxidation of *Arabidopsis* seeds is altered in mutant of the high-affinity nitrate transporter NRT2.7. *J. Exp. Bot.* **2014**, *65*, 885–893. [CrossRef]
10. Chopin, F.; Orsel, M.; Dorbe, M.F.; Chardon, F.; Truong, H.N.; Miller, A.J.; Krapp, A.; Daniel-Vedele, F. The *Arabidopsis* ATNRT2.7 nitrate transporter controls nitrate content in seeds. *Plant Cell* **2007**, *19*, 1590–1602. [CrossRef]
11. Kotur, Z.; Glass, A.D. A 150 kDa plasma membrane complex of AtNRT2.5 and AtNAR2.1 is the major contributor to constitutive high-affinity nitrate influx in *Arabidopsis thaliana*. *Plant Cell Environ.* **2015**, *38*, 1490–1502. [CrossRef]
12. Lezhneva, L.; Kiba, T.; Feria-Bourrellier, A.B.; Lafouge, F.; Boutet-Mercey, S.; Zoufan, P.; Sakakibara, H.; Daniel-Vedele, F.; Krapp, A. The *Arabidopsis* nitrate transporter NRT2.5 plays a role in nitrate acquisition and remobilization in nitrogen-starved plants. *Plant J. Cell Mol. Biol.* **2014**, *80*, 230–241. [CrossRef]
13. Wei, J.; Zheng, Y.; Feng, H.; Qu, H.; Fan, X.; Yamaji, N.; Ma, J.F.; Xu, G. OsNRT2.4 encodes a dual-affinity nitrate transporter and functions in nitrate-regulated root growth and nitrate distribution in rice. *J. Exp. Bot.* **2018**, *69*, 1095–1107. [CrossRef]
14. Fine, K.E.; Smith, M.W.; Cole, J.C. Partitioning of nitrogen, phosphorus, and potassium in redbud trees. *Sci. Hortic.* **2013**, *152*, 1–8. [CrossRef]
15. Fam, R.R.S.; Hiong, K.C.; Choo, C.Y.L.; Wong, W.P.; Chew, S.F.; Ip, Y.K. Molecular characterization of a novel algal glutamine synthetase (GS) and an algal glutamate synthase (GOGAT) from the colorful outer mantle of the giant clam, Tridacna squamosa, and the putative GS-GOGAT cycle in its symbiotic zooxanthellae. *Gene* **2018**, *656*, 40–52. [CrossRef]
16. Gan, H.; Jiao, Y.; Jia, J.; Wang, X.; Li, H.; Shi, W.; Peng, C.; Polle, A.; Luo, Z.B. Phosphorus and nitrogen physiology of two contrasting poplar genotypes when exposed to phosphorus and/or nitrogen starvation. *Tree Physiol.* **2016**, *36*, 22–38. [CrossRef]
17. Okumoto, S.; Koch, W.; Tegeder, M.; Fischer, W.N.; Biehl, A.; Leister, D.; Stierhof, Y.D.; Frommer, W.B. Root phloem-specific expression of the plasma membrane amino acid proton co-transporter AAP3. *J. Exp. Bot.* **2004**, *55*, 2155–2168. [CrossRef]
18. Lu, K.; Wu, B.; Wang, J.; Zhu, W.; Nie, H.; Qian, J.; Huang, W.; Fang, Z. Blocking amino acid transporter OsAAP3 improves grain yield by promoting outgrowth buds and increasing tiller number in rice. *Plant Biotechnol. J.* **2018**, *16*, 1710–1722. [CrossRef]
19. Fäth, J.; Kohlpaintner, M.; Blum, U.; Göttlein, A.; Mellert, K.H. Assessing phosphorus nutrition of the main European tree species by simple soil extraction methods. *For. Ecol. Manag.* **2019**, *432*, 895–901. [CrossRef]

20. Brown, K.R.; Courtin, P.J. Can phosphorus additions increase long-term growth and survival of red alder (*Alnus rubra* Bong.) on periodically dry sites? *For. Ecol. Manag.* **2018**, *430*, 545–557. [CrossRef]
21. Huang, C.Y.; Shirley, N.; Genc, Y.; Shi, B.; Langridge, P. Phosphate utilization efficiency correlates with expression of low-affinity phosphate transporters and noncoding RNA, IPS1, in barley. *Plant Physiol.* **2011**, *156*, 1217–1229. [CrossRef]
22. Shin, H.; Shin, H.S.; Dewbre, G.R.; Harrison, M.J. Phosphate transport in *Arabidopsis*: Pht1;1 and Pht1;4 play a major role in phosphate acquisition from both low- and high-phosphate environments. *Plant J. Cell Mol. Biol.* **2004**, *39*, 629–642. [CrossRef]
23. Kamalanathan, M.; Pierangelini, M.; Shearman, L.A.; Gleadow, R.; Beardall, J. Impacts of nitrogen and phosphorus starvation on the physiology of *Chlamydomonas reinhardtii*. *J. Appl. Phycol.* **2015**, *28*, 1509–1520. [CrossRef]
24. Wang, Y.; Ribot, C.; Rezzonico, E.; Poirier, Y. Structure and expression profile of the *Arabidopsis* PHO1 gene family indicates a broad role in inorganic phosphate homeostasis. *Plant Physiol.* **2004**, *135*, 400–411. [CrossRef]
25. Karthikeyan, A.S.; Varadarajan, D.K.; Mukatira, U.T.; D'Urzo, M.P.; Damsz, B.; Raghothama, K.G. Regulated expression of *Arabidopsis* phosphate transporters. *Plant Physiol.* **2002**, *130*, 221–233. [CrossRef]
26. Versaw, W.K.; Harrison, M.J. A chloroplast phosphate transporter, PHT2;1, influences allocation of phosphate within the plant and phosphate-starvation responses. *Plant Cell* **2002**, *14*, 1751–1766. [CrossRef]
27. Tang, H.; Chen, X.; Gao, Y.; Hong, L.; Chen, Y. Alteration in root morphological and physiological traits of two maize cultivars in response to phosphorus deficiency. *Rhizosphere* **2020**, *14*, 100201. [CrossRef]
28. Fang, X.-M.; Zhang, X.-L.; Chen, F.-S.; Zong, Y.-Y.; Bu, W.-S.; Wan, S.-Z.; Luo, Y.; Wang, H. Phosphorus addition alters the response of soil organic carbon decomposition to nitrogen deposition in a subtropical forest. *Soil Biol. Biochem.* **2019**, *133*, 119–128. [CrossRef]
29. Duff, S.M.G.; Sarath, G.; Plaxton, W.C. The role of acid phosphatases in plant phosphorus metabolism. *Physiol. Plant.* **1994**, *90*, 791–800. [CrossRef]
30. Penaloza, E.; Munoz, G.; Salvo-Garrido, H.; Silva, H.; Corcuera, L.J. Phosphate deficiency regulates phosphoenolpyruvate carboxylase expression in proteoid root clusters of white lupin. *J. Exp. Bot.* **2005**, *56*, 145–153. [CrossRef]
31. Xia, G.; Cheng, L. Foliar Urea Application in the Fall Affects Both Nitrogen and Carbon Storage in Young 'Concord' Grapevines Grown under a Wide Range of Nitrogen Supply. *J. Am. Soc. Hortic. Sci.* **2004**, *129*, 653–659. [CrossRef]
32. Cheng, L.; Fuchigami, L.H. Growth of young apple trees in relation to reserve nitrogen and carbohydrates. *Tree Physiol.* **2002**, *22*, 1297–1303. [CrossRef] [PubMed]
33. Yan, B.; Wu, B.; Gao, Y.; Wu, J.; Niu, J.; Xie, Y.; Cui, Z.; Zhang, Z. Effects of nitrogen and phosphorus on the regulation of nonstructural carbohydrate accumulation, translocation and the yield formation of oilseed flax. *Field Crop. Res.* **2018**, *219*, 229–241. [CrossRef]
34. Sanchez-Calderon, L.; Lopez-Bucio, J.; Chacon-Lopez, A.; Gutierrez-Ortega, A.; Hernandez-Abreu, E.; Herrera-Estrella, L. Characterization of low phosphorus insensitive mutants reveals a crosstalk between low phosphorus-induced determinate root development and the activation of genes involved in the adaptation of Arabidopsis to phosphorus deficiency. *Plant Physiol.* **2006**, *140*, 879–889. [CrossRef]
35. Jia, S.; Wang, Z.; Li, X.; Sun, Y.; Zhang, X.; Liang, A. N fertilization affects on soil respiration, microbial biomass and root respiration in *Larix gmelinii* and *Fraxinus mandshurica* plantations in China. *Plant Soil* **2010**, *333*, 325–336. [CrossRef]
36. Xin, J.S.; Zheng, L.; Huang, Y.R.; Pan, Y.D.; Lu, G.J. *NYT 2419-2013 Determination of Total Nitrogen in Plant, Automatic Kjeldahl Apparatus Method*; Ministry of Agriculture of the People's Republic of China: Beijing, China, 2014.
37. Bao, S.D. (Ed.) *Soil Agrochemical Analysis: Determination of Phosphorus in Plants*; China Agriculture Press: Beijing, China, 1999; pp. 268–270.
38. Yang, J.H. (Ed.) *Soil Agrochemical Analysis and Environmental Monitoring*; China Land Press: Beijing, China, 2008.
39. Wu, Y.-W.; Li, Q.; Jin, R.; Chen, W.; Liu, X.-L.; Kong, F.-L.; Ke, Y.-P.; Shi, H.-C.; Yuan, J.-C. Effect of low-nitrogen stress on photosynthesis and chlorophyll fluorescence characteristics of maize cultivars with different low-nitrogen tolerances. *J. Integr. Agric.* **2019**, *18*, 1246–1256. [CrossRef]
40. Hao, Y.-Q.; Lu, G.-Q.; Wang, L.-H.; Wang, C.-L.; Guo, H.-M.; Li, Y.-F.; Cheng, H.-M. Overexpression of AmDUF1517 enhanced tolerance to salinity, drought, and cold stress in transgenic cotton. *J. Integr. Agric.* **2018**, *17*, 2204–2214. [CrossRef]
41. Cataldo, D.A.; Maroon, M.; Schrader, L.E.; Youngs, V.L. Rapid colorimetric determination of nitrate in plant tissue by nitration of salicylic acid. *Commun. Soil Sci. Plant Anal.* **2008**, *6*, 71–80. [CrossRef]
42. Bradrord, M.M. A rapid and sensitive method for the quantitation of microgram quantities of protein utilizing the principle of protein-dye binding. *Anal. Biochem.* **1976**, *72*, 248–254. [CrossRef]
43. Rosen, H. A Modified Ninhydrin Calorimetric Analysis for Amino Acids. *Arch. Biochem. Biophys.* **1957**, *67*, 10–15. [CrossRef]
44. Turnao, F.J.; Muhitch, M.J. Differential accumulation of ferredoxin- and NADH-dependent glutamate synthase activities, peptides, and transcripts in developing soybean seedlings in response to light, nitrogen, and nodulation. *Physiol. Plant.* **1999**, *107*, 407–418. [CrossRef]
45. Brugière, N.; Dubois, F.; Limami, A.M.; Lelandais, M.; Roux, Y.; Sangwan, R.S.; Hirel, B. Glutamine Synthetase in the Phloem Plays a Major Role in Controlling Proline Production. *Plant Cell* **1999**, *11*, 1995–2011. [CrossRef]
46. Guiz, C.; Hirel, B.; Shedlofsky, G.; Gadal, P. Occurrence and influence of light on the relative proportions of two glutamine sythetases in rice leaves. *Plant Sci. Lett.* **1979**, *15*, 271–277. [CrossRef]

47. Wang, J.; Hui, D.; Lu, H.; Wang, F.; Liu, N.; Sun, Z.; Ren, H. Main and interactive effects of increased precipitation and nitrogen addition on growth, morphology, and nutrition of *Cinnamomum burmanni* seedlings in a tropical forest. *Glob. Ecol. Conserv.* **2019**, *20*, e00734. [CrossRef]
48. Zhao, Q.; Zeng, D.-H. Nitrogen addition effects on tree growth and soil properties mediated by soil phosphorus availability and tree species identity. *For. Ecol. Manag.* **2019**, *449*, 117478. [CrossRef]
49. Luo, J.; Li, H.; Liu, T.; Polle, A.; Peng, C.; Luo, Z.B. Nitrogen metabolism of two contrasting poplar species during acclimation to limiting nitrogen availability. *J. Exp. Bot.* **2013**, *64*, 4207–4224. [CrossRef]
50. Song, H.; Lei, Y.; Zhang, S. Differences in resistance to nitrogen and phosphorus deficiencies explain male-biased populations of poplar in nutrient-deficient habitats. *J. Proteom.* **2018**, *178*, 123–127. [CrossRef]
51. Netzer, F.; Pozzi, L.; Dubbert, D.; Herschbach, C. Improved photosynthesis and growth of poplar during nitrogen fertilization is accompanied by phosphorus depletion that indicates phosphorus remobilization from older stem tissues. *Environ. Exp. Bot.* **2019**, *162*, 421–432. [CrossRef]
52. Zheng, L.-L.; Zhao, Q.; Sun, Q.-Y.; Liu, L.; Zeng, D.-H. Nitrogen addition elevated autumn phosphorus retranslocation of living needles but not resorption in a nutrient-poor *Pinus sylvestris* var. Mongolica plantation. *For. Ecol. Manag.* **2020**, *468*, 118174. [CrossRef]
53. Gu, C.; Song, A.; Zhang, X.; Wang, H.; Li, T.; Chen, Y.; Jiang, J.; Chen, F.; Chen, S. Cloning of chrysanthemum high-affinity nitrate transporter family (CmNRT2) and characterization of CmNRT2.1. *Sci. Rep.* **2016**, *6*, 23462. [CrossRef]
54. Talkner, U.; Meiwes, K.J.; Potočić, N.; Seletković, I.; Cools, N.; De Vos, B.; Rautio, P. Phosphorus nutrition of beech (*Fagus sylvatica* L.) is decreasing in Europe. *Ann. For. Sci.* **2015**, *72*, 919–928. [CrossRef]
55. Clarkson, D.T.; Hanson, J.B. The Mineral Nutrition of Higher Plants. *Ann. Rev. Plant Physol.* **1980**, *31*, 239–298. [CrossRef]
56. Gerke, J. The acquisition of phosphate by higher plants: Effect of carboxylate release by the roots. A critical review. *J. Plant Nutr. Soil Sci.* **2015**, *178*, 351–364. [CrossRef]
57. Chen, Y.; Nguyen, T.H.N.; Qin, J.; Jiao, Y.; Li, Z.; Ding, S.; Lu, Y.; Liu, Q.; Luo, Z.-B. Phosphorus assimilation of Chinese fir from two provenances during acclimation to changing phosphorus availability. *Environ. Exp. Bot.* **2018**, *153*, 21–34. [CrossRef]
58. Stefanovic, A.; Ribot, C.; Rouached, H.; Wang, Y.; Chong, J.; Belbahri, L.; Delessert, S.; Poirier, Y. Members of the PHO1 gene family show limited functional redundancy in phosphate transfer to the shoot, and are regulated by phosphate deficiency via distinct pathways. *Plant J. Cell Mol. Biol.* **2007**, *50*, 982–994. [CrossRef]
59. You, C.; Wu, F.; Yang, W.; Xu, Z.; Tan, B.; Zhang, L.; Yue, K.; Ni, X.; Li, H.; Chang, C.; et al. Does foliar nutrient resorption regulate the coupled relationship between nitrogen and phosphorus in plant leaves in response to nitrogen deposition? *Sci. Total Environ.* **2018**, *645*, 733–742. [CrossRef]
60. Li, J.; Guo, Q.; Zhang, J.; Korpelainen, H.; Li, C. Effects of nitrogen and phosphorus supply on growth and physiological traits of two *Larix* species. *Environ. Exp. Bot.* **2016**, *130*, 206–215. [CrossRef]
61. Almeida, J.P.; Rosenstock, N.P.; Forsmark, B.; Bergh, J.; Wallander, H. Ectomycorrhizal community composition and function in a spruce forest transitioning between nitrogen and phosphorus limitation. *Fungal Ecol.* **2019**, *40*, 20–31. [CrossRef]
62. Chen, H.; Chen, M.; Li, D.; Mao, Q.; Zhang, W.; Mo, J. Responses of soil phosphorus availability to nitrogen addition in a legume and a non-legume plantation. *Geoderma* **2018**, *322*, 12–18. [CrossRef]
63. Du, E.; de Vries, W. Nitrogen-induced new net primary production and carbon sequestration in global forests. *Environ. Pollut.* **2018**, *242 Pt B*, 1476–1487. [CrossRef]
64. Linkohr, B.I.; Williamson, L.C.; Fitter, A.H.; Leyser, H.M. Nitrate and phosphate availability and distribution have different effects on root system architecture of Arabidopsis. *Plant J. Cell Mol. Biol.* **2002**, *29*, 751–760. [CrossRef] [PubMed]
65. Vance, C.P.; Stone, C.U.; Allan, D.L. Phosphorus acquisition and use critical adaptations by plants for securing a nonrenewable resource. *New Phytol.* **2003**, *157*, 423–447. [CrossRef] [PubMed]
66. Li, J.; Wu, G.; Guo, Q.; Korpelainen, H.; Li, C. Fast-growing *Larix kaempferi* suffers under nutrient imbalance caused by phosphorus fertilization in larch plantation soil. *For. Ecol. Manag.* **2018**, *417*, 49–62. [CrossRef]

MDPI

Article

Site-Effects Dominate the Plant Availability of Nutrients under Salix Species during the First Cutting Cycle

Piotr Koczorski [1,2], Bliss Furtado [2], Katarzyna Hrynkiewicz [2], Michelle Breezmann [1], Martin Weih [3] and Christel Baum [1,*]

[1] Soil Science, Faculty of Agricultural and Environmental Sciences, University of Rostock, 18051 Rostock, Germany; p.koczorski@doktorant.umk.pl (P.K.); michelle.breezmann@posteo.net (M.B.)

[2] Department of Microbiology, Faculty of Biological and Veterinary Sciences, Nicolaus Copernicus University, 87-100 Torun, Poland; bliss.furtado@umk.pl (B.F.); hrynk@umk.pl (K.H.)

[3] Department of Crop Production Ecology, Swedish University of Agricultural Sciences, 750 07 Uppsala, Sweden; martin.weih@slu.se

* Correspondence: christel.baum@uni-rostock.de; Tel.: +49-381-498-3100; Fax: +49-381-498-3122

Abstract: Fast-growing willows (*Salix* spp.) provide alternative sources of renewable energy generation, but need an adequate nutrient availability in the soil for high biomass production. In general, species mixtures can be more nutrient-efficient than pure cultures, but this is scarcely known for *Salix* spp. Therefore, this study evaluates the nutrient availability and P mobilization under two willow species, *Salix dasyclados* var. 'Loden' and *S. schwerinii* × *viminalis* var. 'Tora', grown as pure and mixed cultures at non-fertilized former arable sites in Germany (Stagnic Cambisol) and Sweden (Vertic Cambisol). The plant availability of potassium (K), magnesium (Mg) and phosphorus (P) and soil phosphatase activities in the topsoil were measured in spring of the year of planting (initial) and under 4 years-old stocks (one year after the first 3-year cutting cycle). The initial plant availability of the nutrients significantly differed between the sites and the two sampling dates at both sites. The plant availability of K and Mg was optimal to high at both sites and sampling dates, but rather low for P (after 4 years $\leq$5 mg P 100 g^{-1} soil). The plant-available P and K content in soil significantly decreased within the 4 years of willow growth at both sites. The acid and alkaline phosphatase activity in the soil of the German site (Rostock) was significantly lower after 4 years of willow growth, but differed not significantly between the two sampling dates at the Swedish site (Uppsala). Higher activity of acid phosphatase compared to alkaline phosphatase was recorded in the soils at both test sites based on the site-specific soil pH (<7). The slight decrease of plant availability of P after 4 years of *Salix* growth in pure culture differed not significantly between the different species. Mixed growth did not decrease the plant availability of P within this period, although no significant difference in the biomass production of pure and mixed growth was observed. This was valid at both sites, and therefore, seems independent of the site-specific differences in soil and climate conditions. The general validity of the assumptions should be tested also for other species mixtures and soil conditions in the future before site-adapted growth designs can be recommended in biomass production of *Salix*.

Keywords: short rotation coppice; phosphatase activity; nutrient content; growth stages; biomass; willow; *Salix*

Citation: Koczorski, P.; Furtado, B.; Hrynkiewicz, K.; Breezmann, M.; Weih, M.; Baum, C. Site-Effects Dominate the Plant Availability of Nutrients under Salix Species during the First Cutting Cycle. *Forests* **2021**, *12*, 1226. https://doi.org/10.3390/f12091226

Academic Editor: Dirk Landgraf

Received: 29 July 2021
Accepted: 7 September 2021
Published: 9 September 2021

1. Introduction

Short rotation coppices (SRCs) with poplar (*Populus*) and willows (*Salix*) species can be established on many types of land, including marginal lands that are unsuitable for agriculture [1,2]. SRCs offer a promising contribution to fuel wood supply, providing an alternative to fossil fuels and other nonrenewable resources [3]. The harvest from SRCs are usually used as feedstock in combined heat and power plants for energy generation [4,5]. Moreover, SRC plantations have ecological benefits compared to annual crops. They improve the soil water retention, enhance biodiversity in comparison to agricultural monocultures, improve water quality, protect ground water, prevent soil erosion and it is a low-input agricultural practice, thus implying low greenhouse gas emissions due to limited applications of chemicals [6–10] SRCs have been investigated intensively for the last 20 years (reviewed by Rödl [11]). SRC are managed using agricultural techniques, including high-density plantings and a regular cutting every 2 to 6 years without replanting [12]. The choice of tree species for SRC is generally confined to fast-growing tree species, such as those from the genera *Populus* and *Salix* [13]. In Europe, poplar and willow (*Populus* and *Salix*) belong to the natural vegetation of the softwood floodplain forest [14,15]. Poplar is typical for Western and Central European floodplains, where willow is more tolerant to low temperatures and can grow further north and in continental climates in the east of Europe [16].

Willows are deciduous trees or shrubs and comprise 330–500 species around the world [17]. They adapt in cool climates and high altitudes or on wet soils [18]. Willows require sufficient moisture supply during site establishment, while in later plantation stages, they can adapt to dry environments with heat and drought stress [17]. Studies have proven that willows have high tolerance to marginal or contaminated soils [6,8,19]. Willow SRCs are gaining increasing interest, because of their efficient and sustainable land use in combination with a growing demand for biofuel resources [20]. In Sweden, willow coppices are often used for phytoremediation where wastewaters or sewage sludge is applied to plantations in order to reduce pollutants or excess nutrients in the water [21].

Previous studies have reported that the productivity in SRCs is determined mainly by the soil fertility [22], soil pH (usually 5–7.5, but willow and poplar are tolerant to pH outside this range) [23], climatic conditions, nutrient and water availability [24], plant species and plantation density [25]. The need for fertilizers in SRCs is small compared to conventional agricultural crops [26]. SRC yields can be maximized by establishing them at fertile soils or by applying organic fertilizers, such as slurry, digested sewage sludge, manure from biogas plants [27]. However, most of these products contain high levels of nitrogen and phosphorus (P), which is risky for the leaching of nitrate and phosphate. Many studies have shown that willows and poplars have high evapotranspiration rates and are able to uptake large amounts of nutrients present in waste, thus allowing significant wastewater disposal over the growing season [28]. Furthermore, some microorganisms, mainly bacteria, are capable of converting phosphates (through solubilization and mineralization processes), and in turn supplying P to plants [29]. Soil enzymes such as acid phosphatases and alkaline phosphatases aid in enzymatically mineralizing P hydrolytically [29]. These enzymes improve the P supply to plants and strengthen the activity of many beneficial microorganisms in the adjacent soil [29]. Additionally, studies have shown that phosphatases are involved in plant growth promotion, activity against plant pathogens, waste remediation and metal recovery [30–33].

The cultivation area of SRCs is expected to increase in many European countries, such as Sweden [27], Germany [34], Ireland [35] and England [25]. As the land under SRC cultivation increases, information on the potential impact of SRC on soil quality and nutrient use efficiency is needed. Early decreases in the nutrient availability under *Salix purpurea* (cv. Hotel) within the first cutting cycle were described from a Canadian site by Ens et al. [36].

Thus far, pure cultures of one *Salix* species are the common praxis, but mixed growth was tested during the last years [37]. Species mixtures can be more efficient in the nutrient mobilization [38], since they combine e.g., different microbial communities in the rhizosphere [39]. However, the impact of mixed growth on the nutrient availability was scarcely tested for *Salix* spp. thus far. We hypothesize that species mixtures of *Salix* can have a higher nutrient mobilization and soil nutrient availability than pure stands by their higher microbial diversity, and thereby, activity in the rhizosphere.

Therefore, the main aims of this study were: (i) to evaluate the impact of growth of *Salix* species/varieties in pure and mixed cultures on the nutrient availability in the soil at two test sites with different soil and climate conditions; and (ii) to analyze the effects of mixed vs. pure growth on soil enzymatic P mobilization by phosphatases.

2. Material and Methods

2.1. Study Sites and Soil Sampling

The SRCs selected for this study are among two of the three experimental field sites of the ECOLINK-*Salix* project. The goal of this project is to investigate the relationship between genotype diversity, genotype identity, productivity and ecosystem function [37]. The two SRC sites are located in Uppsala in Central Sweden (59°49′ N 17°39′ E) and Rostock in Northern Germany (54°02′ N 12°05′ E). These two field sites with different climatic and soil conditions were selected to test whether the effects of *Salix* species and mixture are common or limited to defined conditions only. The dominating soil type at the site in Uppsala is a Vertic Cambisol and was previously arable farmland. The area of 4147 m^2 is divided into 45 plots (92.16 m^2 in size). In 2014, 6480 trees of four different species/varieties were planted on this site. The species/variety pool of the trial in Uppsala includes four different *Salix* varieties partly belonging to different species [37], of which the pure and mixed culture for two of them was selected for the present study: *Salix dasyclados* var. 'Loden' (acronym L) and *S. schwerinii* × *viminalis* var. 'Tora' (T), based on their significant physiological differences [37]. The mean annual precipitation sum between March and October of 2014 to 2016 was 374 mm and the mean annual temperature between March and October of 2014 to 2016 was 11.0 °C in Uppsala. Samples from the trial in Uppsala were collected in an early stage (initial plantation year, 2014) and late plantation stage (after the first cutting cycle, 2018) of the plantation.

The site in Northern Germany near Rostock was previously used as arable farmland and is established on a Stagnic Cambisol. The area of 829 m^2 is smaller than in Uppsala, due to space and funding restrictions, which strongly compromised the trial size [37]. However, the plot size remained the same as in Uppsala, resulting in nine plots in Rostock. As a consequence of the smaller size of the trial area, the number of trees planted in 2014 was reduced to 1296, comprising only two species/varieties of *Salix* [Tora (T) and Loden (L)] instead of four [39,40]. The mean annual precipitation sum between March and October of 2014 to 2016 was lower in Rostock than Uppsala, with 281 mm, whereas the mean annual temperature between March and October of 2014 to 2016 was higher at 13.3 °C. Samples from the trial in Rostock were collected in an early stage (initial plantation year, 2014) and late plantation stage (after the first cutting cycle, 2018) of the plantation.

2.2. Planting Design and Sample Collection

In preparation for the experiment, both sites were treated with Roundup (glyphosate, 4 L ha^{-1}) in order to kill any existing weeds in the trial areas, which were subsequently cultivated with a rotavator prior to planting [37]. The planting of the 18 cm long stem cuttings was carried out manually [37]. All the cuttings were obtained from the same stock and were soaked in water for two days before being planted in such a way that the tips of the cuttings were flush with the surface [37]. In the beginning, the trial sites were weeded by hand; later, the weeds were controlled by mowing between the rows of plants when necessary [37]. No additional nutrient fertilizers were applied [37].

The planting set up on both the sites was a randomized block design with three replicates (blocks). The blocks in Uppsala have 15 plots each (i.e., four species/varieties and three replicates), whereas the blocks in Rostock consist of three plots [40]. The four (Uppsala) or two (Rostock) species/varieties of *Salix* were planted in every possible combination. Thus, some plots were planted with only one variety (e.g., L or T) pure cultures, some with mixtures of two varieties (e.g., LT) and, in Uppsala, even plots with three or four varieties were planted [37].

The patterns in which the cuttings were planted differed according to the number of species/varieties in the plots: if there were two species/varieties, they were planted in a checker board pattern; if there were three or four species/varieties, their planting positions were randomized, although with the single restriction that no two individuals of the same species/variety should be planted directly next to each other in one row [37]. Twelve rows of twelve plants were fit into the 9.6 m × 9.6 m plots with every other row being set off [37]. This led to a hexagonal planting pattern with 0.8 m between every plant.

The selected two *Salix* species/varieties display contrasting characteristics (Weih and Nordh, 2002). For example, *S. schwerinii* × *viminalis* var. 'Tora' (T) is generally high-performing but less stress resistant, while *S. dasyclados* var. 'Loden' (L) is sturdier and more stress-tolerant.

Soil samples were taken with a soil corer (3 cm diameter) down to 10 cm soil depth with five replicates per plot in spring 2014 and 2018. The early plantation stage was defined as the year of establishment of the short rotation coppice (initial, 2014). The late plantation stage was defined as the year after the first cutting cycle (after four years of growth, 2018).

This soil depth was selected, since the highest fine root density of *Salix* spp. was revealed in this range [41], and therefore, the highest soil ecological impacts were assumed at this depth. For the soil chemical analyses, soil was sieved <2 mm. Soil phosphatases were measured in fresh wet soil. All other soil chemical properties were measured in air-dried soil.

2.3. Biochemical Analyses of Soil

The activity of acid and alkaline phosphatases in the soil were determined colorimetrically according to [42]. The enzyme activities were expressed as µg p nitrophenol (pNP) g^{-1} soil h^{-1} released from the pre-given substrate solution (p-nitrophenyl-phosphate) within one hour of incubation in modified universal buffer with pH 6.5 (for acid phosphatases) and pH 11 (for alkaline phosphatases) in April 2014 and April 2018.

2.4. Chemical Analyses of Soil

The total carbon (TC) and total nitrogen (TN) concentrations of soil samples were determined by dry combustion using a VARIO EL analyzer (Vario EL Fa. Foss Heraeus, Hanau, Germany). The concentration of SOC was valued by deducting the separately determined inorganic C (dissolution with HCl and volumetric CO_2-determination) from the concentration of TC.

The soil pH was measured in a 0.01 M $CaCl_2$ solution using a soil:solution ratio of *w*/*v* 1/2.5. Double lactate-extractable P (P_{dl}), Mg (Mg_{dl}) and K (K_{dl}) were considered to be the plant-available P fractions (e.g., [43]) and were determined by extracting P, Mg and K from 12 g soil with 150 mL lactate solution ($C_6H_{10}CaO_6$ * H_2O + 10 N HCl) according to [44]. Concentrations of P, Mg and K were determined with inductively coupled plasma-optical emission spectroscopy (ICP-OES, Optima 8300, Perkin Elmer, Waltham, MA, USA) at wavelengths of 214.914 nm, 285.213 nm and 766.490 nm, respectively.

2.5. Biomass Measurements

During the late winter of 2016/2017, i.e., three growing seasons after planting, here representing the "late plantation stage", all individual shoots within a central measurement area of 8.0 m × 3.2 m of each plot were cut at 0.1 m above ground and weighed in fresh condition (fresh weight). A stratified sample of 30 shoots per species/variety was done among the plants situated outside the central measurement area of all pure culture plots to determine the relationships between fresh and dry weights of shoots separately for all species/varieties. The dry weights (biomasses) of the stratified sample shoots were determined after oven-drying at 70 °C for 96 h, and the species/variety specific regressions between the fresh and dry weights of the stratified samples were used to estimate the biomasses of all individual shoots sampled within the central measurement area of all plots [37].

2.6. Statistical Analyses

The effect of the site, the growth design and their interactions on the soil properties were analyzed by two-way ANOVA using the software PAST [45]. Statistical analysis was performed using the Statistica software package (version 13.0, StatSoft, Tulsa). Principal Component Analysis (PCA) was performed using R package. The samples used in the PCA were attributes (mean values) measured in two test sites (Rostock and Uppsala), two plantation stages (early and late) and three species identity and culture conditions ('Loden' pure culture, 'Tora' pure culture and a mixture of 'Loden' and 'Tora'). The attributes analyzed in the present study were: alkaline phosphatase activity, acidic phosphatase activity, willow biomass and plant-available K, Mg and P content in the soil.

3. Results

3.1. Plant-Available Nutrient Contents (K, Mg and P) in the Soil

The initial plant-available concentrations of K and Mg in the soil differed significantly between the two test sites and between the early and late plantation stage per test site (Table 1). The plant-available concentration of P was low and at the same level at both test sites (Figure 1). The plant-available concentrations of P and K in soil significantly decreased with the progressing willow growth (initial vs. 4 years of growth) at the test site Rostock (Figure 1 and Table 1).

Table 1. Nutrient contents (K, Mg and P) in mg/100 g soil in in sampling sites (a) Rostock and (b) Uppsala initally and after 4 years for each species identity and cultivation condition (Loden, Tora and their mixture). The biomass of the willow species was measured after harvest (kg dry matter per plot).

(a) Rostock							
Species	**Initial**			**After 4 Years**			**Biomass (First Harvest)**
	K (mg/100 g)	Mg (mg/100 g)	P (mg/100 g)	K (mg/100 g)	Mg (mg/100 g)	P (mg/100 g)	(kg dry matter/plot)
Loden [L]	13.4 ± 6.1	25.2 ± 9.5	6.8 ± 1.6 ^{A}	10 ± 2.2	20.4 ± 3.8	4.4 ± 0.01 ^{B}	16.14 ± 4.94
Tora [T]	32.9 ± 10.7 ^{A}	27.5 ± 5.9	5.8 ± 0.9 ^{A}	10.7 ± 3 ^{B}	22.0 ± 0.3	4 ± 0.1 ^{B}	30.71 ± 11.94
Loden, Tora [LT]	23.5 ± 14.3	21.4 ± 9.4	5.2 ± 1.9	10.2 ± 3.6	21.1 ± 4.3	4.4 ± 1.1	19.57 ± 5.21
(b) Uppsala							
Species	**Initial**			**After 4 Years**			**Biomass (First Harvest)**
	K (mg/100 g)	Mg (mg/100 g)	P (mg/100 g)	K (mg/100 g)	Mg (mg/100 g)	P (mg/100 g)	(kg dry matter/plot)
Loden [L]	17.6 ± 3.3	25.9 ± 1.6 ^{B}	5.1 ± 0.3 ^{A}	22.2 ± 2 *	32.1 ± 1.8 ^{A*}	4.3 ± 0.2 ^{B}	11.19 ± 5.14
Tora [T]	16.6 ± 2	27.7 ± 3.3	4.5 ± 0.6	19.9 ± 3.6 *	28.8 ± 3.3 *	3.6 ± 0.4	13.67 ± 2.89
Loden, Tora [LT]	17.8 ± 4.1	30.8 ± 12.3	6.0 ± 2.4	23.6 ± 4.6 *	27.9 ± 6.9	5.0 ± 1.8	13.76 ± 4.74

Soil properties were compared by site, growth stages (early and late plantation stage) and species (plots: L—Loden, T—Tora and LT—mixture). Values are means ± SDs (n = 3). The significant differences with $p < 0.05$ are marked by the following symbols: *—differences between sites, small letters—differences between species within one site, capital letters—differences between growth stages within one site.

Conversely, the amount of plant-available Mg in Uppsala soils increased from early to late plantation stages mainly under Loden (Figure 1). The plant-available concentration of K at the test site Uppsala was significantly higher than at the test site Rostock (see Tables 1 and 2). Likewise, the plant-available concentration of Mg in soil was higher in pure culture (Loden and Tora) in Uppsala in comparison to Rostock after 4 years of growth.

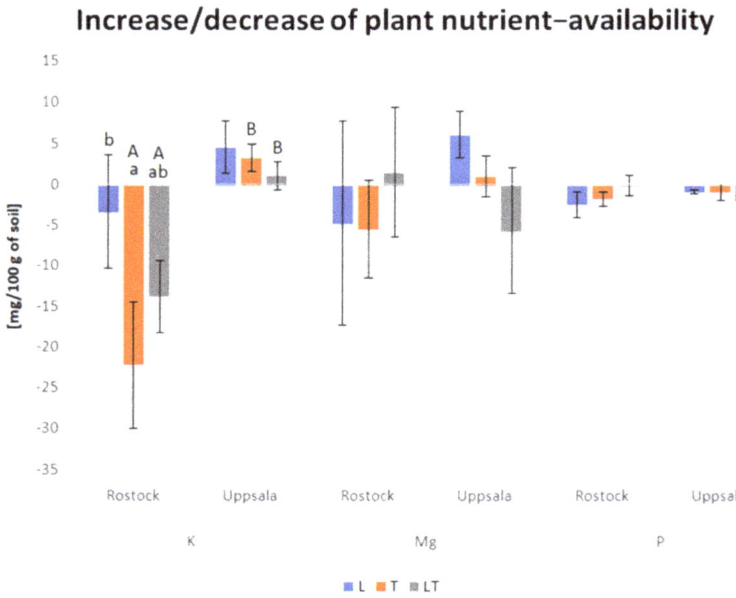

Figure 1. Nutrient levels increase/decrease between the years 2014 and 2018 in the soil at the two sites (Rostock and Uppsala). Small letters represent significant differences between the *Salix* species or the growth design on the same sampling site and capital letters represent the significant difference between sampling sites.

Table 2. Results of the two-way analysis of variance (ANOVA) on the effect of site, the growth design (with different host plant diversity; pure vs. mixture) and their interactions (site × growth design) on soil properties under *Salix* in spring 2014 (initial) and spring 2018 (after 4 years of growth).

Parameter		Site Initial	Site 4 Years	Growth Design Initial	Growth Design 4 Years	Site × Growth Design Initial	Site × Growth Design 4 Years
Plant-available P content	p	0.304	0.950	0.603	0.249	0.335	0.590
	F	1.15	0.004	0.53	1.56	1.20	0.335
Plant-available Mg content	p	0.092	<0.001	0.612	0.747	0.101	0.495
	F	3.36	20.48	0.51	0.29	2.79	0.74
Plant-available K content	p	0.020	<0.001	0.053	0.702	0.039	0.536
	F	7.15	606.01	3.79	3.94	4.28	7.13
Alkaline phosphatase activity	p	0.378	<0.001	0.928	0.065	0.344	0.184
	F	0.84	95.30	0.07	3.45	1.17	1.95
Acid phosphatase activity	p	0.200	<0.001	0.016	0.513	0.147	0.297
	F	1.84	198.20	5.92	0.71	2.26	1.34

3.2. Acid and Alkaline Phosphatase Activity in the Soil

The site and the growth design affected the activities of acid and alkaline phosphatases significantly (Table 2). The alkaline phosphatase activity in the soil Rostock was significantly higher at the early plantation stages, and a significance between species (mainly in Loden) was observed (Figure 2). In the late plantation stage, the soils showed very low activity, specifically in the soil under Loden and under the mixture. Initially, under Loden, significantly higher alkaline phosphatase activity in the soil was revealed at the test site Rostock (Figure 2), while after 4 years under Tora and Loden, significantly higher activities were measured than under the mixture at the test site Uppsala.

The activity of acid phosphatases significantly increased in the early plantation stages of willow species than in the late plantation stages in Rostock (Figure 2). The highest activity in soil was observed in plots with the mixed culture plots (LT) in comparison to the plots with monocultures Tora and Loden. No significant differences were observed among the monocultures and mixed culture plots in the early nor in the late plantation stages in Uppsala. Only plots with Loden displayed a statistical significance, with the highest activity in the late plantation stage. Overall, the acid phosphatase activity was the lowest at the late plantation stages of species in Rostock. On average, the activity in the early plantation stage was approximately 320 μg p-nitrophenol g^{-1} DM h^{-1}, whereas in the late plantation stage, it lowered to an average of approximately 130 μg p-nitrophenol g^{-1} DM h^{-1}. A significantly high phosphatase activity was seen in the late plantation stages of willow species in Uppsala. Similarly, this activity was the lowest in Rostock.

Overall, a significantly higher acid phosphatase production was recorded at both the investigated sites compared to alkaline phosphatase production (Figure 2).

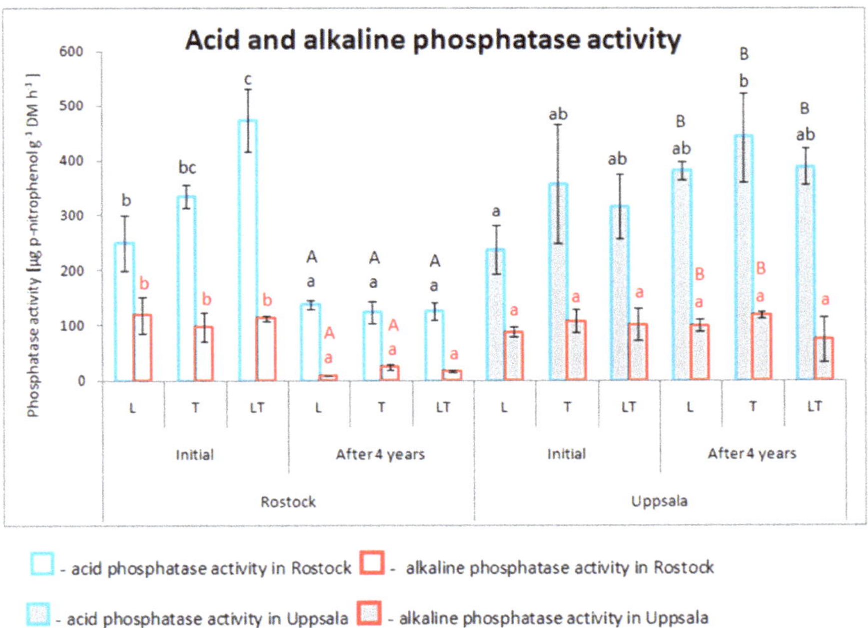

Figure 2. The alkaline phosphatase and acid phosphatase activity [μg p-nitrophenol g^{-1} DM h^{-1}] in soils obtained from Rostock and Uppsala. The data present comparisons between two sites, Rostock (on the left site) and Uppsala (on the right), between growth stages (initial and after 4 years) and species (plots: L—Loden, T—Tora and LT—mixture). Values are means ± SDs (n = 3). The significant differences with $p < 0.05$ are marked by the following symbols: small letters—differences between species within one site, capital letters—differences between initial and after four years. DM—dry matter.

Regardless of the tested parameters, approximately 55% of the total variance was explained by the first two components in the PCA analysis (Figure 3). The PCA analysis revealed that the samples were differentiated mainly based on the test sites, i.e., Rostock and Uppsala. A positive tendency towards the increase in willow biomass production was observed for Rostock samples with higher acid and alkaline phosphatase activity.

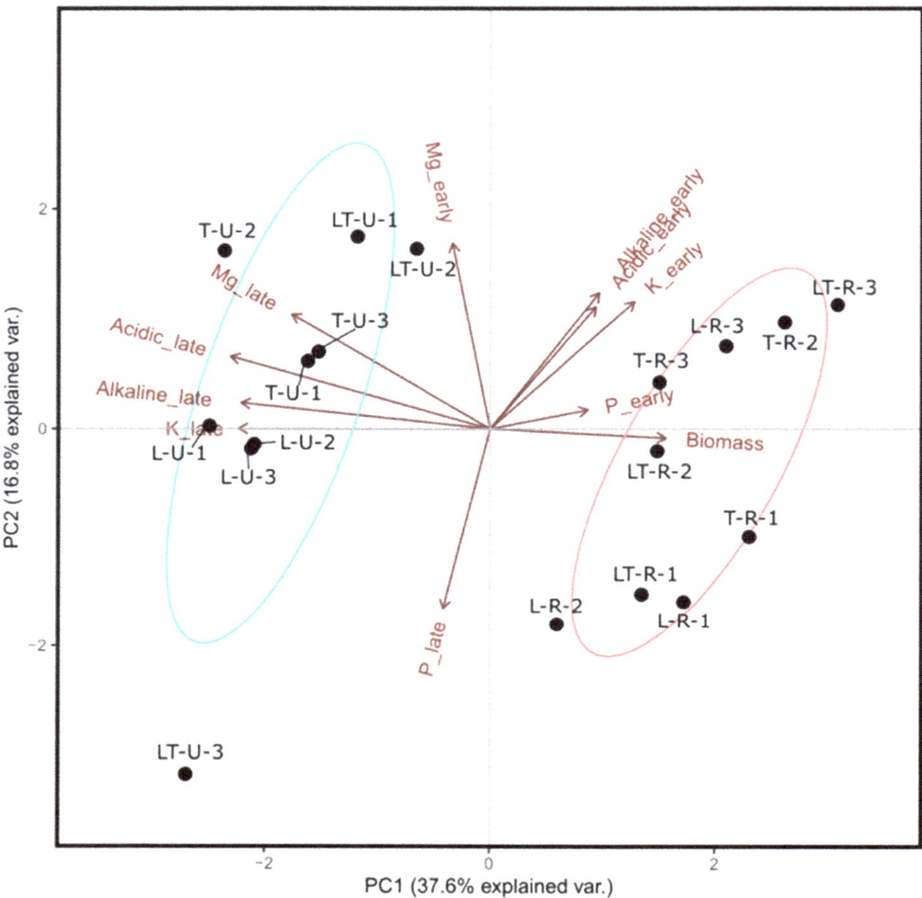

Figure 3. Grouping of samples and corresponding variable component loadings according to the Principal Component Analysis (PCA). Samples are attributes (mean values of the attributes indicated by arrows) measured in two test sites (Rostock—R and Uppsala—U), two plantation stages (initial (early) and after 4 years (late)) and three species identity and culture conditions (L—Loden pure culture, T—Tora pure culture and LT—mixture of Loden and Tora). Attributes were soil phosphatase activity (Alkaline and Acid), willow biomass and soil nutrient content (K, Mg and P in early and late plantation stages). PC1 explained 37.6% of total variance; PC2 explained 18.8% of total variance.

4. Discussion

Short rotation coppices (SRCs) generate crops used in renewable energy generation in Europe. The success of SRC establishment can be assessed by studying the adaptability of crops by monitoring their growth, climate and site conditions. Factors such as climate, soil nutrient availability, plant species and growth design may significantly influence plant nutrient cycling and overall biomass production. This study investigated the Ecolink *SALIX* SRC plantations located in Germany (Rostock) and Sweden (Uppsala). Both test sites were maintained in a similar way and planted with same two *Salix* species and their mixture. We analyzed the effect of the growth design on the plant availability and hydrolytic mobilization of P at the year of planting and after 4 years of growth, including the first harvest. The test cultivars Loden and Tora were selected for our experiment because they are both phenotypically and genotypically very distinct *Salix* species. Loden is rather slow-growing but more stress-tolerant, while Tora is generally high-performing and less stress-tolerant [37].

The initial concentrations of plant-available nutrients (Mg, K and P) at both test sites in the present study were high compared to the recommended level for arable crops [44] for Mg and K, but below the recommended level (10–18 mg P/100 g soil) for an optimal P supply. The plant-available concentration of Mg in the soils of Rostock and Uppsala was even higher than the recommended value of >19, i.e., to 32 mg/100 g soil for an optimal plant supply [44]. Moreover, the soil of the test site Rostock displayed high initial levels of plant-available K concentrations (see Table 1).

Overall, the biomass production at the test site Rostock was higher than at the test site Uppsala; however, no significant differences were observed between species and pure vs. mixed culture plots [37]. The biomass production was not correlated with the plant-availability of one of the tested nutrients (Mg, K, P); however, a correlation between the phosphatase activities and the biomass production was indicated (see Figure 3).

In agreement with Ens et al. [36], we measured significantly decreased plant-available P concentrations in the soil under the pure stands after the first cutting period, and a significant site effect on the P cycling (phosphatase activities, see Table 2) was observed. However, significant differences were observed for soil nutrient concentrations (mainly P and K) between the initial stage early and after four years of growth at both test sites (see Table 1). The acid and alkaline phosphatase activities decreased strongly from the initial level to the 4-year growth stage, which might be caused by the former grassland vegetation with a higher fine root density [46].

A general effect of the species and the growth design on the phosphatase activity under *Salix* is in agreement with the results of the mycorrhizosphere observation by Baum et al. [47]. However, only two *Salix* species were investigated in the present study due to the limitation of available plant variants at the test site Rostock in Germany (only Loden and Tora were present). Increased number of plant species and a higher amount of diversity in the mixtures (three or more species) might have increased the validity of the present information. Furthermore, a joint impact of the P and N supply might be assumed [36] and was not investigated in the present study.

The initial nutrient surplus at the arable test sites in Sweden and Germany agrees with results of former investigations of SRC [48] and underlines no need for fertilization in the first cutting period at such sites. This is because formerly arable farmland was usually regularly fertilized, which often results in high nutrient contents [49]. The nutrient concentrations in the soil changed significantly within the first four years (see Table 1). The cultivar Loden, which was included in the present study, is frequently reported with great potential in nutrient acquisition from soil, especially nitrogen and increased biomass production when paired with other *Salix* species [40]. The analysis of the P content in soil of both the sampling sites showed a significant decrease from early plantation to late plantation stage, which might suggest that Loden is efficient in P uptake. Since both experimental sites were not fertilized after willows were planted, most of the P present is in organic or low soluble form, which is not easily accessible to plants. P depletion in Uppsala and Rostock was also paired with a significant increase and decrease of acid phosphatase activity, respectively (see Figures 1 and 2). This inconsistency might suggest that acid phosphatase activity is not strongly connected with P supply in soil, but may be connected with other factors. Study performed by Criquet et al. [50] revealed that increased leaf litter moisture is positively correlated with acid phosphatase activity. Additionally, experiments performed in three forest ecosystems in China showed that increased precipitation during the dry season had a positive effect on enzyme activity [51]. The cultivar Tora showed slightly fewer prominent differences from the initial nutrient availability to the level after the first cutting cycle; only in Rostock site it differed in both P and Mg between these two sampling dates. Although P depletion was not as severe as for Loden, the plant-available Mg content decreased by almost two-fold compared to the initial content.

Of note, differences were observed in willow biomass production at the two test sites, which may be due to the varying nutrient concentrations in soils. Most of the plant-available nutrient concentrations (Mg and P) were decreased in the soil within the first cutting period, although the level of K varied at both the test site and growth stages. The PCA analysis revealed that biomass was positively correlated with P and K during the early plantation stage and negatively with late plantation stage concentration of P and K (Figure 3). The correlation was prominent, although we observed K depletion in one experimental site and an increase in the other site. Willows are known to efficiently uptake organic P when paired with ectomycorrhizal fungi [52]. This suggests that the presence of these two compounds during early plantation stages is key to reaching higher biomass production efficiency in later plantation stages. Additionally, the K content was in direct correlation with alkaline and acid phosphatase activity in the early and late plantation stages, respectively, thus indicating that changes in phosphatase activity are not bound to soil P concentration but to other, more complex sets of factors. High correlation of plant-available K with phosphatase activity is probably connected with its important role as a co-factor of many enzymes. Tabaldi et al. [53] investigated the effect of various metals on *Cucumis sativus* L., e.g., Zn, K and Na. As a result, they observed increased acid phosphatase activity in higher presence of K ions.

Acid phosphatase activity was about three-fold higher than alkaline phosphatase. According to measurements performed in our previous study, pH on both sampling site is around 6 (measurements done at 2018 and 2019) [39], which promotes the activity of acid phosphatases. pH in which acid phosphatase is active is between 4.5–6, whereas for alkaline, this is 8–11 [54]. Additionally, the pH value in SRCs is known to drop slowly with time, which further promotes the activity of acid over alkaline phosphatase [55]. Another very important factor in acid phosphatase activity is the presence of arbuscular and ectomycorrhizal fungi. Baum et al. [47] pointed out the impact of both mycorrhiza types on various factors, including acid phosphatase activity. Loden was mostly colonized by ectomycorrhizal fungi and showed higher acid phosphatase activity in pure cultures, while Tora was colonized by arbuscular mycorrhizal fungi with slightly lower activity. Additionally, they reported that mixed growth of *Salix* possessed higher phosphatase activity than monocultures [47]. The abovementioned factors contribute to increased P mobilization and were in agreement with the insignificant decrease of the plant-available P concentrations in the soil under mixed growth of *Salix* species within the first cutting cycle in the present study.

5. Conclusions

The site impacts are the main controls of the changes in the concentrations of plant-available nutrients under *Salix*. The changes of the plant availability of P within one cutting cycle are generally low, independently of the site and growth design. Mixed growth of *Salix* species promotes the activity of alkaline phosphatases in P-deficient soil conditions. The impact of the growth design on the nutrient cycling differs significantly and site-specifically in the direction and amplitude. The future challenge will be to select a site-specific optimized growth design.

Author Contributions: Conceptualization, M.W. and C.B.; data curation, M.B.; supervision, K.H. and C.B.; Writing original draft, P.K.; Writing review & editing, B.F., M.W. and C.B. All authors have read and agreed to the published version of the manuscript.

Funding: Part of the research in the Swedish trial was funded by The Swedish Research Council Formas, project no. 2020-02339. Part of the soil analyses was funded by Deutsche Forschungsgesellschaft (DFG), project no. BA 1494/9-1.

Institutional Review Board Statement: Not applicable.

Informed Consent Statement: Not applicable.

Data Availability Statement: The data were presented in this article.

Conflicts of Interest: The authors declare no conflict of interest.

References

1. Baum, C.; Leinweber, P.; Weih, M.; Lamersdorf, N.; Dimitriou, L. Effects of short rotation coppice with willows and poplar on soil ecology. *Landbauforsch. Volkenrode* **2009**, *59*, 183–196.
2. Dimitriou, L.; Baum, C.; Baum, S.; Busch, G.; Schulz, U.; Köhn, J.; Lamersdorf, N.; Leinweber, P.; Aronsson, P.; Weih, M.; et al. The impact of Short Rotation Coppice (SRC) cultivation on the environment. *Landbauforsch. Volkenrode* **2009**, *59*, 159–162.
3. Dimitriou, I.; Rutz, D. *Sustainable Short Rotation Coppice: A Handbook*; WIP Renewable Energies: Munich, Germany, 2015; ISBN 978-3-936338-36-2.
4. Jørgensen, U.; Dalgaard, T.; Kristensen, E.S. Biomass energy in organic farming—The potential role of short rotation coppice. *Biomass Bioenergy* **2005**, *28*, 237–248. [CrossRef]
5. Berhongaray, G.; Verlinden, M.S.; Broeckx, L.S.; Janssens, I.A.; Ceulemans, R. Soil carbon and belowground carbon balance of a short-rotation coppice: Assessments from three different approaches. *GCB Bioenergy* **2017**, *9*, 299–313. [CrossRef]
6. Padoan, E.; Passarella, I.; Prati, M.; Bergante, S.; Facciotto, G.; Ajmone-Marsan, F. The suitability of short rotation coppice crops for phytoremediation of urban soils. *Appl. Sci.* **2020**, *10*, 307. [CrossRef]
7. Vanbeveren, S.P.P.; Ceulemans, R. Biodiversity in short-rotation coppice. Renew. *Sustain. Energy Rev.* **2019**, *111*, 34–43. [CrossRef]
8. Dimitriou, I.; Mola-Yudego, B.; Aronsson, P. Impact of Willow Short Rotation Coppice on Water Quality. *Bioenergy Res.* **2012**, *5*, 537–545. [CrossRef]
9. Beringer, T.; Lucht, W.; Schaphoff, S. Bioenergy production potential of global biomass plantations under environmental and agricultural constraints. *GCB Bioenergy* **2011**, *3*, 299–312. [CrossRef]
10. Kahle, P.; Baum, C.; Boelcke, B. Effect of afforestation on soil properties and mycorrhizal formation. *Pedosphere* **2005**, *15*, 754–760.
11. Rödl, A. Short Rotation Coppice: Status and Prospects. In *Meyers*; Springer: Berlin/Heidelberg, Germany; New York, NY, USA, 2019.
12. Eppler, U.; Petersen, J.E.; Couturier, C. *Short Rotation Forestry, Short Rotation Coppice and Perennial Grasses in the European Union: Agro-Environmental Aspects, Present Use and Perspectives. Short Rotation Forestry, Short Rotation Coppice and Perennial Grasses in the European Union: Agro-Environmental Aspects, Present Use and Perspectives*; JRC Scientific and Technical Reports: Ispra, Italy, 2008; pp. 95–132.
13. Wolf, H.; Schildbach, M.; Hartmann, K.-U. *Plantagenbaumarten und Deren Züchtung*; Verlag, W., Ed.; Weißensee Verlag: Berlin, Germany, 2010.
14. Tullus, H.; Tullus, A.; Rytter, L. *Short-Rotation Forestry for Supplying Biomass for Energy Production*; Forest BioEnergy Production; Springer: New York, NY, USA, 2013; pp. 39–56.
15. Pleguezuelo, C.R.R.; Zuazo, V.H.D.; Bielders, C.; Bocanegra, J.A.J.; PereaTorres, F.; Martínez, J.R.F. Bioenergy farming using woody crops. A review. *Agron. Sustain. Dev.* **2014**, *35*, 95–119. [CrossRef]
16. Hughes, F.; Richards, K.; Girel, J.; Moss, T.; Muller, E.; Nilsson, C.; Rood, S. *The Flooded Forest: Guidance for Policy Makers and River Managers in Europe on the Restoration of Floodplain Forests*; The FLOBAR2 Project: Cambridge, UK, 2003.
17. Dickmann, D.I.; Kuzovkina, J. Poplars and willows of the world, with emphasis on silviculturally important species. *Poplars Willows Trees Soc. Environ.* **2014**, *22*, 8.
18. Castaño-Díaz, M.; Barrio-Anta, M.; Afif-Khouri, E.; Cámara-Obregón, A. Willow short rotation coppice trial in a former mining area in northern spain: Effects of clone, fertilization and planting density on yield after five years. *Forests* **2018**, *9*, 154. [CrossRef]
19. Ruttens, A.; Boulet, J.; Weyens, N.; Smeets, K.; Adriaensen, K.; Meers, E.; van Slycken, S.; Tack, F.; Meiresonne, L.; Thewys, T.; et al. Short rotation coppice culture of willows and poplars as energy crops on metal contaminated agricultural soils. *Int. J. Phytoremediation* **2011**, *13*, 194–207. [CrossRef] [PubMed]
20. Weih, M.; Hansson, P.-A.; Ohlsson, J.A.; Sandgren, M.; Schnürer, A.; Rönnberg-Wästljung, A.-C. *Sustainable Production of Willow for Biofuel Use*; Saffron, C., Ed.; Burleigh Dodds Science Publishing Limited: Cambridge, UK, 2020.
21. Dimitriou, I.; Aronsson, P. Willows for Energy and Phytoremediation in Sweden. *Unasylva-FAO* **2005**, *56*, 47.
22. Stolarski, M.; Szczukowski, S.; Tworkowski, J.; Klasa, A. Productivity of seven clones of willow coppice in annual and quadrennial cutting cycles. *Biomass Bioenergy* **2008**, *32*, 1227–1234. [CrossRef]
23. Caslin, B.; Teagasc, O.P. Energy Crops Agronomy–lessons to date. In *Energy Crops Business Contacts*; IrBEA: Dublin, Ireland, 2010; Volume 4.
24. Langeveld, H.; Quist-Wessel, F.; Dimitriou, I.; Aronsson, P.; Baum, C.; Schulz, U.; Bolte, A.; Baum, S.; Köhn, J.; Weih, M.; et al. Assessing Environmental Impacts of Short Rotation Coppice (SRC) Expansion: Model Definition and Preliminary Results. *BioEnergy Res.* **2012**, *5*, 621–635. [CrossRef]
25. Wilkinson, J.M.; Evans, E.J.; Bilsborrow, P.E.; Wright, C.; Hewison, W.O.; Pilbeam, D.J. Yield of willow cultivars at different planting densities in a commercial short rotation coppice in the north of England. *Biomass Bioenergy* **2007**, *31*, 469–474. [CrossRef]
26. Dimitriou, I.; Rutz, D. *Sustainability Criteria and Recommendations for Short Rotation Woody Crops*; IEE Project SRCplus; WIP Renewable Energies: Munich, Germany, 2014.
27. Dimitriou, I.; Rosenqvist, H.; Berndes, G. Slow expansion and low yields of willow short rotation coppice in Sweden; implications for future strategies. *Biomass Bioenergy* **2011**, *35*, 4613–4618. [CrossRef]

28. Mirck, J.; Isebrands, J.G.; Verwijst, T.; Ledin, S. Development of short-rotation willow coppice systems for environmental purposes in Sweden. *Biomass Bioenergy* **2005**, *28*, 219–228. [CrossRef]
29. Richardson, A.E.; Simpson, R.J. Soil microorganisms mediating phosphorus availability update on microbial phosphorus. *Plant Physiol.* **2011**, *156*, 989–996. [CrossRef]
30. Zhu, J.; Qu, B.; Li, M. Phosphorus mobilization in the Yeyahu Wetland: Phosphatase enzyme activities and organic phosphorus fractions in the rhizosphere soils. *Int. Biodeterior. Biodegrad.* **2017**, *124*, 304–313. [CrossRef]
31. Macaskie, L.E.; Bonthrone, K.M.; Yong, P.; Goddard, D.T. Enzymically mediated bioprecipitation of uranium by a Citrobacter sp.: A concerted role for exocellular lipopolysaccharide and associated phosphatase in biomineral formation. *Microbiology* **2000**, *146*, 1855–1867. [CrossRef]
32. Anzuay, M.S.; Ludueña, L.M.; Angelini, J.G.; Fabra, A.; Taurian, T. Beneficial effects of native phosphate solubilizing bacteria on peanut (*Arachis hypogaea* L.) growth and phosphorus acquisition. *Symbiosis* **2015**, *66*, 89–97. [CrossRef]
33. Nasto, M.K.; Osborne, B.B.; Lekberg, Y.; Asner, G.P.; Balzotti, C.S.; Porder, S.; Taylor, P.G.; Townsend, A.R.; Cleveland, C.C. Nutrient acquisition, soil phosphorus partitioning and competition among trees in a lowland tropical rain forest. *New Phytol.* **2017**, *214*, 1506–1517. [CrossRef]
34. Hauk, S.; Wittkopf, S.; Knoke, T. Analysis of commercial short rotation coppices in Bavaria, southern Germany. *Biomass Bioenergy* **2014**, *67*, 401–412. [CrossRef]
35. Wickham, J.; Rice, B.; Finnan, J.; McConnon, R. *A Review of Past and Current Research on Short Rotation Coppice in Ireland and Abroad*; Coford: Dublin, Ireland, 2010.
36. Ens, J.; Farrell, R.E.; Bélanger, N. Early effects of afforestation with willow (Salix purpurea,"Hotel") on soil carbon and nutrient availability. *Forests* **2013**, *4*, 137–154. [CrossRef]
37. Hoeber, S.; Arranz, C.; Nordh, N.E.; Baum, C.; Low, M.; Nock, C.; Scherer-Lorenzen, M.; Weih, M. Genotype identity has a more important influence than genotype diversity on shoot biomass productivity in willow short-rotation coppices. *GCB Bioenergy* **2018**, *10*, 534–547. [CrossRef]
38. Liu, C.L.C.; Kuchma, O.; Krutovsky, K.V. Mixed-species versus monocultures in plantation forestry: Development, benefits, ecosystem services and perspectives for the future. *Glob. Ecol. Conserv.* **2018**, *15*, e00419. [CrossRef]
39. Koczorski, P.; Furtado, B.U.; Gołębiewski, M.; Hulisz, P.; Baum, C.; Weih, M.; Hrynkiewicz, K. The effects of host plant genotype and environmental conditions on fungal community composition and phosphorus solubilization in willow short rotation coppice. *Front. Plant Sci.* **2021**, *12*, 647709. [CrossRef] [PubMed]
40. Hoeber, S.; Fransson, P.; Prieto-Ruiz, I.; Manzoni, S.; Weih, M. Two Salix genotypes differ in productivity and nitrogen economy when grown in monoculture and mixture. *Front. Plant Sci.* **2017**, *8*, 231. [CrossRef] [PubMed]
41. Rytter, R.M.; Hansson, A.C. Seasonal amount, growth and depth distribution of fine roots in an irrigated and fertilized Salix viminalis L. plantation. *Biomass Bioenergy* **1996**, *11*, 129–137. [CrossRef]
42. Tabatabai, M.A.; Bremner, J.M. Use of p-nitrophenyl phosphate for assay of soil phosphatase activity. *Soil Biol. Biochem.* **1969**, *1*, 301–307. [CrossRef]
43. Kerschberger, M.; Hege, U.; Jungk, A. Phosphordüngung nach Bodenuntersuchung und Pflanzenbedarf. *VDLUFA Standpkt.* **1997**, *8*, 1–14.
44. Hoffmann, G. Bestimmung von Phosphor und Kalium im Calcium-Acetat-Lactat (CAL)-Auszug. In *Die Untersuchung von Boden*; VDLUFA-Methodenbuch Bd. I, 4. Auflage, A 6.2.1.1, VDLUFA Verlag D Darmstadt; VDLUFA: Speyer, Germany, 1991.
45. Hammer, D.A.T.; Ryan, P.D.; Hammer, Ø.; Harper, D.A.T. Past: Paleontological statistics software package for education and data analysis. *Palaeontol. Electron.* **2001**, *4*, 178.
46. Lugli, L.F.; Andersen, K.M.; Aragão, L.E.O.C.; Cordeiro, A.L.; Cunha, H.F.V.; Fuchslueger, L.; Meir, P.; Mercado, L.M.; Oblitas, E.; Quesada, C.A.; et al. Multiple phosphorus acquisition strategies adopted by fine roots in low-fertility soils in Central Amazonia. *Plant Soil* **2019**, *450*, 49–63. [CrossRef]
47. Baum, C.; Hrynkiewicz, K.; Szymanska, S.; Vitow, N.; Hoeber, S.; Fransson, P.M.A.; Weih, M. Mixture of Salix genotypes promotes root colonization with dark septate endophytes and changes P cycling in the mycorrhizosphere. *Front. Microbiol.* **2018**, *9*, 1012. [CrossRef] [PubMed]
48. Kahle, P.; Baum, C.; Boelcke, B.; Kohl, J.; Ulrich, R. Vertical distribution of soil properties under short-rotation forestry in Northern Germany. *J. Plant Nutr. Soil Sci.* **2010**, *173*, 737–746. [CrossRef]
49. Lutter, R.; Tullus, A.; Kanal, A.; Tullus, T.; Tullus, H. The impact of short-rotation hybrid aspen (*Populus tremula* L.× *P. tremuloides* Michx.) plantations on nutritional status of former arable soils. *For. Ecol. Manag.* **2016**, *362*, 184–193. [CrossRef]
50. Criquet, S.; Ferre, E.; Farnet, A.M.; Le Petit, J. Annual dynamics of phosphatase activities in an evergreen oak litter: Influence of biotic and abiotic factors. *Soil Biol. Biochem.* **2004**, *36*, 1111–1118. [CrossRef]
51. Huang, W.; Liu, J.; Zhou, G.; Zhang, D.; Deng, Q. Effects of precipitation on soil acid phosphatase activity in three successional forests in southern China. *Biogeosciences* **2011**, *8*, 1901–1910. [CrossRef]
52. Rennenberg, H.; Herschbach, C. Phosphorus nutrition of woody plants: Many questions-few answers. *Plant Biol.* **2013**, *15*, 785–788. [CrossRef] [PubMed]
53. Tabaldi, L.A.; Ruppenthal, R.; Cargnelutti, D.; Morsch, V.M.; Pereira, L.B.; Schetinger, M.R.C. Effects of metal elements on acid phosphatase activity in cucumber (Cucumis sativus L.) seedlings. *Environ. Exp. Bot.* **2007**, *59*, 43–48. [CrossRef]

54. Nannipieri, P.; Giagnoni, L.; Landi, L.; Renella, G. *Role of Phosphatase Enzymes in Soil*; Springer: Berlin/Heidelberg, Germany, 2011; pp. 215–243.
55. Pellegrino, E.; Di Bene, C.; Tozzini, C.; Bonari, E. Impact on soil quality of a 10-year-old short-rotation coppice poplar stand compared with intensive agricultural and uncultivated systems in a Mediterranean area. *Agric. Ecosyst. Environ.* **2011**, *140*, 245–254. [CrossRef]

MDPI

Article

Physiological Performance and Biomass Growth of Different Black Locust Origins Growing on a Post-Mining Reclamation Site in Eastern Germany

Christian A. Lange [1,*], Dirk Knoche [1], Robin Hanschke [1], Sonja Löffler [2] and Volker Schneck [3]

[1] Forschungsinstitut für Bergbaufolgelandschaften e.V. (FIB), Brauhausweg 2, 03238 Finsterwalde, Germany; d.knoche@fib-ev.de (D.K.); hanschkerobin@gmail.com (R.H.)
[2] Landeskompetenzzentrum Forst Eberswalde (LFE), Alfred-Möller-Straße 1, 16225 Eberswalde, Germany; sonja.loeffler@lfb.brandenburg.de
[3] Thuenen Institute of Forest Genetics, Eberswalder Chaussee 3a, 15377 Waldsieversdorf, Germany; volker.schneck@thuenen.de
* Correspondence: c.lange@fib-ev.de; Tel.: +49-3531-7907-31

Citation: Lange, C.A.; Knoche, D.; Hanschke, R.; Löffler, S.; Schneck, V. Physiological Performance and Biomass Growth of Different Black Locust Origins Growing on a Post-Mining Reclamation Site in Eastern Germany. *Forests* **2022**, *13*, 315. https://doi.org/10.3390/f13020315

Academic Editor: Dirk Landgraf

Received: 13 January 2022
Accepted: 11 February 2022
Published: 15 February 2022

Publisher's Note: MDPI stays neutral with regard to jurisdictional claims in published maps and institutional affiliations.

Abstract: Black Locust/Robinia can play an important role in land reclamation due to its pronounced nitrogen fixation capability, fast initial growth and relative high drought tolerance. Hence, we set up a trial to test 12 Black Locust clones and three provenances growing on sandy overburden material within the open cast lignite mine *Welzow-Süd* (South Brandenburg) in March 2014. Since then, biomass growth of the Black Locust trees was examined and physiological performance was studied on several occasions using chlorophyll a fluorescence and Dualex® measuring technique. Plant physiological measurements revealed differences in photosynthetic vitality (PI_{ABS}), although the PI_{ABS} values followed a similar pattern and sequences across the plot. While the genotypes *Fra3* and *Roy* show the highest photosynthetic vitality, the clones *Rog* and *Rob* display the lowest PI_{ABS} mean values. Chlorophyll and phenol content as well as the nutrition supply of the test trees vary depending on their origin and site conditions. The annual biomass growth rate corresponds to photosynthetic vitality and both depend on weather conditions during the growing season. After six years, the growing biomass amounts to 14.7 Mg d.m. ha^{-1} for clone *Rob* and 44.8 Mg d.m. ha^{-1} for clone *Fra3*, i.e., 2.5 to 7.5 Mg d.m. ha^{-1} $year^{-1}$. Our data demonstrate a good correlation between biophysical parameters and biomass growth. We, thus, infer that physiological measuring methods can be combined to strengthen predictions regarding the physiological performance of Black Locust origins.

Keywords: *Robinia pseudoacacia* L.; photosynthetic vitality; chlorophyll and phenol content; nutrition supply; dry matter yield; land reclamation

1. Introduction

Black Locust (*Robinia pseudoacacia* L.)—once introduced to central Europe on the strength of its remarkable flowering—has established itself as a common tree species. Since the time of its introduction, the prolific Robinia has demonstrated its remarkable ability to spread [1,2], though it often remains neglected by silviculture. Current estimates suggest that Black Locust covers 34,000 hectares of German forest cultivation area, with over two thirds of this area situated in the north-east German lowlands [3]. Black Locust is considered a fast-growing tree species and stands out due to its superior wood characteristics which make it suitable for wide-ranging and high-quality usage. Furthermore, in times of climate change, Robinia is gaining in importance by virtue of its outstanding tolerance to drought and heat [4] as well as its good adaptability to climate change. Hence, Black Locust is predestined to contribute to the sustainable productivity of forests, even during critical weather situations such as drought and frost. In addition, Robinia plays an important role in the rehabilitation/reclamation of nutrient-poor lignite mining and abandoned land on account of its ability to fix atmospheric nitrogen. Black Locust stands currently cover 4.9% (i.e., 102 hectares) of the total reclamation area managed by the lignite mining company LEAG.

The declared aim of breeding Black Locust genotypes is to provide vital plus trees prior to material utilisation. However, to date, Robinia breeding initiatives have predominantly focused on external characteristics such as the stem form and improving the biomass growth performance. Nonetheless, criteria such as plants' physiological performance and tolerance to abiotic stressors are becoming increasingly decisive for the successful establishment of forest stands as well as for short rotation coppices in agriculture. Despite this, surprisingly little is known about the physiological performance of the Black Locust clones and the Black Locust provenances available on the market, and the research on cultivating Robinia genotypes which meet both criteria—high biomass growth as well as promising stress tolerance—is still in its infancy [5]. There is evidence showing significant genetic variation with regard to drought tolerance [6–9].

Within the framework of the FastWOOD subprojects 6 and 7, we investigated biomass growth as well as individual physiological performance of different Black Locust origins (i.e., clones and provenances). Our goal was to determine specific reaction patterns of promising origins to abiotic stressors. We used so-called biomarkers as major indicators to identify and classify Robinia clones and provenances with regard to their climate adaptability [8,10]. We relied on an experimental blend consisting of plot experiments [10] as well as field trials [8] to test selected Black Locust origins. In this context, we applied physiological marker techniques to reveal the individual adaptive capacity of the Robinia clones and provenances to drought, nutrient deficiencies and late frost.

Given that Robinia clones and provenances differ in their physiological performance and stress tolerance, insights into a genotype's suitability for cultivation in short coppice rotations and on forest land are valuable. The earlier a measurement procedure can deliver reliable results predicting the physiological performance, the faster research into relevant tree breeding can progress. Hence, our hypothesis is that there is a relationship between biophysical parameters and biomass growth which can be used to deliver physiological performance predictions of different Robinia origins at an early stage.

2. Materials and Methods

2.1. Experimental Site

Our test plot (51°36′49″ N, 14°14′05″ E) is situated in the northern part of *Welzow-Süd* open-cast lignite mining site, in the south of the Federal State of Brandenburg (Germany), Figure 1. Our experimental plot covers 0.4 hectares in total. We also run a weather station approximately 5 km east of our experimental plot which records standard meteorological data (i.e., precipitation, temperature, etc.).

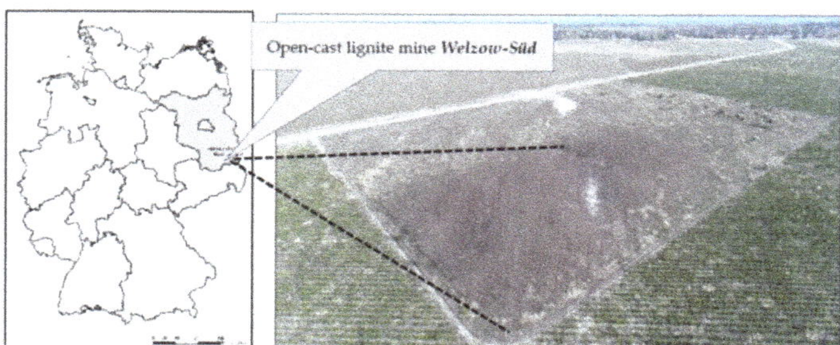

Figure 1. Map and aerial image showing our test plot in the north of lignite surface mine *Welzow-Süd*, photo taken on 12 April 2018.

2.2. Soil Conditions

The test area, situated in the northern part of the lignite mining pit *Welzow-Süd*, has been undergoing rehabilitation since 2012. Investigations on local dump soil substrate by Hanschke [11] revealed the prevalence of two different dumped materials:

(1) Loamy Sand with few gravels;
(2) Loamy Sand with few gravels and finely divided lignite material.

Note that the evaluation of the post-mining soil substrate by Hanschke [11] was carried out prior to lime application on our test site. Both dump substrates units are characterised by low pH value, low nutrition level and low water holding capacity [11], Table 1.

Table 1. Soil chemical composition of two identified mining substrates [11] related to the Robinia clonal test plot.

							10% HCl Extraction			
Substrate No.	pH ($CaCl_2$)	C_{tot} [%]	N_{tot} [%]	C/N	S_{tot} [%]	$CaCO_3$ [%]	CaO [mg kg^{-1}]	MgO [mg kg^{-1}]	K_2O [mg kg^{-1}]	P_2O_5 [mg kg^{-1}]
1	3.8	0.62	0.016	38.8	0.07	<0.1	865	443	430	94
2	4.1	2.07	0.027	76.7	0.04	<0.1	1617	753	634	144

C_{tot}: total carbon content; N_{tot}: total nitrogen content; S_{tot}: total sulphur content; C/N: C/N ratio, total carbon content divided by total nitrogen content.

2.3. Test Design

The trial was designed to investigate 12 selected clones of Black Locust along with 3 provenances (Table 2). The original plus trees of the clones (ortets) were selected on the basis of their outstanding stem form and growth performance in 1990 [12]. The trial was established with one-year-old plants from tissue culture (clones) and provenances in spring 2014. All plants were cut after planting 5–7 cm above ground. Potential competitive shoots were removed during the first growing season. A complete randomised subplot design was used with 12 replications and 9 trees per plot. The spacing was 1.5 m by 1.5 m.

Table 2. Experimental set up for testing different Black Locust clones and provenances. The background colour is highlighting the 10 test units actually investigated in the study (10 of 15 in total).

No.	*Name*	*Test Unit*	*Origin*	*Remarks*
1	Bendida	Ben	Bulgaria	selection by company *Lignum*
2	Tangra	Tan	Bulgaria	selection by company *Lignum*
3	Fraport 1	Fra1	Hesse (GER)	identical to *"Nyrségi"*
4	Fraport 2	Fra2	Hesse (GER)	identical to *"Jázkiséri"*
5	Fraport 3	Fra3	Hesse (GER)	
6	Langen	Lan	Forest district *Langen* (GER)	
7	Habichtborn	Hab	Arboretum Habichtborn (GER)	clone "Appalachia 4183"
8	Robert	Rob	Brandenburg (GER)	Forest *Hasenholz*
9	Roger	Rog	Brandenburg (GER)	Forest *Hasenholz*
10	Romy	Rom	Brandenburg (GER)	Forest *Buckow*
11	Rowena	Row	Brandenburg (GER)	Forest *Bollersdorf*
12	Roy	Roy	Brandenburg (GER)	Forest *Waldsieversdorf*
15	Kiskunsagi	Kis	Hungary	origin: seed stocks
22	Schöneiche	Schö	Brandenburg (GER)	origin: seed stocks
23	Cuci	Cuc	Romania	origin: seed stocks

However, in order to reduce costs and delay we only examine 7 clones and the 3 provenances in this study (grey shaded in Table 2). Five of the investigated clones have their origins in the Federal State of Brandenburg (*Rob, Rog, Romy, Row* and *Roy*, respectively) and two in the Federal State of Hesse (*Fra3, Lan*).

According to the principle described in [8], we picked five out of nine test plants per patch for biophysical measurement (Figure 2).

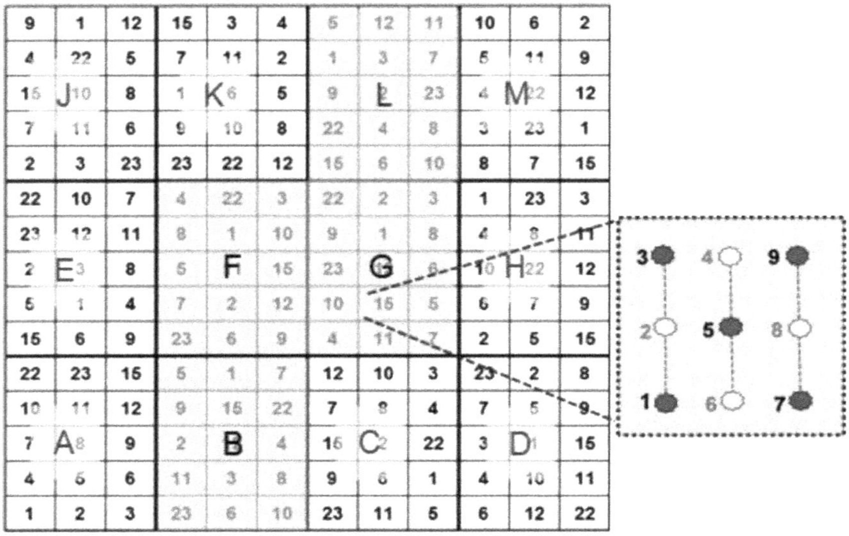

Figure 2. Test design, comprising 4 (B, F, G, L) out of a total of 12 test subplots. Per test unit, 5 out of 9 Robinia test plants were used in our investigations (see [8]).

2.4. *Chlorophyll a Fluorescence Measurements and JIP-Test*

A portable Plant Efficiency Analyser (Pocket PEA, Hansatech, King's Lynn, UK) was used to non-invasively measure chlorophyll fluorescence on the leaves' surface of Black Locust test plants in order to assess their physiological status (i.e., vitality). Lange et al. [13] successfully employed this methodology to determine plant physiological effects of different soil ameliorants applied to young sessile oaks (*Quercus petraea* (Matt.) Liebl.) growing on a uranium tailings dump in *Schlema* (Ore Mountains, Germany). Before taking the actual measurement, leaves of the test plants were dark-adapted for at least one hour before the chlorophyll a fluorescence measurements were performed. The fast phase fluorescence transients were quantified by means of the JIP-test [14,15] and using the Biolyzer software [15]. The JIP-test, developed and tested under both laboratory and practical conditions, is well accepted amongst experts to detect, describe and quantify the dynamic capacities of the photosynthetic sample. It has been widely and successfully used for the investigation of photosystem II behaviour in various photosynthetic organisms under different stress conditions and enables the study of synergistic and antagonistic effects of different co-stressors [10,13,16–18].

In order to quantify and compare the individual physiological performance of the Robinia test trees both under normal and stress conditions we chose to use the Performance Index (PI_{ABS}) as the JIP-test parameter. PI_{ABS} is a multiparametric expression which incorporates the independent parameters contributing to photosynthesis, namely absorption (RC/ABS), the quantum efficiency of trapping ($\varphi Po/(1-\varphi Po)$) and efficiency of conversion of trapped excitation energy to electron transport ($\psi o/(1-\psi o)$), see Strasser et al. [19]. The Performance index (PI_{ABS}) is presented below on absorption basis, Equation (1) [19]:

$$PI_{ABS} = \frac{\gamma RC}{1-\gamma RC} \cdot \frac{\varphi Po}{1-\varphi Po} \cdot \frac{\psi o}{1-\psi o} = \frac{RC}{ABS} \cdot \frac{\varphi Po}{1-\varphi Po} \cdot \frac{\psi o}{1-\psi o} \quad (1)$$

where γRC is the fraction of reaction center chlorophylls relative to the total chlorophyll: $\gamma RC = Chl_{RC}/Chl_{total}$. Since $Chl_{tot} = Chl_{antenna} + Chl_{RC}$, we get: $\gamma RC/(1 - \gamma RC) = Chl_{RC}/Chl_{antenna} = RC/ABS$.

2.5. *Dualex® Scientific+™ for Determining Chlorophyll and Flavonols in Leaves*

According to the manufacturer's specification, the Dualex® Scientific+™ system was developed on the basis of research conducted by the CNRS (Centre national de la recherche scientifique, Paris, France) and Force-A (University of Paris-Sud, Orsay, France). Using a photometric measurement principle, the Dualex® is able to perform a non-destructive and rapid measurement of the chlorophyll content in leaves as well as flavonol and anthocyanin contents in the epidermis with sufficient accuracy and in real time [20]. There is some evidence that polyphenols, especially antioxidative flavonols such as anthocyanin, are reliable indicators for plants' vitality. Under abiotic stress and/or as a result of nutrition deficiency, biosynthesis of chlorophyll decreases whereas the production of secondary plant substances such as flavonols increases [21]. The Nitrogen Balance Index (NBI), calculated from the ratio of chlorophyll content and flavonol concentration, indicates the nitrogen supply status of the tested plant [22,23]. Ultimately, an efficient and field-suited optical sensor is available to screen large datasets of leaf samples in a relatively short period of time.

2.6. *Biophysical Measurements*

Throughout the growing seasons in 2015 and 2016, we carried out in vivo chlorophyll a fluorescence and Dualex® measurements (usually five measurements per tree) on five out of nine test plants from seven different Robinia clones and three provenances located in the subplots B, F, G and L (Figure 2). Altogether, physiological test results were collected for 200 Black Locust trees over seven measuring dates.

2.7. *Plants' Leaf Analysis*

After sampling at the beginning of August 2015 and 2016, the leaf tissue was analysed to determine the (N, P, K, Ca, Mg and S) content. Leaves were dried at 80 °C for 48 h and then finely ground using a vibrating sample mill. The total nitrogen content was derived by combustion according to the Dumas principle using an element analyser. The other major elements (K, Ca, Mg, P, S) were measured with an inductively coupled plasma atomic emission spectrometer (ICP-AES). Prior to ICP-AES measurement, dried leaf powder was digested in a microwave pressure digestion system with HNO_3. Determination methods are listed in Table 3.

Table 3. Determination methods used for leaf sample analysis.

Parameter	Test Method	Detection Limit	Dimension
Sample preparation	VDLUFA Bd.III 2.1.1 (1983)		
Sample preparation	VDLUFA Bd. III 2.2.1–2.2.4 (1976)		
N_{tot}	DIN ISO 13878 (1998-11)	0.005	%
C_{tot}	DIN ISO 10694 (1996-08)	0.02	%
Nitric acid pressure digestion (microwave)	EPA Method 3052 (1996-12)		
Ca	DIN EN ISO 11885 (E22, 2009-09)	0.01	g kg^{-1}
Mg	DIN EN ISO 11885 (E22, 2009-09)	0.002	g kg^{-1}
K	DIN EN ISO 11885 (E22, 2009-09)	0.01	g kg^{-1}
P	DIN EN ISO 11885 (E22, 2009-09)	0.005	g kg^{-1}
Water/moisture content	VDLUFA Bd. III 3.1 (1976)	0.1	%

2.8. *Biomass Growth Measurements*

Towards the end of the particular growing season, we recorded plant height and diameter at breast height (i.e., 1.3 m above ground level) for 200 test specimens. Subsequently, we calculated the mean annual height and diameter growth of the clones and provenances. In addition, we estimated the individual biomass growth rate (aboveground annual woody biomass) using an allometric equation [24–26]. The biometric investigations focusing on 265 Black Locust test trees in the joint project FastWOOD resulted in the allometric equation (Equation (2)) which is tailored toward young Robinia forests on reclamation sites [27,28].

$$BM_{dry\ matter}\ [kg] = 0.00059909 \cdot d_{1.3}{}^{2.356} \quad (2)$$

2.9. Quality Assessment

Parallel to growth measurements, the stem form and crown formation were also evaluated. The number of trees with multiple stems, forks and/or ramicorns was assessed as well as trees with stem and bark injuries, and branch and crown fractures.

3. Results and Discussions

3.1. Weather Conditions from 2014–2019

Weather recordings from the time period 2014–2019 presented in Table 4 help to characterise the growth conditions for our Robinia test plants.

Table 4. Monthly sum of rainfall [mm] and average air temperature [°C] throughout the years 2014–2019. Data recorded at FIB weather station *Welzow*, Brandenburg, Germany. The background colour is a stylistic instruments.

	Precipitation **[mm]**						***Air Temperature*** **[°C]**					
	2014	*2015*	*2016*	*2017*	*2018*	*2019*	*2014*	*2015*	*2016*	*2017*	*2018*	*2019*
January	18.4	60.6	40.4	23.6	40.9	56.3	0.8	2.8	−0.3	−2.4	3.2	0.6
February	13.5	5.2	46.3	32.4	4.7	30.6	4.6	1.3	3.6	2.0	−2.6	3.5
March	24.9	38.0	30.6	37.8	27.1	47.2	7.4	5.7	4.3	6.9	1.4	7.0
April	20.1	28.9	35.6	24.2	31.3	12.3	11.3	8.9	8.5	7.9	13.5	10.4
May	102.3	23.9	29.3	29.0	31.6	61.0	13.3	13.5	14.8	14.7	16.7	12.3
June	41.5	57.2	97.9	106.4	11.5	7.6	17.0	16.3	18.6	18.5	19.0	22.5
July	100.7	62.2	84.9	112.3	33.0	36.9	20.7	20.1	19.2	19.0	20.7	20.0
August	124.9	21.6	39.3	74.4	26.8	24.9	17.1	22.1	17.9	19.0	21.4	20.5
September	79.6	68.7	8.2	27.0	35.0	31.3	15.6	14.0	16.8	13.3	16.0	14.8
October	55.8	42.2	71.0	64.2	20.5	34.3	12.0	8.0	8.6	11.7	11.1	11.2
November	6.5	68.2	35.4	47.3	13.6	20.6	6.7	7.6	3.6	5.5	5.4	6.2
December	36.6	16.8	39.6	41.7	69.1	26.9	2.6	6.6	2.1	2.9	4.3	4.2
Year (Jan–Dec)	**624.8**	**493.5**	**558.5**	**620.3**	**345.1**	**389.9**	**10.8**	**10.6**	**9.8**	**9.9**	**10.9**	**11.1**
Veg. (Apr–Oct)	**524.9**	**304.7**	**366.2**	**437.5**	**189.7**	**208.3**	**15.3**	**14.7**	**14.9**	**14.9**	**16.9**	**15.9**

As shown in Table 4, the total precipitation was considerably lower in 2018 and 2019 than in the years from 2014 to 2017, especially during the vegetation period (Apr–Oct). Indeed, the average precipitation during the vegetation period in 2018 and 2019 was, respectively, 219 mm (−53.5%) and 200 mm (−49.0%) lower than the 2014–2017 average. Compared to the average air temperature in the years 2014–2017, the mean air temperature also increased by 2° K (growth year 2018) and by 1° K (2019). Therefore, in 2018, the young Black Locust trees were faced with the most extreme conditions since their planting in March 2014.

3.2. Chlorophyll a Fluorescence and Photosynthetic Vitality

Results determined from the measurements taken during growth season 2015 and 2016 are presented in the following section. PI_{ABS}, a significant JIP-test parameter which enables the quantification of plants' vitality status at a specific time point, will be addressed in greater detail. PI_{ABS} average values vary across the Black Locust test units, but follow similar sequences and patterns over time (Figures 3 and 4). After starting on a relative low level in June of both of the test years, average PI_{ABS} values of all the test trees were found to have increased on subsequent occasions.

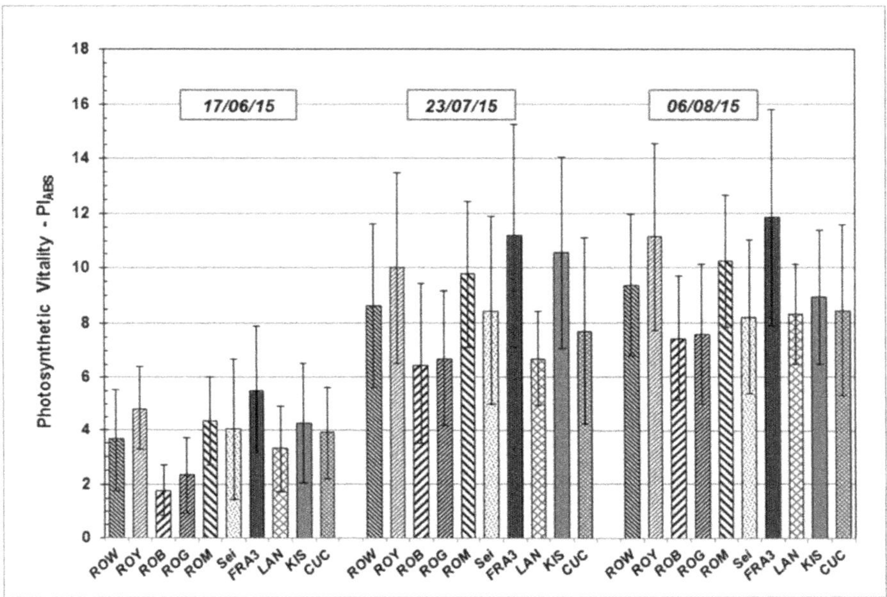

Figure 3. PI_{ABS} mean values and standard deviation of Black Locust clones and provenances during the growth season 2015 at clonal test plot *Welzow*.

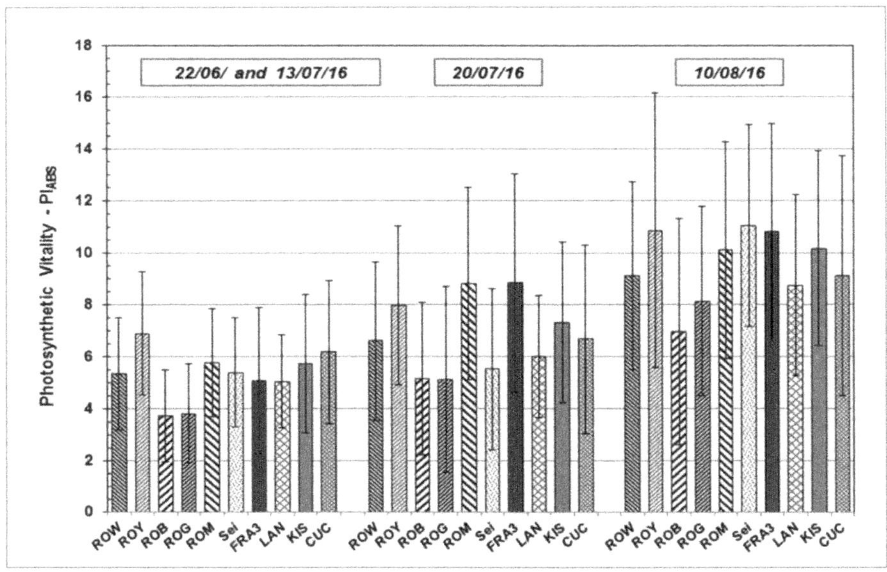

Figure 4. PI_{ABS} mean values and standard deviation of Black Locust clones and provenances during the growth season 2016 at clonal test plot *Welzow*.

As shown, individual PI_{ABS} mean values of each particular Robinia clone and provenance remain more or less the same in their relationship to each other, independent of measuring date and time. The highest average PI_{ABS} values were observed in the test clones *Fra3* and *Roy*. Test trees of *Rog* and *Rob* clones show the lowest PI_{ABS} mean values, indicating a weak photosynthetic vitality and low physiological performance.

3.3. Chlorophyll and Phenol Content

In addition to the chlorophyll a fluorescence described above, we also performed Dualex® measurements on exactly the same Robinia leaves. The results of these measurements enable us to check for correlations between parameters originating from different measuring methods. Table 5 contains average values of chlorophyll, flavonol and anthocyanin as well as the above-mentioned NBI collected over 7 measuring dates in 2015 and 2016.

Table 5. Mean values of Dualex® parameter determined on Robinia leaves of different clones and provenances at clonal test plot *Welzow* in 2015 and 2016.

		Chlorophyll [μg cm^{-2}]		*Flavonol* [μg cm^{-2}]		*Anthocyanin* [μg cm^{-2}]		*NBI*	
		2015	*2016*	*2015*	*2016*	*2015*	*2016*	*2015*	*2016*
Row	Avg	**37.6**	**34.3**	**1.58**	**1.22**	**0.025**	**0.004**	**24.1**	**28.6**
	SD	*2.8*	*3.6*	*0.14*	*0.20*	*0.017*	*0.004*	*2.9*	*3.7*
Roy	Avg	**37.5**	**35.3**	**1.56**	**1.22**	**0.031**	**0.008**	**24.3**	**29.5**
	SD	*2.9*	*3.8*	*0.10*	*0.20*	*0.016*	*0.009*	*2.6*	*4.3*
Rob	Avg	**37.0**	**37.2**	**1.68**	**1.49**	**0.049**	**0.023**	**22.2**	**25.7**
	SD	*3.5*	*3.1*	*0.13*	*0.25*	*0.017*	*0.023*	*2.5*	*4.4*
Rog	Avg	**34.8**	**33.8**	**1.70**	**1.35**	**0.050**	**0.016**	**20.6**	**25.9**
	SD	*2.5*	*3.8*	*0.12*	*0.27*	*0.023*	*0.017*	*2.2*	*4.2*
Romy	Avg	**39.5**	**38.3**	**1.71**	**1.48**	**0.025**	**0.007**	**23.1**	**26.5**
	SD	*2.2*	*4.0*	*0.08*	*0.26*	*0.015*	*0.011*	*1.6*	*3.5*
Schö	Avg	**37.8**	**37.3**	**1.58**	**1.26**	**0.032**	**0.004**	**24.2**	**30.3**
	SD	*3.3*	*3.7*	*0.12*	*0.20*	*0.018*	*0.008*	*2.8*	*4.0*
Fra3	Avg	**35.5**	**32.5**	**1.56**	**1.33**	**0.038**	**0.017**	**23.0**	**25.3**
	SD	*2.1*	*2.9*	*0.13*	*0.25*	*0.017*	*0.019*	*2.8*	*3.5*
Lan	Avg	**40.5**	**38.0**	**1.40**	**1.09**	**0.017**	**0.005**	**29.3**	**36.0**
	SD	*3.5*	*4.5*	*0.11*	*0.20*	*0.012*	*0.008*	*4.1*	*6.4*
Kis	Avg	**38.0**	**35.3**	**1.53**	**1.11**	**0.026**	**0.003**	**25.1**	**32.6**
	SD	*5.0*	*5.1*	*0.12*	*0.18*	*0.027*	*0.006*	*4.3*	*4.7*
Cuc	Avg	**38.2**	**36.8**	**1.63**	**1.31**	**0.033**	**0.010**	**23.7**	**29.3**
	SD	*3.6*	*4.3*	*0.13*	*0.25*	*0.018*	*0.013*	*3.2*	*6.5*
All	Avg	**37.6**	**35.9**	**1.59**	**1.29**	**0.033**	**0.010**	**24.0**	**29.0**
	SD	*1.7*	*1.9*	*0.09*	*0.13*	*0.011*	*0.007*	*2.3*	*3.4*

When comparing the data collected in 2015 and in 2016, it becomes apparent that both the mean chlorophyll content as well as average flavonol and anthocyanin concentrations were higher in 2015, although the average NBI was found to be lower in 2015. The increase in the plant's own secondary substances in the leaves during growing season 2015 and the decrease in NBI can likely be attributed to the weather conditions (relatively dry and warm August 2015, Table 4).

3.4. Plant Nutrition

As mentioned above, the NBI indicates the nitrogen supply status of the test plant. The nutrient supply of test trees belonging to Robinia clones and provenances *Row*, *Roy*, *Fra3*, *Lan*, *Kis* and *Cuc* is therefore discussed in the next section (Table 6).

Table 6. Results of the determination of plant nutrients in the leaf tissue of Robinia test trees taken on 10 August 2015 and 10 August 2016 at clonal test plot *Welzow*. The font colour and italics are stylistic instruments.

	N_{tot} [w.-%]		P [w.-%]		K [w.-%]		Ca [w.-%]		Mg [w.-%]		S_{tot} [w.-%]	
	2015	**2016**	2015	**2016**	2015	**2016**	2015	**2016**	2015	**2016**	2015	**2016**
Row	2.87	3.76	0.12	0.15	1.34	1.35	2.38	1.84	0.24	0.16	0.18	0.19
Roy	3.29	3.70	0.15	0.17	1.25	1.18	1.95	1.55	0.19	0.13	0.20	0.21
Fra3	2.67	3.14	0.11	0.14	0.98	1.22	3.33	2.17	0.30	0.17	0.18	0.21
Lan	3.02	3.97	0.12	0.18	1.33	1.35	2.08	1.40	0.27	0.15	0.19	0.23
Kis	3.21	3.93	0.14	0.17	1.35	1.35	2.02	1.74	0.21	0.15	0.21	0.21
Cuci	3.16	3.55	0.13	0.16	1.26	1.17	2.13	1.76	0.27	0.15	0.25	0.23
Mean	***3.04***	***3.67***	***0.13***	***0.16***	***1.25***	***1.27***	***2.32***	***1.74***	***0.25***	***0.15***	***0.20***	***0.21***
SD	*0.23*	*0.30*	*0.01*	*0.02*	*0.14*	*0.09*	*0.52*	*0.27*	*0.04*	*0.01*	*0.03*	*0.02*

The data presented in Table 6 indicate that with a few exceptions, the uptake of nitrogen, phosphor, potassium and sulphur in Robinia leaves was slightly higher during the growing season in 2016 than in 2015. In contrast, calcium and magnesium contents were found to have decreased from 2015 to 2016, regardless of plant origin. Test trees of *Kis* and *Roy* displayed the highest N and P concentrations (valid for 2015), whereas *Fra3* and *Row* exhibited the highest Ca and Mg contents in general.

Bearing in mind that the growing season in 2015 was warmer and drier than in 2016 (Table 4), it is highly likely that drought, monitored in late summer 2015, limited nitrogen fixation and resulted in lower N and P content in the Robinia leaf tissue.

Our measured leaf nutrient concentrations correspond closely with previously published data, e.g., [29,30]. Heinsdorf [30] investigated nutrient uptake and nutrient supply of a seven-year-old Black Locust stand growing on a former open cast lignite mining site in East Germany and reported the follow leaf analysis data (mean values): N_{tot} 3.64%, $N_{soluble}$ 0.92%, P 0.18%, K 1.19% and Mg 0.16%. Our findings also showed a K-to-Ca-antagonism, as reported by Heinsdorf [30], where high K concentrations in Robinia leaves correspond to low Ca concentrations (Figure 5).

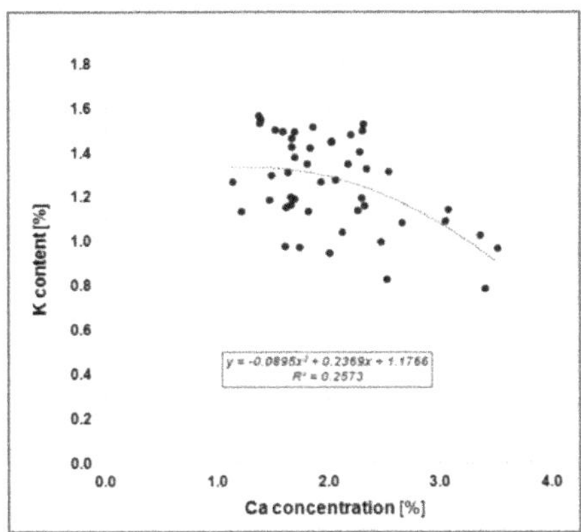

Figure 5. Bivariate scatter plot illustrating the K-to-Ca-antagonism, i.e., a negative relationship between K and Ca concentration in leaves of Black Locust trees growing on clonal test plot *Welzow*.

Furthermore, we examined our data for a systematic relationship between Dualex®-derived NBI values and N_{tot} content as well as NBI and the P concentration, respectively (Figure 6).

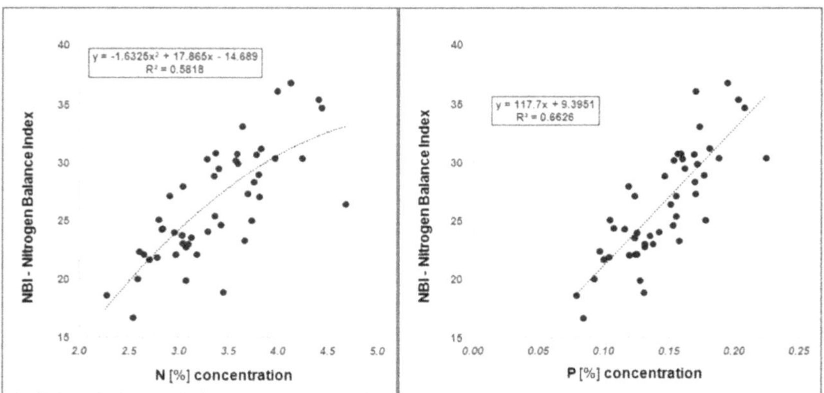

Figure 6. Regression analysis between NBI and N content (**left**) and between NBI and P concentration (**right**) in Robinia leaves.

The regression analyses show a good correlation between the NBI, determined using the Dualex®, and N_{tot} content in Robinia leaves as well as NBI and P concentration, respectively (Figure 6). Hence, Dualex® measurements offer a more cost-effective and time-efficient method for determining the nutrient status of plants.

3.5. Biomass Growth

As already mentioned in the previous section, PI_{ABS} mean values vary depending on the origin of the plant sampled. Generally, it can be stated that higher PI_{ABS} values, indicating greater efficiency in primary photosynthetic processes, are likely to result in greater plant growth. Clear and significant differences can be observed between the clones and provenances with regard to the mean annual height growth (Figure 7).

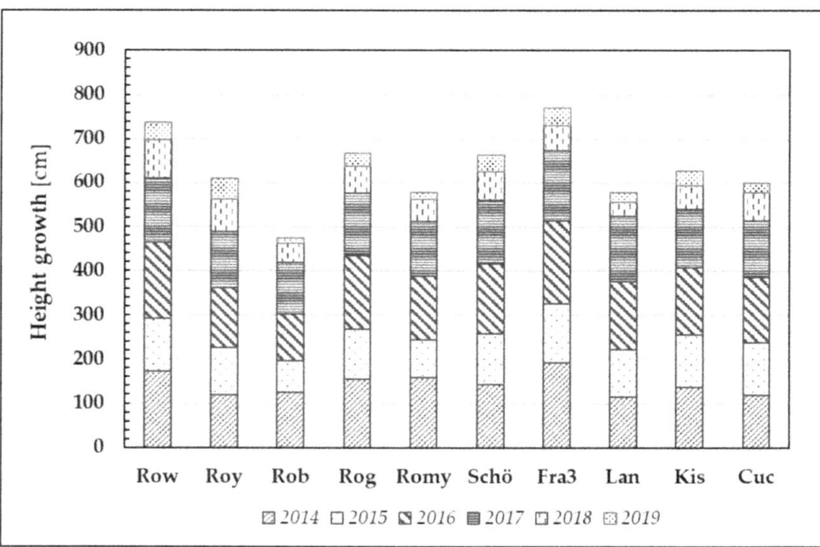

Figure 7. Plant height growth through the years 2014–2019 of ten Robinia test units. Note that one bar usually comprised 20 Robinia test trees per origin and per year.

Clones *Fra3* and *Row* demonstrated the best height growth (mean values) after six years of growth (Ø 772 cm and Ø 738 cm, respectively). In contrast, the clone *Rob* has the lowest performance (Ø 474 cm).

After the first growing season, Robinia test trees gained a height increment of 144 cm (averaged over all trees). Despite the relatively dry year that followed (2015), the trees nonetheless produced 110 cm growth on average. To date, even the 2018 and 2019 growing seasons, namely the driest and hottest vegetation periods, resulted in a minimum average height increment of 59 cm (2018) and 30 cm (2019).

Furthermore, we examined biomass formation during the growth periods of 2014 to 2019 (Figure 8). Annual biomass growth rates vary depending on the weather conditions during the growing season and corresponding to origin. After six years, the mean biomass yield averaged over all tested Robinia clones and provenances was 28.20 Mg d.m. ha^{-1}, whereas the mean annual increment was 4.70 Mg d.m. ha^{-1} year^{-1}.

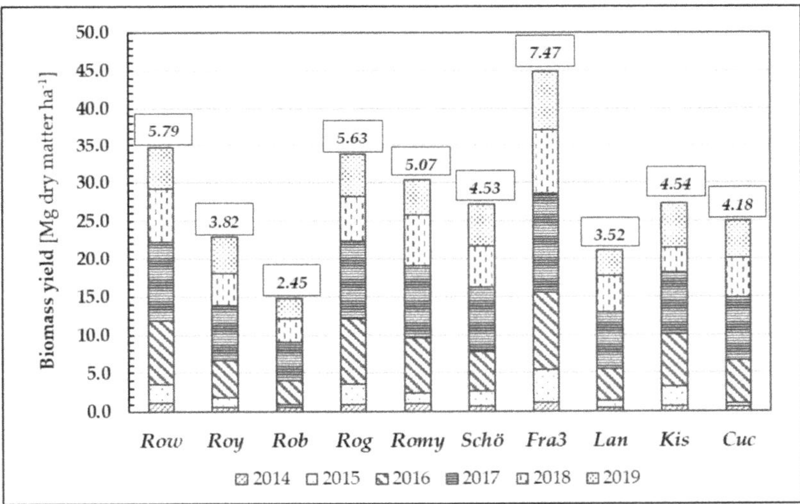

Figure 8. Biomass growth [Mg dry matter ha^{-1}] through the years 2014–2019 of different Robinia test units; data given in small boxes represent individual average increment of biomass [Mg dry matter ha^{-1}] in 6 years of time.

Similar to their performance in height growth, the clones *Fra3* and *Row* achieved the highest mean biomass growth values (7.47 Mg d.m. ha^{-1} year^{-1} and 5.79 Mg d.m. ha^{-1} year^{-1}, respectively) whilst the low-performing clone *Rob* yielded only 2.45 Mg d.m. ha^{-1} year^{-1}. The differences in biomass yield between the test units were statistically significant.

3.6. Quality Assessment

A quality assessment of all the test trees revealed a frequent occurrence of crooked stem forms and unfavourable crown formation. Altogether, 66.1% of all trees tested showed multiple stem formation with varying severity between individuals. Only 30% of the test trees belonging to clone *Romy* were found to have forks, whereas more than two thirds (83.3%) of the plants of the provenance *Kis* exhibited forks. These findings raise the question of whether it is possible to generate a sufficient number of straight-boled trunks for material utilisation.

4. Discussion

Our results show that the drought, monitored in late summer 2015, is very likely to have caused limitations of nitrogen fixation and resulted in lower N and P content in the Robinia leaves' tissue (Table 6). Such findings have been confirmed by other authors. Mantovani et al. [31], for example, studied carbon allocation, nodulation, and biological nitrogen fixation of two-year-old Black Locust (*Robinia pseudoacacia* L.) saplings under soil water limitation. The authors used stable isotopic composition of C (δ 13C) and N (δ 15N) of the leaves to investigate adverse effects of drought as well as to identify the portion N accrued from the atmosphere by biological nitrogen fixation. They also found that drought stress significantly reduces total aboveground biomass production of the test plants as well as increases the nodule biomass of Black Locust in order to maintain biological nitrogen fixation and counteract the lower soil nitrogen availability.

Regarding adaptation to climate change, Mantovani et al. [32] showed that Black Locust plants can adapt to prolonged drought conditions by lessening water loss through both reduced transpiration and leaf size. However, under well-watered conditions, Robinia does not regulate its transpiration. It, therefore, cannot be considered a water-saving tree species. Veste and Kriebitzsch [7] carried out pot experiments in order to evaluate the growth and ecophysiological performance of Black Locust under drought stress. They demonstrated that when Black Locust is exposed to drought, it drastically reduces leaf area in order to minimise transpiration. Moreover, their test plants showed different adaptations and a high plasticity of the ecophysiological processes to cope with long-term drought stress and high temperatures, which also enables them to grow in drier regions [7]. Bhusal et al. [33] have shown that drought resistance is indicated by leaf mass per area, photosynthetic rate, leaf water potential and further factors. While drought resistance was not concretely explored in this study, it opens avenues to combine our results with investigations into drought response of Black Locust origins in the future.

Furthermore, Seserman et al. [34] pointed out that tree yields in Black Locust short rotation coppices were positively impacted by air temperature increase and negatively by decreasing precipitation.

The reduction in biomass growth of Black Locust during the drought years (Figure 8) is in accordance with previous studies [34,35]. Hence, Mantovani et al. [35] investigated spatial and temporal variation of drought impact on Black Locust's water status and growth. They conducted their study at two different sites: one site with fertile agricultural soil (site 1) and a reclaimed post-mining site with heterogeneous unstructured soil (site 2). They found that stem growth was drastically reduced during a period of summer drought, particularly in the post-mining area, as a result of the adverse edaphic conditions (below the critical pre-dawn water potential value of -0.5 MPa). However, the trees could cope with the extreme soil and weather conditions in the post-mining site without perishing.

Our biomass growth rates are in range with average annual values gained from other sites in the Lusatian post-mining area reported by Knoche et al. [28]. The latter reported biomass yields after six growth years for study site *Drebkau 1* = 37.7 Mg d.m. ha^{-1} (Ø 6.3 Mg d.m. ha^{-1} year^{-1}) and for site *Drebkau 2* = 40.7 Mg d.m. ha^{-1} (Ø 6.8 Mg d.m. ha−1 year^{-1}) as well as 19.3 Mg d.m. ha^{-1} (Ø 3.2 Mg d.m. ha^{-1} year^{-1}) for site *Senftenberg*.

To give a comprehensive evaluation of all clones and provenances studied, plant physiological performance, biomass growth and quality data were individually evaluated and ranked (Table 7). Note that parameters given in the table are not weighted and deep balanced but may nonetheless assist in identifying the best Robinia plant material for land users in post-mining areas.

Table 7. Ranking of major plant-physiological, biomass growth and quality parameters of ten Black Locust clones and provenances. The background colour is a stylistic instrument.

	Row	*Roy*	*Rob*	*Rog*	*Romy*	*Schö*	*Fra3*	*Lan*	*Kis*	*Cuc*
PI_{ABS}—*photosynthetic vitality*	5	2	10	9	3	6	1	8	4	7
Chlorophyll content	8	7	5	9	2	3	10	1	6	4
Phenol content	3	4	10	9	6	5	6	1	2	6
Nutrition supply (only N, P, K)	4	2	-	-	-	-	6	3	1	5
NBI—*Nitrogen Balance Index*	6	4	9	10	7	3	8	1	2	5
Height growth (2014–2019)	2	6	10	3	9	4	1	8	5	7
Height increment (2018–2019)	1	2	9	5	8	3	4	10	6	7
Total biomass formation (2014–2019)	2	8	10	3	4	6	1	9	5	7
Biomass increment (2018–2019)	2	7	10	3	4	5	1	9	8	6
Multiple stem formation 2019	3	6	5	2	1	6	6	9	10	4
Evaluation stem quality 2019	7	3	9	5	2	4	1	8	10	5
Final Rank	*1*	*5*	*10*	*6*	*4*	*2*	*2*	*9*	*7*	*8*

5. Conclusions

Our study reveals pronounced differences in the physiological performance, biomass growth and stem quality of the Black Locust clones and provenances studied. Under the challenging climatic and edaphic conditions of our test site—considering photosynthetic vitality, chlorophyll and phenol content, nutrition state, biomass growth and stem quality—Robinia genotypes *Rowena, Fra3* and *Romy* as well as the Brandenburg provenance *Schöneiche* show the most promise with regard to growth performance, especially for cultivation in short coppice rotation and on forest land.

Biophysical measurements using the Pocket PEA and Dualex®, especially when combined and used in parallel, are reliable indicators for detecting abiotic stress already in an early stage. We were able to show that while mean PI_{ABS} values vary across Black Locust test units, they still follow similar sequences/patterns. We found correlations between biophysical parameters resulting from chlorophyll a fluorescence and Dualex® measurements. Correlation analysis revealed good accordance of Dualex®-derived NBI (Nitrogen Balance Index) and N_{tot} content as well as P content, detected in Robinia leaves. Hence, using Dualex® measurements can provide insights into the nutrient status (especially nitrogen) of plants in a cost and time efficient manner.

We conclude that biophysical measurements have the potential to shorten otherwise long-lasting research plans in tree breeding. In addition, biophysical measurements enable the early assessment of the physiological performance and stress tolerance of different Robinia clones and provenances.

Author Contributions: Conceptualization, C.A.L., D.K. and V.S.; methodology, C.A.L. and S.L.; validation, D.K., C.A.L. and V.S.; formal analysis, C.A.L., R.H. and S.L.; investigation, D.K., C.A.L. and V.S.; resources, V.S., S.L. and C.A.L.; data curation, C.A.L., S.L. and R.H.; writing—original draft preparation, C.A.L.; writing—review and editing, D.K., R.H., C.A.L. and V.S.; visualization, C.A.L., S.L. and R.H.; supervision, D.K.; project administration, D.K.; funding acquisition, V.S. and D.K. All authors have read and agreed to the published version of the manuscript.

Funding: This work was kindly funded by the German Fachagentur für Nachwachsende Rohstoffe with support by the Federal Ministry of Food and Agriculture (joint project FastWOOD, Project No. 22001014). We would also like to thank the mining company LEAG for financial support of perennial data acquisition.

Data Availability Statement: The raw data sets generated during the time period 2014–2019 are available from the corresponding author on reasonable request.

Conflicts of Interest: The authors declare no conflict of interest. The funders had no role in the design of the study; in the collection, analyses, or interpretation of data; in the writing of the manuscript, or in the decision to publish the results.

References

1. Kleinbauer, I.; Dullinger, S.; Peterseil, J.; Essl, F. Climate change might drive the invasive tree *Robinia pseudoacacia* into nature reserves and endangered habitats. *Biol. Conserv.* **2010**, *143*, 382–390. [CrossRef]
2. Meyer-Münzer, B.; Grotehusmann, H.; Vor, T. Robinie (*Robinia pseudoacacia* L.). In *Potenziale und Risiken eingeführter Baumarten —Baumartenportraits mit naturschutzfachlicher Bewertung*; Vor, T., Spellmann, H., Bolte, A., Ammer, C., Eds.; Göttinger Forstwissenschaften, Universitätsverlag Göttingen: Göttingen, Germany, 2015; Volume 7, pp. 277–296.
3. Knoche, D.; Lange, C.; Engel, J. Entwicklung von innovativen Bewirtschaftungsverfahren für Robinienbestände (*Robinia pseudoacacia* L.). In *FastWOOD II: Züchtung schnellwachsender Baumarten für die Produktion nachwachsender Rohstoffe im Kurzumtrieb—Erkenntnisse aus 6 Jahren FastWOOD*; Thünen Report; Liesebach, M., Ed.; Johann Heinrich von Thünen-Institut: Braunschweig, Germany, 2015; Volume 26, pp. 71–84.
4. Roloff, A.; Grundmann, B. Waldbaumarten und ihre Verwendung im Klimawandel. *Arch. Forstwes. Landsch. Ökol.* **2008**, *42*, 97–109.
5. Schneck, V. Robinie—Züchtungsansätze und Begründungsverfahren. In *Proceedings of the Beiträge—Agrarholz 2010, Symposium: Berlin, Germany, 18–19 May 2010*; Deutschland/Bundesministerium für Ernährung, Landwirtschaft und Verbraucherschutz, Ed.; BMELV: Bonn, Germany, 2010; pp. 1–8. Available online: https://literatur.thuenen.de/digbib_extern/dn054837.pdf (accessed on 12 January 2022).
6. Guse, T.; Schneck, V.; Liesebach, M.; von Wuehlisch, G. Improving growth performance and drought tolerance of *Robinia pseudoacacia* L.—Analysis of seedlings of European progenies. *Indian J. Ecol.* **2011**, *38*, 95–98.
7. Veste, M.; Kriebitzsch, W.U. Einfluss von Trockenstress auf Photosynthese, Transpiration und Wachstum junger Robinien (*Robinia pseudoacacia* L.). *Forstarchiv* **2013**, *84*, 35–42.
8. Löffler, S.; Lange, C.; Kätzel, R. Der kleine Unterschied macht's—Klonprüfung zur Trockenstresstoleranz von Robinien (Ergebnisse des FNR-Projektes—FastWOOD 3). *Eberswalder Forstl. Schr.* **2017**, *64*, 37–44.
9. Lange, C.A.; Knoche, D.; Schneck, V.; Anders, A. Die Robinie—Anbauversuch unter schwierigen Bedingungen. *AFZ-DerWald* **2021**, *2*, 24–29.
10. Lange, C.A.; Knoche, D.; Hanschke, R. Biophysikalische Untersuchungen zur Stresstoleranz von Robinien unterschiedlicher Herkunft. In *Wasser—Lebensgrundlage für Landschaften. 25 Jahre FIB e.V.*; Haubold-Rosar, M., Knoche, D., Hildmann, C., Eds.; Shaker: Düren, Germany, 2020; Volume 3, pp. 167–178.
11. Hanschke, L. *Bodengeologischer Kartierungsbericht. Tagebaubereich Welzow/Süd Fläche WW 0 70—H 52. Kartierungsbericht des Büros für Planung, Beratung und Standorterkundung*; Vattenfall Europe Mining AG: Turnow-Preilack, Germany, 2011.
12. Naujoks, G.; Schneck, V.; Hutter, I.; Grotkass, C. Die Mikrovermehrung ausgelesener Elite-Robinienklone—Eine große Chance für erfolgreichen Robinienanbau. *Forst Holz* **2005**, *60*, 458–460.
13. Lange, C.A.; Böcker, L.; Katzur, J. Revegetation of a uranium mine dump by using fertilizer treated sessile oaks. *Int. J. Phytoremediation* **2011**, *13*, 18–34. [CrossRef]
14. Strasser, B.J.; Strasser, R.J. Measuring fast fluorescence transients to address environmental questions: The JIP-test. In *Photosynthesis: From Light to Biosphere*; Mathis, P., Ed.; Kluwer Academic Publisher: Alphen aan den Rijn, The Netherlands, 1995; Volume 5, pp. 977–980.
15. Strasser, R.J.; Srivastava, A.; Tsimilli-Michael, M. Analysis of the chlorophyll a fluorescence transient. In *Chlorophyll Fluorescence A Signature of Photosynthesis, Advances in Photosynthesis and Respiration*; Papageorgiou, G., Govindjee, Eds.; Springer: Dordrecht, The Netherlands, 2004; Volume 19, pp. 321–362.
16. Lange, C.A.; Weissflog, L.; Strasser, R.J.; Krueger, G.H.J.; Pfennigsdorff, A. Phytotoxic effects of trichloroacetic acid on Scots pine and birch determined by chl a fluorescence and the JIP-test. *S. Afr. J. Bot.* **2004**, *70*, 683–694. [CrossRef]
17. Smit, M.F.; van Heerden, P.D.R.; Pienaar, J.J.; Weissflog, L.; Strasser, R.J.; Krüger, G.H.J. Effect of trifluoroacetate, a persistent degradation product of fluorinated hydrocarbons, on Phaseolus vulgaris and Zea mays. *Plant Physiol. Biochem.* **2009**, *47*, 623–634. [CrossRef]
18. Lange, C.A.; Kotte, K.; Smit, M.; van Deventer, P.W.; van Rensburg, L. Effects of different soil ameliorants on Karee trees (*Searsia lancea*) growing on mine tailings dump soil—Part I: Pot trials. *Int. J. Phytoremediat.* **2012**, *14*, 908–924. [CrossRef] [PubMed]
19. Strasser, R.J.; Srivastava, A.; Tsimilli-Michael, M. Screening the vitality and phytosynthetic activity of plants by fluorescence transient. In *Crop Improvement for Food Security*; Behl, R.K., Punia, M.S., Lather, B.P.S., Eds.; SSARM: Hisar, India, 1999; pp. 72–115.
20. Goulas, Y.; Cerovic, Z.G.; Cartelat, A.; MOYA, I. Dualex: A new instrument for field measurements of epidermal ultraviolet absorbance by chlorophyll fluorescence. *Appl. Opt.* **2004**, *43*, 4488–4496. [CrossRef] [PubMed]
21. Blackmer, T.M.; Schepers, J.S. Use of a chlorophyll meter to monitor N status and schedule fertigation of corn. *J. Prod. Agric.* **1995**, *8*, 56–60. [CrossRef]
22. Meyer, S.; Cerovic, Z.G.; Goulas, Y.; Montpied, P.; Desmotes-Mainard, S.; Bidel, L.; Moya, I.; Dreyer, E. Relationships between optically assessed polyphenols and chlorophyll content and the dry mass per leaf area of woody plants: A signature of the carbon and nitrogen balance within leaf? *Plant Cell Environ.* **2006**, *29*, 1338–1348. [CrossRef]

23. Tremblay, N.; Bélec, C.; Jenni, S.; Fortier, E.; Mellgren, R. The Dualex—A new tool to determine nitrogen sufficiency in broccoli. *Acta Hort.* **2009**, *824*, 121–132. [CrossRef]
24. Ketterings, Q.M.; Coe, R.; van Noordwijk, M.; Ambagu, Y.; Palm, C.A. Reducing uncertainty in the use of allometric biomass equations for predicting above-ground tree biomass in mixed secondary forests. *For. Ecol. Manag.* **2001**, *146*, 199–209. [CrossRef]
25. Röhle, H.; Hartmann, K.-U.; Steinke, C.; Murach, D. Leistungsvermögen und Leistungserfassung von Kurzumtriebsbeständen. In *Anbau und Nutzung von Bäumen auf landwirtschaftlichen Flächen*; Reeg, T., Bemmann, A., Emmann, A., Konold, W., Murach, D., Spiecker, H., Eds.; Wiley-VCH: Weinheim, Germany, 2009; pp. 41–55.
26. Carl, C.; Biber, P.; Landgraf, D.; Buras, A.; Pretzsch, H. Allometric models to predict aboveground woody biomass of black locust (*Robinia pseudoacacia* L.) in short rotation coppice in previous mining and agricultural areas in Germany. *Forests* **2017**, *8*, 328. [CrossRef]
27. Knoche, D.; Lange, C. *Verbundvorhaben: Züchtung schnell wachsender Baumarten der Gattungen Populus, Robinia und Salix für die Produktion nachwachsender Rohstoffe im Kurzumtrieb (FastWOOD III). Teilvorhaben 7: Frühdiagnose der ökophysiologischen Leistungsfähigkeit von Robinien (Robinia pseudoacacia L.) heimischer Bestände (FIB-Forschungsinstitut für Bergbaufolgelandschaften e.V.)*; Förderkennzeichen 22001014, Schlussbericht 2018; FIB-Forschungsinstitut für Bergbaufolgelandschaften e.V.: Finsterwalde, Germany, 2018; pp. 1–109. [CrossRef]
28. Knoche, D.; Lange, C.; Engel, J. Short-rotation management of black locust (*Robinia pseudoacacia* L.) in the federal state of Brandenburg, Germany. In *Novel Methods and Results of Landscape Research in Europe, Central Asia and Siberia*; Sychev, V.G., Mueller, L., Eds.; Landscape Planning, Management and Rehabilitation: FGBNU "VNII agrochimii": Moscow, Russia, 2018; Volume 5, pp. 101–104.
29. Kanzler, M.; Böhm, C.; Freese, D. Impact of P fertilisation on the growth performance of black locust (*Robinia pseudoacacia* L.) in a lignite post-mining area in Germany. *Ann. For. Res.* **2015**, *58*, 39–54. [CrossRef]
30. Heinsdorf, D. Ergebnisse eines Nährstoffmangelversuches zur Robinie (*Robinia pseudoacacia* L.) auf Kipprohboden. *Beiträge Forstwirtsch.* **1987**, *21*, 13–17.
31. Mantovani, D.; Veste, M.; Boldt-Burisch, K.; Fritsch, S.; Koning, L.; Freese, D. Carbon allocation, nodulation, and biological nitrogen fixation of black locust (*Robinia pseudoacacia* L.) under soil water limitation. *Ann. For. Res.* **2015**, *58*, 259–274. [CrossRef]
32. Mantovani, D.; Veste, M.; Freese, D. Black locust (*Robinia pseudoacacia* L.) ecophysiological and morphological adaptations to drought and their consequence on biomass production and water use efficiency. *N. Z. J. For. Sci.* **2014**, *44*, 29. [CrossRef]
33. Bhusal, N.; Lee, M.; Lee, H.; Adhikari, A.; Han, A.R.; Han, A.; Kim, H.S. Evaluation of morphological, physiological, and biochemical traits for assessing drought resistance in eleven tree species. *Sci. Total Environ.* **2021**, *779*, 146466. [CrossRef] [PubMed]
34. Seserman, D.-M.; Pohle, I.; Veste, M.; Freese, D. Simulating Climate Change Impacts on Hybrid-Poplar and Black Locust Short Rotation Coppices. *Forests* **2018**, *9*, 419. [CrossRef]
35. Mantovani, D.; Veste, M.; Böhm, C.; Vignudelli, M.; Freese, D. Spatial and temporal variation of drought impact on black locust (*Robinia pseudoacacia* L.) water status and growth. *iForest* **2015**, *8*, 743–747. [CrossRef]

 MDPI

Article

Soil Carbon Modelling in *Salix* Biomass Plantations: Variety Determines Carbon Sequestration and Climate Impacts

Saurav Kalita [1], Hanna Karlsson Potter [1], Martin Weih [2], Christel Baum [3,*], Åke Nordberg [1] and Per-Anders Hansson [1]

1 Department of Energy and Technology, Swedish University of Agricultural Sciences, 750 07 Uppsala, Sweden; saurav.kalita@slu.se (S.K.); hanna.e.karlsson@slu.se (H.K.P.); ake.nordberg@slu.se (Å.N.); per-anders.hansson@slu.se (P.-A.H.)
2 Department of Crop Production Ecology, Swedish University of Agricultural Sciences, 750 07 Uppsala, Sweden; martin.weih@slu.se
3 Soil Science, Faculty of Agricultural and Environmental Sciences, University of Rostock, 18051 Rostock, Germany
* Correspondence: christel.baum@uni-rostock.de; Tel.: +49-381-498-3100; Fax: +49-381-498-3122

Abstract: Short-rotation coppice (SRC) *Salix* plantations have the potential to provide fast-growing biomass feedstock with significant soil and climate mitigation benefits. *Salix* varieties exhibit significant variation in their physiological traits, growth patterns and soil ecology—but the effects of these variations have rarely been studied from a systems perspective. This study analyses the influence of variety on soil organic carbon (SOC) dynamics and climate impacts from *Salix* cultivation for heat production for a Swedish site with specific conditions. Soil carbon modelling was combined with a life cycle assessment (LCA) approach to quantify SOC sequestration and climate impacts over a 50-year period. The analysis used data from a Swedish field trial of six *Salix* varieties grown under fertilized and unfertilized treatments on Vertic Cambisols during 2001–2018. The *Salix* systems were compared with a reference case where heat is produced from natural gas and green fallow was the land use alternative. Climate impacts were determined using time-dependent LCA methodology—on a land-use (per hectare) and delivered energy unit (per MJ_{heat}) basis. All *Salix* varieties and treatments increased SOC, but the magnitude depended on the variety. Fertilization led to lower carbon sequestration than the equivalent unfertilized case. There was no clear relationship between biomass yield and SOC increase. In comparison with reference cases, all *Salix* varieties had significant potential for climate change mitigation. From a land-use perspective, high yield was the most important determining factor, followed by SOC sequestration, therefore high-yielding fertilized varieties such as 'Tordis', 'Tora' and 'Björn' performed best. On an energy-delivered basis, SOC sequestration potential was the determining factor for the climate change mitigation effect, with unfertilized 'Jorr' and 'Loden' outperforming the other varieties. These results show that *Salix* variety has a strong influence on SOC sequestration potential, biomass yield, growth pattern, response to fertilization and, ultimately, climate impact.

Keywords: biomass production; life cycle assessment; climate impact; soil organic carbon; *Salix*; willow; short rotation coppice; genotypic difference

Citation: Kalita, S.; Karlsson Potter, H.; Weih, M.; Baum, C.; Nordberg, Å.; Hansson, P.-A. Soil Carbon Modelling in *Salix* Biomass Plantations: Variety Determines Carbon Sequestration and Climate Impacts. *Forests* **2021**, *12*, 1529. https://doi.org/10.3390/f12111529

Academic Editor: Dirk Landgraf

Received: 20 September 2021
Accepted: 4 November 2021
Published: 6 November 2021

1. Introduction

It has been established that the current atmospheric concentrations of three major greenhouse gases (GHGs)—carbon dioxide (CO_2), methane (CH_4) and nitrous oxide (N_2O), are at the highest levels estimated for the past 800,000 years [1]. Most of this increase has happened post 1750, which was the beginning of the Industrial Revolution. The most alarming trend is that the decadal rate of increase in atmospheric CO_2 was highest in 2002–2011 since direct measurements began in 1958 [2]. There is consensus among the scientific community that the principal cause of this rapid increase is use of fossil fuels and land use change associated with the start of the Industrial Age. The increased atmospheric concentration of GHGs has enhanced radiative forcing, leading to higher average global temperatures and climate change.

Countries and organizations worldwide have set certain regulations and targets to limit the increase in average global temperatures to avoid the negative impacts of climate change. The European Commission has set targets to cut GHG emissions by at least 40% below 1990 levels and to increase renewable energy share to at least 32% by 2030 [3]. The long-term strategy is to reach a climate-neutral EU by 2050 [4]. Sweden has made an ambitious commitment to phase out all GHG emissions completely by 2045 [5,6]. The climate crisis induced by increased GHG emissions has led to a quest for different strategies to mitigate the problem. Bioenergy from sustainable biomass can be part of a viable climate mitigation strategy by replacing fossil fuels for heat and electricity generation. At the global scale biomass accounted for 9% of renewable electricity generation and 96% of renewable heat generation in 2018 [7].

The cultivation of plant species such as *Salix* (willow) and *Populus* (poplar) in short-rotation coppice (SRC) systems has emerged as an interesting approach to sustainably produce renewable biomass [8,9]. *Salix* SRC systems are characterized by short growth cycles of 2–5 years, after which the stems are harvested, and shoots regrow from the stumps left in the soil [10]. SRC plantations can have a positive effect on soil organic carbon (SOC) sequestration, because of the addition of large amounts of root and leaf litter to the soil, which are better incorporated into the soil due to minimal soil disturbance compared with annual crops [11]. *Salix* propagates easily via cuttings and is well suited to growth in temperate and Arctic climatological conditions. Commercial plantations of *Salix* are gaining interest worldwide for use as a biomass crop, with the largest cultivated areas (as of 2015) in China and Argentina, followed by North America and Europe [12]. There is high interest in European countries such as Sweden, where commercial *Salix* plantations were established in the 1990s, with policies proposed to increase energy crop cultivation to 40,000 hectares by 2030 [13].

In Sweden, the area under SRC plantations reached a peak of about 18,000 hectares in the mid-1990s, which decreased to about 12,000 hectares by 2015 [14,15]. This was attributed to a combination of factors such as poor management, inefficient policy and low prices—which meant that the practical results did not meet the high expectations [16,17]. New varieties and better management practices adapted to Swedish conditions have emerged in the past two decades. These, combined with the ambitious Swedish emission reduction targets, make SRC *Salix* an interesting prospect for biomass feedstock in the Swedish context.

The SOC sequestration potential of SRC plantations is gaining attention among researchers for its climate mitigation effects. Multiple studies [18–22] have found that SRC *Salix* systems sequester more carbon than conventional cropping systems. However, the SOC sequestration of *Salix* established on grasslands is more uncertain and can be lower [23,24]. The magnitude and potential for SOC change depend on previous land use, soil and climate conditions [18,24,25]. This, combined with the different soil profile depths considered in different studies leads to variation in reported SOC stock change rates. Long-term field data, and especially those on belowground biomass production rates, are necessary to validate and improve the accuracy of SOC sequestration estimates for SRC *Salix* plantations under different growth and soil conditions.

Biomass for bioenergy utilization can be considered carbon neutral as CO_2 emitted from its conversion phase is recaptured by new growth. However, there is a need to assess the climate impact in a system perspective including changes in SOC and land use, and impacts from site preparation, production of inputs, machinery operations, transports and energy conversion. Quantification of the potential effects and impacts of biomass use over spatial and temporal horizons is needed to ensure its sustainability.

There are several tools for environmental impact evaluation, and one of the most commonly used is Life Cycle Assessment (LCA). LCA is a well-established and standardized tool for estimation of potential environmental impacts from a product or service over its whole lifespan. The LCA methodology was originally designed for industrial processes and products but has been expanded in recent decades to evaluate and compare agricultural, forestry and bioenergy processes and products [26,27]. In the context of bioenergy production system evaluation, LCA helps by expanding the perspective beyond the production system itself. This is important as the environmental consequences of a bioenergy production system frequently depend more on the impacts on other parts of the value chain than on the production system itself. Thus, the broad system perspective makes LCA a suitable tool for planning of bioenergy systems and policymaking, especially in the context of the potential effects of bioenergy production systems on climate change mitigation. However, when modelling large and often complicated systems in LCA studies, parts of the data are often more uncertain and some subjective aspects may be handled in order to reach the broad system perspectives [28,29]. These limitations are not unique to LCA, and similar problems occur even in other methods for environmental systems analysis. The decisions on data quality requirements play an important role in the results of the assessment. Ambitions about completeness of data must be balanced against availability of resources and workload. These are intrinsic and accepted aspects of LCA studies, as long as the relevance, data quality and relevant major assumptions are appropriately described [30]. The LCA methodology is constantly evolving as understanding of climate and environmental impacts develops.

The most common climate impact metric used in LCA is global warming potential (GWP_{100}), which is based on radiative forcing and captures the integrated impacts over a single time horizon of 100 years [31]. It does not capture the effect of timing and persistence of GHG fluxes and temporal changes in SOC [32]. It does not represent the actual impacts on ecosystems such as temperature change, sea level change or biodiversity loss.

Using a time-dependent method can counter this by expressing the climate metric as a function of time. Several studies have developed such alternative methods and applied them in LCA to capture the emissions and fluxes of carbon flows between the atmosphere, biomass and soil [17,32–34]. An absolute time-dependent climate metric such as the absolute global temperature change potential (ΔTs) developed by Ericsson et al. [35] represents the impact on global mean surface temperature from emission or removal of a GHG at a particular point in time. This can aid in better understanding of climate impacts of bioenergy as biomass systems capture and emit carbon at different points in time. Several LCA studies have assessed *Salix* cultivation for bioenergy utilization [17,32,33,36–40]. However, studies looking at the magnitude of impact of differences between *Salix* varieties on the overall bioenergy system are rare.

Differences between *Salix* varieties can have a significant impact on physiological traits, biomass quality, growth patterns and soil ecology. Weih and Nordh [41] showed that key traits and shoot biomass production are variety-specific and that there is a need to account for these variety differences at the field level. Adegbidi et al. [42] found that biomass production, nutrient use efficiency and nutrient removal are strongly influenced by variety in *Salix* plantations. Cunniff et al. [43] observed significant differences in allocation between aboveground and belowground biomass in different varieties and at different locations in the UK. Data from *Salix* field trials in Sweden have demonstrated that the effects of fertilization on soil ecology are also affected by variety [44]. *Salix* varieties have been found to differ significantly in their response to fertilization and in carbon storage potential in shoots and soil [44].

Despite of the many plant-and field-scale reports indicating significant impacts of *Salix* varieties on plant traits that potentially affect their environmental performance, there is a lack of systems-scale research (such as LCAs) accounting for these differences. Thus, there is a need to address the differences between *Salix* varieties regarding the impact on soil carbon sequestration and climate impact when assessing bioenergy systems in a life cycle perspective

This study aimed to analyze the effects of *Salix* variety and fertilization treatment on SOC dynamics, and subsequent effects on climate impacts of *Salix* cultivation for bioenergy on a commercial scale, with a 50-year time horizon. A field trial established in 2001 is the source of the harvest and SOC data for the selected *Salix* varieties in this study [44,45]. Unfortunately, root biomass data over time from field-grown trees were not available from the trials used here, and we therefore used indirect methods to estimate root biomass allocation over time from published reports using pot and lysimeter experiments, in which root biomass can be assessed more easily. Other data are either taken from literature and studies on *Salix* systems where available, or based on assumptions derived from other biomass systems.

Specific objectives of the study were to:

1. Estimate the potential for soil carbon sequestration for the selected *Salix* varieties under the specific site conditions of Vertic Cambisols by using soil carbon modelling;
2. Assess the climate impact from utilizing *Salix* grown on existing fallow land as feedstock in an incineration plant using two metrics—GWP_{100} and a time-dependent climate metric (ΔTs);
3. Calculate the energy balance and performance (in terms of energy ratio) for the selected *Salix* varieties.

It is expected that quantification of the magnitude of varietal effects will highlight the importance of their inclusion in systems analysis studies of bioenergy. The intention was to provide a basis for comparison of *Salix* varieties in terms of energy and climate performance, which can aid in the consideration of optimal *Salix* variety selection for a particular purpose, e.g., maximized carbon sequestration potential.

We believe that studies like this investigation will motivate the need for variety- and location-specific root and belowground data to make realistic, accurate and detailed assessments of the environmental performance of bioenergy systems.

2. Materials and Methods

The effect of *Salix* variety on the climate impact and energy performance of a *Salix*-biomass production system under Swedish conditions (Uppsala region) was analyzed using LCA methodology. Two functional units (FU) of 1 MJ of heat and 1 hectare of land were chosen to describe the two different functions of the system—generation of heat and use of land as a resource for mitigating climate impacts. The energy FU compares the relative impact of using the *Salix* varieties as an energy source, while the land FU unit compares the different impacts from a land use perspective considering land as a restricted resource.

The climate impact calculation considers three major GHGs (CO_2, N_2O and CH_4) and is expressed in terms of two metrics—global warming potential (GWP_{100}) and a time-dependent climate impact (ΔTs) as defined in [35], with a one-year time step. The flux of carbon in the soil due to addition and decomposition of biomass was modelled with the carbon model ICBM developed by Andrén and Kätterer [46]. Annual net flux of the selected GHGs was estimated for each source and sink, and the associated emission impulses were based on the timing of the emissions.

2.1. Plant Material, Field Trial and Data Collection

The analysis was based on data collected from a field trial during the period 2001−2017 at Pustnäs, near Uppsala in central Sweden by Weih and Nordh [45]. The following six commercial *Salix* varieties were part of the study: 'Björn' (*Salix schwerinii* E. Wolf. × *S. viminalis* L.), 'Gudrun' (*S. burjatica* Nasarow × *S. dasyclados* Wimm.), 'Jorr' (*S. viminalis*), 'Loden' (*S. dasyclados*), 'Tora' (*S. schwerinii* × *S. viminalis*) and 'Tordis' ((*S. schwerinii* × *S. viminalis*) × *S. viminalis*). There were two experimental treatments—fertilized (approx. 100 kg N, 14 kg P, 47 kg K ha^{-1} yr^{-1}) and unfertilized. Plots were 6.75 m × 7.00 m in size and contained 84 plants each, corresponding to a planting density of about 18,000 plants per hectare. Each variety and treatment had four replicate plots. The dominating soil type was a vertic cambisol with a sandy loam as topsoil (0–20 cm soil depth) with 66% sand, 16% silt and 18% clay. Initial SOC content at 0–10 cm soil depth was 11.1 g kg^{-1}, with a bulk density of 1.3 g cm^{-3}. Further details of the field trial can be found in Weih and Nordh [45].

After establishment of the plantation in 2001, the plantation was managed in three-year cutting cycles with shoots harvested during winter in 2004, 2007, 2010, 2013 and 2016. Mean air temperature during the growing season (April to October) in the years relevant to this study was 12.5 °C, and the corresponding mean annual precipitation sum was 841 mm [44].

For the present analysis, the average yield for the first harvest (2004) and for subsequent harvests (average value for 2007–2016 harvests) were calculated. The first yield after planting is usually lower, as the plant root system is still establishing. The shoot growth and biomass yield figures after the field measurement period (post-2017) were assumed to follow the average values calculated from the field trial data. Table 1 presents the average harvest values from the field study used as input to the modelling work.

Table 1. Average harvested biomass yield (dry weight, DW) and standard deviation (SD) of the six commercial *Salix* varieties grown under two fertilization regimes in central Sweden from 2001 to 2018. F0 and F+ refer to the unfertilized and fertilized treatment, respectively.

Variety and Treatment	1st Harvest		2nd–5th Harvest	
	Average (DW Mg ha^{-1})	SD	Average (DW Mg ha^{-1})	SD
Björn F0	7.4	3.8	31.9	8.3
Björn F+	15.5	4.3	42.7	16.0
Gudrun F0	8.8	4.1	20.8	5.9
Gudrun F+	11.6	1.2	20.6	5.2
Jorr F0	4.5	1.3	14.4	7.6
Jorr F+	16.9	0.9	36.9	6.0
Loden F0	3.9	1.2	14.4	4.1
Loden F+	10.4	4.9	18.3	10.2
Tora F0	6.7	4.7	18.2	8.7
Tora F+	16.6	6.1	38.3	11.1
Tordis F0	10.8	5.1	28.5	13.0
Tordis F+	19.8	6.4	48.5	9.0

The field site was ploughed shortly before planting of the *Salix* stem cuttings. The soil in each plot was sampled (five replicates per plot) with a soil corer (3 cm diameter), to a depth of 10 cm in spring 2001 and to a depth of 20 cm in 2018. The initial soil sampling was performed prior to laying out the plots. The field site is characterized by a flat surface without relief-promoted erosion, which contributed to the lack of significant differences in soil properties between the different plots. An additional follow-up soil sampling in 2002 showed no significant differences in the bulk density and SOC content among the plots. As such, the ploughing did not cause a measurable difference between the first (2001) and second year (2002). The SOC content in the 0–10 cm layer was recorded and is reported by Baum et al. [44], who provide full details of the soil sampling and analysis procedures.

As the plough depth was about 25 cm during the year of establishment of the field trial, the topsoil (0–20 cm soil depth) was assumed to be homogenous and to have similar characteristics. Hence, the initial SOC stock in the 10–20 cm soil layer in 2001 was assumed similar to that in the 0–10 cm soil layer. The bulk density in 2018 had not changed significantly from the initial value of 1.3 g cm^{-3}, which can be expected as consequence of combined lack of loosening by tillage under the perennial crops, but improved aeration of the soil by increased SOC content. The SOC stock in the 10–20 cm soil layer from 2018 was analyzed following the same methodology as was described by Baum et al. [44] for the 0–10 cm soil layer. The resulting SOC stocks in the 0–10 cm, 10–20 cm and 0–20 cm layers in 2001 and 2018 are displayed in Table 2. The reduction in SOC content in the 10–20 cm layer for some of the *Salix* varieties is not unexpected under SRC as evidenced by similar results reported by Kahle et al. [47].

Table 2. Soil organic carbon stock (Mg ha^{-1}) in the 0–10 and 10–20 cm soil layers measured in field trials on six *Salix* varieties at Pustnäs, Sweden, in 2001 (pre-establishment) and in 2018. F0 and F+ refer to the unfertilized and fertilized treatments respectively.

Variety and Treatment	Soil Carbon Stock 2001 (Mg ha^{-1})			Soil Carbon Stock 2018 (Mg ha^{-1})			Increase in SOC Stock (Mg ha^{-1})
	0–10 cm	10–20 cm	0–20 cm	0–10 cm	10–20 cm	0–20 cm	0–20 cm
Björn F0	14.4	14.4	28.9	24.3	15.3	39.7	10.8
Björn F+	14.4	14.4	28.9	20.9	12.1	33.0	4.2
Gudrun F0	14.4	14.4	28.9	22.3	21.3	43.6	14.7
Gudrun F+	14.4	14.4	28.9	21.9	12.8	34.8	5.9
Jorr F0	14.4	14.4	28.9	31.7	18.6	50.3	21.5
Jorr F+	14.4	14.4	28.9	27.5	14.3	41.8	12.9
Loden F0	14.4	14.4	28.9	26.8	17.3	44.2	15.3
Loden F+	14.4	14.4	28.9	20.5	12.7	33.2	4.4
Tora F0	14.4	14.4	28.9	25.4	16.1	41.6	12.7
Tora F+	14.4	14.4	28.9	25.8	14.2	40.0	11.2
Tordis F0	14.4	14.4	28.9	26.9	16.9	43.9	15.0
Tordis F+	14.4	14.4	28.9	23.9	14.6	38.5	9.6

2.2. System Boundaries

The system studied comprised the steps from preparation of the field site for *Salix* cultivation to production of heat in a boiler in a heating plant (Figure 1). Energy flows and emissions from field operations, production of inputs, biomass transportation and thermochemical conversion were included within the system boundaries. Downstream losses and emissions after production of heat and ash at the incineration plant were considered as outside the system boundaries. Belowground changes and biomass inputs (from leaf, stumps, fine roots and coarse roots) to 20 cm depth were within the system boundaries as the SOC values from the field studies were determined with accuracy within the 0–20 cm soil layer. Highest litter input from fine roots and leaf litter are within this soil profile [48]. As the SOC changes in the sub-20-cm-profile are not part of the study system, a higher total carbon sequestration in the complete soil profile can be assumed.

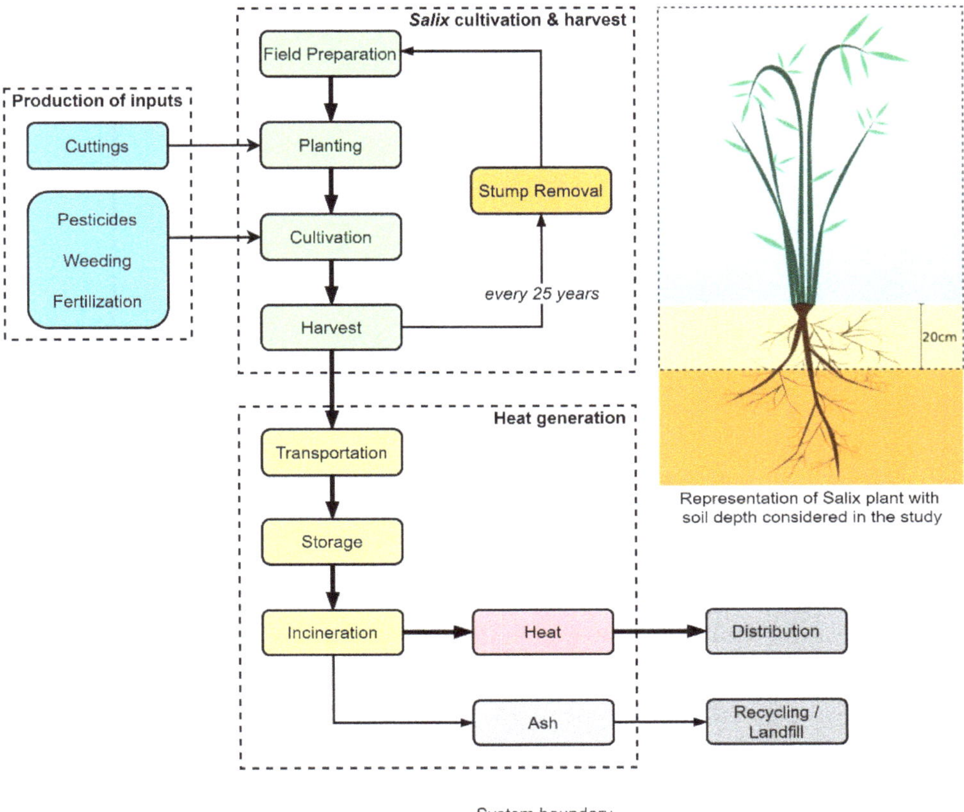

Figure 1. System boundaries (dotted lines) showing the processes considered within the study. Greenhouse gas and energy fluxes associated with the processes within the system boundaries were included in the analysis.

2.3. Field Operations and Management

The SRC *Salix* system followed a typical three-year cutting cycle, with the *Salix* harvested and chipped on-site at the end of every third growth cycle. The *Salix* then regrew from the stumps left in the field. According to current practical recommendations [49], one rotation period was assumed to last 25 years, after which the stumps would be broken up and removed and a new rotation would be established with new cuttings. The study period for the system was set to 50 years, which resulted in two rotation cycles. Technologies and management practices were assumed unchanged during this period. The data and assumptions used to calculate energy and emissions associated with the production of inputs and processes can be found in the Supplementary Material (Tables S6 and S7).

The harvest period for SRC systems is usually during winter months because the biomass is drier, the plant is dormant and the hard frozen soil provides a higher machinery carrying capacity [49,50]. It was assumed that the conventional method of harvesting and direct chipping was followed. Thereafter, the chips were transported to a heating plant for production of heat. The average road transportation distance was set as 40 km in this study.

2.4. Thermochemical Conversion

The higher heating value (HHV) of the *Salix* chips was considered to be 19.9 GJ/Mg DM (dry and ash-free), based on which the lower heating value (LHV) adjusted for moisture content was calculated [51,52]. The average storage period of the chips was 30 days, during which 3% dry matter loss occurred. The heating plant produces heat from biomass incineration and is equipped with flue gas condensation, which raises the overall efficiency. The energy efficiency for heat and flue gas condensation is 84% and 10% respectively (LHV basis), which gives an overall energy efficiency of 94%. The ash produced from biomass incineration was assumed to be transported by road for an average distance of 100 km. Calculation of ash quantities was on ash content of 3% in the *Salix* biomass [53]. The downstream processing and end-use of the ash were deemed outside the system boundaries.

2.5. Reference System

The reference energy system in this study was a fossil fuel-based energy generation system. A natural-gas-powered incineration plant supplied heat equivalent to the amount generated in the same year from the SRC *Salix* system. The alternative land use scenario was green fallow. The modelled SOC increase and use of fossil fuel for topping the land annually were included in the LCA. Assumptions concerning emissions and energy modelling are included in the Supplementary Material (Table S8).

2.6. Energy Performance Indicator

Energy performance was quantified by the indicator energy ratio (ER), which is the ratio between the delivered usable energy (thermal energy in this case) and the total primary energy input to the system [54,55]:

$$ER = \frac{Delivered\ energy\ (E_{out})}{Energy\ Inputs\ (E_{in})} \tag{1}$$

The delivered energy (E_{out}) is the energy produced (as heat) from the heating plant. Energy inputs (E_{in}) is the sum of all primary energy inputs associated with field processes and management, machinery operation, and production of inputs (fertilizers, pesticides and cuttings). E_{in} excludes the energy contained in the *Salix* biomass produced by cultivation.

This means that the losses in the thermochemical conversion process are excluded, but they indirectly reduce the delivered energy (E_{out}). The ER metric is dimensionless and describes the useful energy produced per unit of energy consumed.

2.7. Mineral Fertiliser

Addition of nitrogen in the form of mineral fertilizers and biomass entering the soil lead to direct and indirect emissions of N_2O. The amount of fertilizer was set according to the levels used in the field studies, where all fertilized treatment plots received 100 kg N, 14 kg P, 47 kg K per hectare annually, excluding the year of establishment [44].

The direct (N_2O_{direct}) and indirect ($N_2O_{indirect}$) emissions were calculated as:

$$N_2O_{direct} = EF_N \cdot \left(N_{applied} + N_{litter} + N_{roots}\right) \cdot \frac{44}{28} \tag{2}$$

$$N_2O_{indirect} = N_{applied} \cdot (F_A \cdot EF_D + N_{leached} \cdot EF_L) \cdot \frac{44}{28} \tag{3}$$

where $N_{applied}$ is the nitrogen applied by mineral fertilizer, N_{litter} and N_{roots} is the nitrogen contained in aboveground litter and roots respectively, and $N_{leached}$ is the nitrogen lost by leaching. EF_N, EF_D and EF_L are emission factors for direct emissions from applied nitrogen, indirect emissions from volatilization and re-deposition, and leaching respectively. F_A represents the fraction of applied nitrogen emitted as ammonia. The fraction $\frac{44}{28}$ converts nitrogen to N_2O. The emissions are calculated using default parameter values from IPCC [56], and are presented in Table A1, Appendix A. The same methodology was followed to calculate emissions from the fallow reference case.

N_2O emissions from biomass residues were based on the nitrogen content in *Salix* leaf litter reported for the selected varieties by Weih and Nordh [41] (details in Table S4 in Supplementary Material) and for stems as 0.43% (of total solids) [41]. Root nitrogen content was calculated from a dataset by Manzoni et al [57]. The estimated mean nitrogen content of roots from plants with low and high fertilization was 0.83% and 1.76% (of total solids) respectively (Table S5 in Supplementary Material). There are few studies on nitrogen content between different plant components, especially among different *Salix* varieties.

2.8. Soil Carbon

Soil carbon balances were calculated using the regional Introductory Carbon Balance Model (ICBMr) [46,58]. While the field trials provide measured SOC change for the first 17 years, the soil carbon modelling was used to estimate the SOC sequestration over the study period of 50 years. The model calculates the carbon flux based on variable annual inputs and regional differences. The ICBM model compartmentalizes the soil carbon into two pools, a young pool (Y) and an old pool (O), and the dynamics are governed by five parameters (i, k_y, k_o, h and r_e). The annual carbon input, denoted i, enters the young pool primarily in the form of leaf litter and dead roots. Both the young and old carbon pools undergo decomposition according to first-order kinetics as determined by decay constants k_y and k_o, respectively. The humification coefficient h denotes the fraction of the young pool that enters the old pool, while the remainder returns to the atmosphere as CO_2 emissions. The variable r_e represents the effect of external factors (mostly climatic and edaphic) on the decomposition rates. The initial calibration of the model was carried out using data from the Ultuna long-term field trial [59]. The ICBM parameters from the long-term trials are the basis of the parameters used in our study for SOC modelling as the long-term field trials are in the same region as our study.

The humification factor (h) varies depending on biomass quality and studies have indicated that roots can contribute more to SOC than aboveground residues [60]. *Salix* fine roots specifically have been shown to have higher turnover rates [48]. Therefore, the model was modified to represent the two different input biomass types—aboveground inputs (i_a) and belowground inputs (i_b), with separate humification coefficients (h_a and h_b). Hence, there were two parallel young pools, a young pool representing the aboveground biomass input (Y_a) and a young pool representing the belowground input (Y_b). Equations (4) and (5) were used to calculate the SOC stock with an annual time step:

$$Y_{[a,b]}(t) = \left(Y_{[a,b]_{t-1}} + i_{[a,b]_{t-1}}\right) * exp^{-k_y r_e} \tag{4}$$

$$\begin{aligned} O(t) = &\left(O_{t-1} - \left(\frac{h_a.k_y}{(k_o-k_y)}\left(Y_{a_{t-1}} + i_{a_{t-1}}\right) + \frac{h_b.k_y}{(k_o-k_y)}\left(Y_{b_{t-1}} + i_{b_{t-1}}\right)\right)\right) \cdot exp^{-k_o r_e} \\ &+ \left(\frac{h_a.k_y}{(k_o-k_y)}\left(Y_{a_{t-1}} + i_{a_{t-1}}\right) + \frac{h_b.k_y}{(k_o-k_y)}\left(Y_{b_{t-1}} + i_{b_{t-1}}\right)\right) \cdot exp^{-k_y r_e} \end{aligned} \tag{5}$$

The aboveground input, i_a, consists of the leaf litter. The belowground input, i_b, consists of the yearly fine root turnover and the accumulated coarse roots and stumps broken up and added to the soil after each 25-year rotation. The sum of the young and old pools represents the total SOC content at the specific point in time. Based on Kätterer et al. [60], h_b was assumed to be 2.3 times the value of h_a. The parameters were estimated from previous SOC studies [17,33,35,38] on *Salix* using the same methodology. The parameter details of the ICBM model are included in supplementary material (Tables S1 and S2).

2.9. Biomass Production Allocation

The standing biomass in *Salix* plants was divided into two major pools, aboveground and underground. The aboveground pool consisted of the stems (S) and leaves (L), while the underground pool consisted of the fine roots (F) and coarse roots (C). The stump material was included in the coarse root pool. The biomass growth allocation for these pools in a 3-year growing cycle are included in the Supplementary Material (Table S3). The ratio of 3-year accumulated net primary production (NPP) of aboveground biomass to belowground biomass, denoted as η was calculated as:

$$\eta = \frac{S + L}{F + C} = \frac{(1 + a)S}{(1 + b)F} \tag{6}$$

where S, L, F and C are the net production of stems, leaves, fine roots and coarse roots (including stumps), respectively over the 3-year cutting cycle period, a is the ratio of leaves to stems and b is the ratio of coarse roots to fine roots.

The differences in growth patterns between the various *Salix* varieties and treatments can be expected to lead to variation in values of η between them. Thus, varying the ratio η would lead to different input parameters (i_a and i_b), resulting in different SOC values calculated by the ICBM model. This would lead to differences in biomass input between the varieties and variations in SOC accumulation. The ratios a and b were determined from lysimeter studies on *Salix* growth by Rytter [61] to be 0.244 and 0.238, respectively, and are considered to remain unchanged between the different *Salix* varieties. Introduction of the factor η was an attempt to represent the impact of genetic differences between *Salix* varieties on plant growth and biomass allocation.

Rytter and Hansson [62] found that around 70% of total fine root biomass lies in the upper 20 cm of the soil profile. Based on this, annual root biomass input in the 0–20 cm soil layer was set to 70% of annual root NPP. For the equivalent green fallow reference case, the root biomass was 60% of the root NPP in the 0–20 cm layer [63].

The ICBM model was used to calculate the SOC change in the 0–20 cm soil layer for the 17-year period. The above-to-below ground accumulation ratio (η) was adjusted until the calculated SOC values from the ICBM model matched the measured SOC values from the field trials for all six varieties and treatments. The η values obtained by this method are presented in Table 3.

Table 3. Ratio of aboveground to belowground biomass accumulation (η) over 3-years for the different *Salix* varieties and treatments obtained from optimization of the ICBM soil carbon model with field-based soil organic carbon measurements. F0 and F+ refer to the unfertilized and fertilized treatments respectively.

Parameter	Treatment	Björn	Gudrun	Jorr	Loden	Tora	Tordis
η	F0	1.80	0.85	0.40	0.55	0.80	1.20
	F+	8.00	1.85	1.85	2.00	2.30	3.75

2.10. Climate Impact

In the normalized GWP_{100} metric, the cumulative warming potential of a GHG emission is represented relative to that of CO_2 for a 100-year period [64] and expressed in CO_2-equivalents. The emissions of CO_2, CH_4 and N_2O are multiplied by their respective characterization factors and summed to arrive at the total GWP_{100}. While this is a simplified and popular metric for representation of climate impacts, GWP_{100} does not capture the effects of timing of the emissions and their absolute impacts on the ecosystem [30,54].

Absolute global temperature change potential (AGTP), also referred to as ΔTs, is a metric that takes into account the timing of emissions and represents the climate impact as a change in temporal global mean surface temperature [65]. Using an absolute metric like AGTP displays the climate impact from a GHG emission as change in temperature (ΔTs), which approaches the actual physical effect on global temperature but increases uncertainty. This time-dependent LCA methodology, developed by Ericsson et al. [35], was used here as a climate impact indicator in addition to GWP_{100}.

Emission of a GHG at a particular point in time leads to a change in its atmospheric concentration which affects the radiative forcing (RF). This leads to a change in the energy balance on Earth, which results in an increase or decrease in temperature represented as ΔTs [35,56]. GHGs vary in their radiative efficiency and atmospheric residence time, e.g.,

N_2O and CH_4 have atmospheric residence times of 12.1 and 12.4 years, respectively, while CO_2 stays in the atmosphere until it is absorbed by the ocean or biosphere [66]. The lifetime of CO_2 is modelled based on the Bern carbon cycle. The temperature response of a GHG ($AGTP_x$) is defined as:

$$AGTP_x(H) = \int_0^H RF_x(t)R_T(H-t)dt \left(\mathrm{K\,kg_{gas}}^{-1}\right) \tag{7}$$

which represents the complex interaction between radiative forcing (*RF*) and the temperature response function (R_T) caused by a unit change in RF due to a pulse emission of a GHG '*x*' at a specific time interval (*t*), and '*H*' is the timeframe of the study. The parameter R_T captures the change in temperature due to the change in RF because of emission or uptake of a GHG (*x*) from the atmosphere at time interval (*t*). Integrating over the studied period '*H*' gives the temperature response for a particular GHG ($AGTP_x$) in terms of $\mathrm{K\,kg_{gas}}^{-1}$. The overall temperature response (ΔTs, measured in K) is the summation of the AGTP of the individual GHG emissions over the study timeframe '*H*'. A detailed explanation of the methodology is given in Ericsson et al. [35].

The time-dependent climate impact methodology requires the creation of an inventory of GHG emissions and uptakes distributed over time of the study. Individual temperature responses of each emission are calculated from this inventory. The total system response (ΔTs) is obtained by summing the individual responses and can be plotted as the change in temperature over time.

2.11. Sensitivity Analysis

Even with accurate data collection and standardized methods, uncertainties are unavoidable due to the multiple assumptions and variability involved in modelling and LCA approaches. Sensitivity analysis makes it possible to understand how different factors influence the final results of the analysis [67].

The setting of the system boundary to 20 cm of soil depth is a source of uncertainty. This is a type of parameter uncertainty and model uncertainty, as change in depth of soil profile changes the system boundaries of the model and related parameters such as SOC values and inputs from BGB. To assess how a greater soil profile would influence the SOC modelling and climate impacts from the different *Salix* varieties, a one-at-a-time sensitivity analysis was performed. The system boundary was adjusted to include a soil depth of 25 cm and related parameter of below ground input (i_b) and initial and final SOC values were changed, while other parameters in the analysis remained constant. The average plough depth of 20–25 cm was the motivation for limiting the soil profile depth, as the subsoil characteristics at the site (both before and after establishment of *Salix*) were not known.

In soil carbon modelling, the average SOC stock in the 20–25 cm layer was estimated to be half of the stock in the 10–20 cm layer for each of the varieties described in the previous sections. The root biomass input for *Salix* and the reference fallow case was 80% and 65% of the annual belowground NPP, respectively based on studies of root distribution for *Salix* [48] and grasses [63,68]. The root distribution is subject to variability due to factors such as soil and climate, and hence is a potential source of uncertainty.

3. Results

3.1. Energy Use and Efficiency

Regarding energy performance, the fertilized treatments of varieties 'Tordis', 'Björn', 'Tora' and 'Jorr' performed best in the ambient conditions, with ERs ($\mathrm{GJ_{out}\,GJ_{in}}^{-1}$) of 28.2, 26.5, 25.1 and 24.7, respectively (Table 4). Among the unfertilized varieties, 'Tordis' and 'Björn' gave the best energy performance, with ERs of 47.7 and 48.2, respectively. Average annual net heat output varied from 69 to 234 GJ $\mathrm{ha^{-1}\,year^{-1}}$ between the different *Salix* varieties and treatments. Fertilized 'Tordis' had the highest primary energy input of all the varieties as it had the highest yield levels, leading to high biomass and heat output

(234 GJ ha^{-1} year^{-1}). Fertilization of 'Gudrun' and 'Loden' did not lead to major improvement in their yield over the unfertilized treatment, which resulted in relatively poor energy performance of the fertilized treatment of these two varieties. Among the unfertilized treatments, the variety 'Björn' had the highest annual heat output, 150 GJ ha^{-1} year^{-1}. The energy output from the heating plant is directly proportional to the biomass yield, which was higher when the plots were fertilized. Hence, the energy outputs in the form of heat were consistently higher for the fertilized treatment compared with the unfertilized treatment. The primary energy input for the fertilized treatment of each variety was about 2.5–4.6 higher than in the equivalent non-fertilized treatment. Consequently, ERs for the unfertilized cases was much higher than in the fertilized cases.

Table 4. Primary energy input, heat output, energy in biomass and energy ratio for the six SRC *Salix* varieties in fertilized and unfertilized treatments during two rotation periods (years 0–50). F0 and F+ refer to the unfertilized and fertilized treatments, respectively.

Variety and Treatment	Energy in Biomass (GJ ha^{-1})		Primary Energy Input (GJ ha^{-1})		Net Heat Output (GJ ha^{-1})		Energy Ratio (GJ$_{out}$ GJ$_{in}$$^{-1}$)
	Total	Annual Average	Total	Annual Average	Total	Annual Average	
Björn F0	8245	165	156	3	7518	150	48.2
Björn F+	11252	225	388	8	10259	205	26.5
Gudrun F0	5525	111	110	2	5038	101	46.0
Gudrun F+	5562	111	290	6	5071	101	17.5
Jorr F0	3769	75	79	2	3437	69	43.3
Jorr F+	9834	197	364	7	8967	179	24.7
Loden F0	3757	75	79	2	3426	69	43.2
Loden F+	4956	99	280	6	4519	90	16.1
Tora F0	4789	96	97	2	4366	87	45.1
Tora F+	10177	204	369	7	9279	186	25.1
Tordis F0	7526	151	144	3	6863	137	47.7
Tordis F+	12859	257	415	8	11725	234	28.2

The contribution of the individual cultivation, transportation and handling processes to the total primary energy input over the study period are described in Table 5. The primary energy associated with pesticides, field preparation, production and planting of seedlings, and stump removal were the same for all six *Salix* varieties and treatments, as these processes are independent of variety type and fertilization. These are presented on a per hectare basis. The processes of harvesting, chipping, forwarding and transportation are directly proportional to the amount of shoot biomass produced, and hence are presented on basis of per GJ of energy in biomass. Production and spreading of fertilizers were the greatest contributor to primary energy input for fertilized cases, while it was zero for non-fertilized cases.

Table 5. Primary energy inputs by process category associated with the bioenergy system of six *Salix* varieties in fertilized and unfertilized treatments over the 50-year study period.

Process	Energy (GJ ha^{-1})	Unit
Pesticides [a]	4	GJ ha^{-1}
Field preparation [a]	6	GJ ha^{-1}
Planting & seedlings [a]	4	GJ ha^{-1}
Stump removal [a]	1	GJ ha^{-1}
Fertilizer [b]	0 or 180	GJ ha^{-1}
Harvest & chipping [c]	7.33×10^{-3}	GJ GJ$_{biomass}$$^{-1}$
Forwarding (field transport) [c]	3.58×10^{-3}	GJ GJ$_{biomass}$$^{-1}$
Road transport [c]	6.22×10^{-3}	GJ GJ$_{biomass}$$^{-1}$

[a] Processes which are equal for all varieties. [b] Primary energy associated with fertilization is zero for the unfertilized treatment. [c] These processes are proportional to the amount of biomass produced in the field.

3.2. Soil Organic Carbon

Soil carbon modelling results showed that all varieties and treatments led to an increase in SOC over the initial level in the topsoil (0–20 cm) during the study period (50 years) consisting of two rotation periods (Table 6). The SOC stock calculated by the ICBM model at the end of both the first rotation period (after 25 years) and second rotation period (after 50 years) are shown in Table 6. Fertilized 'Loden' and 'Björn' showed the lowest net increase in SOC during the 50-year period, 15.8 and 13.3 Mg ha^{-1}, respectively. These values were only slightly greater than the SOC increase for the fallow reference case (9.5 Mg C ha^{-1}).

Table 6. Initial, total and net soil organic carbon increase in the 0–20 cm soil layer after two rotation periods (50 years), as calculated by the ICBM soil carbon model. F0 and F+ refer to the unfertilized and fertilized treatments, respectively.

Variety and Treatment	Initial SOC Stock (0–20 cm) (Mg ha^{-1})	SOC Stock after 25 Years (0–20 cm) (Mg ha^{-1})	Total SOC Stock after 50 Years (0–20 cm) (Mg ha^{-1})	Net SOC Increase (0–20 cm) (Mg ha^{-1})	Annual SOC Increase (0–20 cm) (Mg ha^{-1} yr^{-1})	Change in SOC after 50 Years (%)
Björn F0	28.9	52.0	66.1	37.2	0.74	129
Björn F+	28.9	37.2	42.2	13.3	0.27	46
Gudrun F0	28.9	60.4	79.4	50.5	1.01	175
Gudrun F+	28.9	42.1	49.8	20.9	0.42	73
Jorr F0	28.9	74.6	102.4	73.6	1.47	255
Jorr F+	28.9	56.7	73.7	44.9	0.90	155
Loden F0	28.9	61.1	80.4	51.5	1.03	178
Loden F+	28.9	38.9	44.6	15.8	0.32	55
Tora F0	28.9	57.3	74.3	45.5	0.91	158
Tora F+	28.9	52.3	66.6	37.7	0.75	131
Tordis F0	28.9	60.4	79.6	50.7	1.01	176
Tordis F+	28.9	48.7	60.9	32.1	0.64	111
Reference-Fallow	28.9	34.3	38.4	9.5	0.19	33

The carbon modelling results also showed that the unfertilized treatment for each variety was able to sequester about 1.6 to 3.3 times more SOC than the fertilized case, except for 'Tora'. Both treatments of 'Tora' led to a similar increase in SOC stock in the topsoil.

The low-yielding variety 'Jorr' showed the greatest potential for net carbon sequestration, capturing 73.6 Mg C ha^{-1} and 44.9 Mg C ha^{-1} over 50 years in the unfertilized and fertilized treatments, respectively. The variety 'Gudrun' had similar biomass yields for both the fertilized and unfertilized treatments (Table 1), but net SOC increase in the unfertilized case was almost double that in the fertilized case. 'Björn' had high biomass yields, but the SOC increase was at the lower end of the spectrum. Thus, no clear correlation between biomass yield and net SOC increase was established. These results indicate that the impacts on SOC are variety-specific, and that fertilization in general leads to lower net SOC increase.

3.3. Time-Dependent Climate Impact

3.3.1. Impact Per Hectare of Land (Including Substitution Effects)

All *Salix* varieties and treatments gave a negative temperature response (ΔTs) over the study period, which equated to a lowering of the global mean temperature when substituting reference fossil energy (natural gas) and reference land use (fallow) (Figure 2). There was great variation in temperature response between the varieties, from -2.15×10^{-10} K ha^{-1} for fertilized 'Loden' to -5.99×10^{-10} K ha^{-1} for fertilized 'Tordis'. Fertilized 'Tordis', 'Björn', 'Tora' and 'Jorr' had the greatest negative ΔTs per hectare of land, which is explained by the high levels of yield combined with an increase in SOC stocks. These cases represent the best use of land area under the study conditions for climate change mitigation.

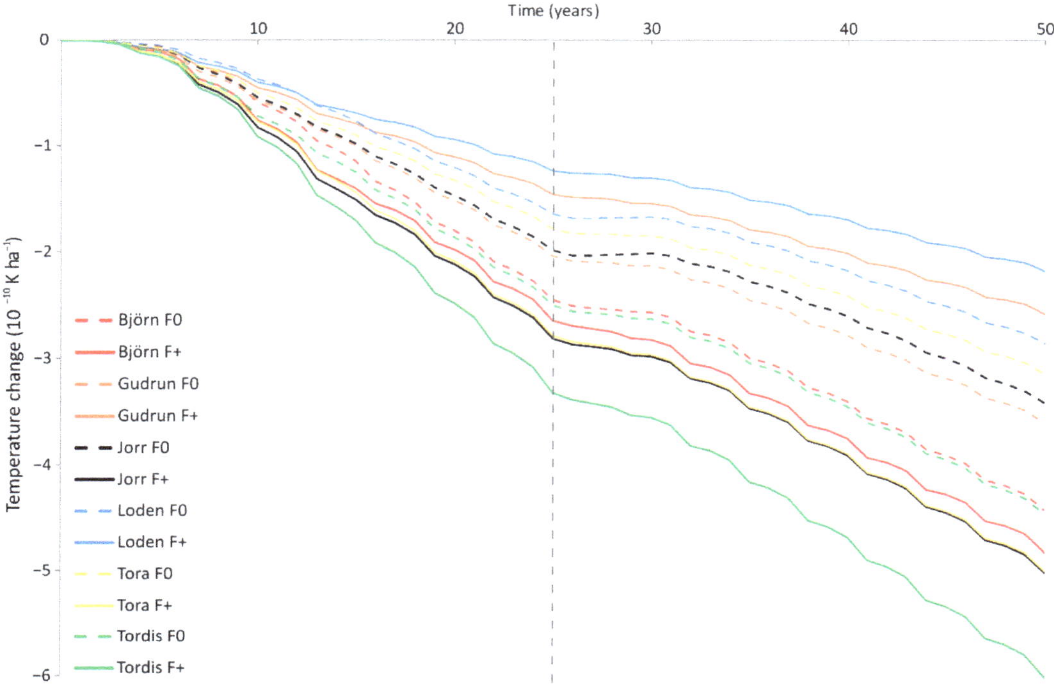

Figure 2. Time-dependent temperature response of the *Salix* SRC systems with substitution effects included. The vertical dashed line represents the end of the first rotation and start of the second (at 25 years). F0 and F+ refer to the unfertilized and fertilized treatments, respectively.

Although the unfertilized treatment of each variety had greater CO_2 sequestration potential, the increase in biomass output achieved by fertilization led to higher replacement of fossil energy. As a result, fertilized cases had lower ΔTs values. 'Loden' and 'Gudrun' were exceptions, as their fertilized cases showed a greater temperature response than the unfertilized cases. These two varieties gained little to no improvement in their yield from fertilization, so the additional energy and material input through fertilization led to a lower climate mitigation potential.

3.3.2. Impact Per Unit of Heat Output (Including Substitution Effects)

A different picture emerges when the climate impacts from all cases were expressed based on their function of delivering energy services (per MJ_{heat}) and replacing fossil-generated heat (Figure 3). Unfertilized 'Jorr' showed the greatest climate mitigation effect (-5.11×10^{-15} K MJ^{-1}), while fertilized 'Björn' (-2.39×10^{-15} K MJ^{-1}) had the lowest. The non-fertilized varieties showed a greater negative temperature response (per MJ_{heat}) than the fertilized varieties. This can be attributed to the higher primary energy demand for the fertilized treatments, combined with the greater SOC increase for the unfertilized cases.

The unfertilized cases were more favorable for climate change mitigation on comparing when the climate impacts per unit of energy delivered (MJ_{heat}) by the biomass systems. This is relevant when comparing energy generation systems and land is not a restricted resource. Unfertilized 'Jorr' and 'Loden' were the best-performing varieties in terms of potential for temperature reduction per unit of energy, although they had the lowest biomass yield. Fertilized 'Loden', 'Gudrun' and 'Björn' had the lowest temperature decrease (ΔTs per MJ_{heat}) over the study period. Those cases also had the lowest SOC increase over the study period.

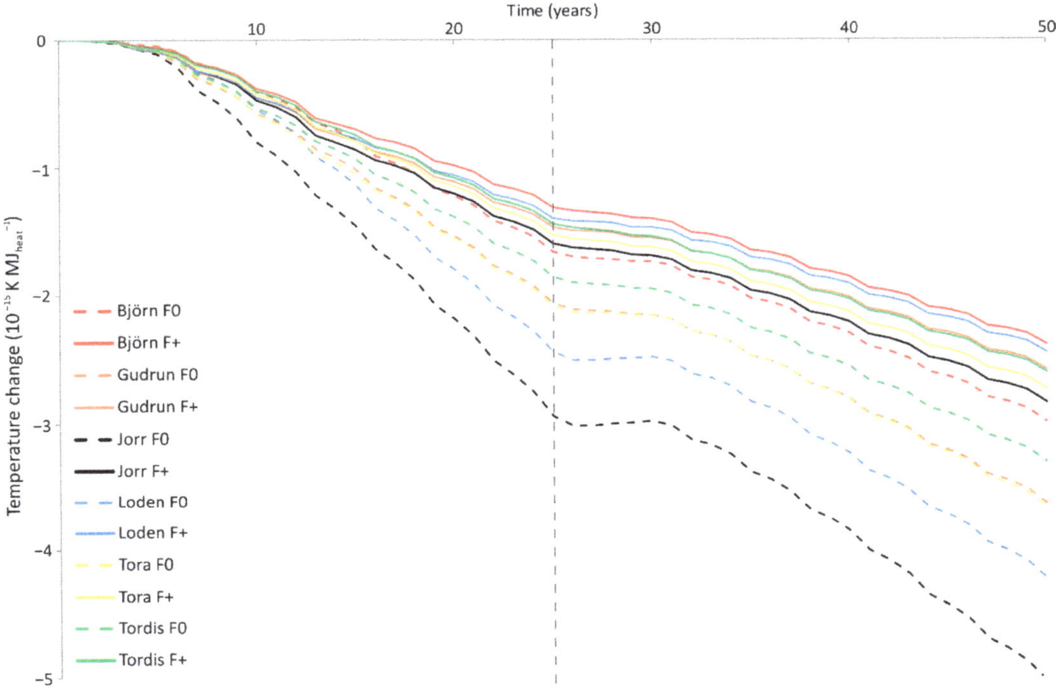

Figure 3. Temperature response per MJ of heat for the *Salix* SRC systems, with substitution effects included. The vertical dashed line represents the end of the first rotation and start of the second (at 25 years). F0 and F+ refer to the unfertilized and fertilized treatments, respectively.

3.4. Global Warming Potential

The life cycle impact assessment of the different varieties under the two fertilization regimes showed varying climate impacts. A negative value of the GWP_{100} metric means that there is a net reduction of atmospheric GHG concentration, leading to a climate mitigation effect. In absolute terms (not including the effect of substituting the reference case), unfertilized 'Jorr' had the lowest GWP_{100} (−333 Mg CO_2-eq. ha^{-1}), while fertilized 'Björn' had the highest total GWP_{100} (30 Mg CO_2-eq. ha^{-1}) (Table 7).

Among the fertilized varieties, 'Björn' and 'Loden' were the worst performing in terms of climate mitigation effects per hectare over 50 years. These varieties had the lowest increase in SOC among the fertilized varieties, which contributed to their poorer climate performance. Fertilized 'Tora' and 'Jorr', which had the highest increase in SOC among fertilized varieties, showed the greatest reduction in GWP, indicating the importance of soil carbon sequestration for achieving a climate change-mitigating effect.

Considering the effects of substitution of a natural gas-based reference system for the SRC *Salix*, all varieties showed a climate-mitigating effect during the study period. The magnitude of the mitigation effect ranged from −312 Mg CO_2-eq.ha^{-1} for fertilized 'Loden' to −858 Mg CO_2-eq. ha^{-1} for fertilized 'Tordis'. On considering the substitution effects, the yield level influenced GWP. High yields contributed to a greater climate mitigation effect, as seen for fertilized 'Tordis', 'Björn', 'Jorr' and 'Tora'. This is a result of avoided equivalent emissions from heat produced in the fossil reference system.

Table 7. Global warming potential (GWP_{100}) for the *Salix* cropping systems and fossil-powered reference system and effect of substitution when *Salix* was assumed to replace the reference system. The GWP is expressed in both Mg CO_2-eq per hectare and g CO_2-eq per MJ of heat during the 50-year study period. F0 and F+ refer to the unfertilized and fertilized treatments, respectively. A positive value indicates emissions to atmosphere, and a negative value indicates reduction.

Variety and Treatment	Global Warming Potential (GWP_{100})					
	SRC System [a]		Reference System [b]		Substitution Effect [c]	
	Mg ha^{-1}	g MJ^{-1}	Mg ha^{-1}	g MJ^{-1}	Mg ha^{-1}	g MJ^{-1}
Björn F0	−143	−19	514	68	−657	−87
Björn F+	30	3	707	94	−677	−91
Gudrun F0	−220	−44	339	67	−558	−111
Gudrun F+	−31	−6	340	68	−372	−74
Jorr F0	−333	−97	225	65	−558	−162
Jorr F+	−114	−13	616	179	−730	−192
Loden F0	−231	−68	225	66	−456	−133
Loden F+	−10	−2	301	88	−312	−90
Tora F0	−198	−45	291	67	−489	−112
Tora F+	−84	−9	638	146	−722	−155
Tordis F0	−210	−31	467	68	−678	−99
Tordis F+	−47	−4	811	118	−858	−122

[a] Climate impact of SRC *Salix* system without substitution effect. [b] Climate impact of reference system—heat from natural gas and green fallow land use. [c] Climate impact of SRC *Salix* system including substitution effects of reference system.

From the perspective of heat delivered with substitution effects, fertilized 'Jorr' had the highest climate mitigation effect, −192 g CO_2-eq.$MJ_{heat}{}^{-1}$ produced, while fertilized 'Gudrun' was at the other end of the spectrum, with −74 g CO_2-eq.$MJ_{heat}{}^{-1}$ produced.

The contribution of the *Salix* production chain emissions, SOC sequestration and substitution effects to the overall net GWP_{100} per hectare for the different *Salix* varieties are presented in Figure 4. The production chain leads to GHG emissions while SOC sequestration and substitution effects remove or replace GHG emissions. Emissions from the production chain (field operations, transportation, fertilizer and soil emissions) are higher for fertilized varieties due to fertilizer production and greater soil N_2O emissions. The substitution effects are the main contributor to the overall negative GWP_{100} for all *Salix* varieties, except for unfertilized Loden and Jorr. These two varieties showed a greater potential of SOC sequestration relative to harvest yields in comparison to the other *Salix* varieties. Alternatively fertilized Gudrun and Loden have a higher GWP_{100} compared to their unfertilized counterparts due to relatively lower improvement in yield.

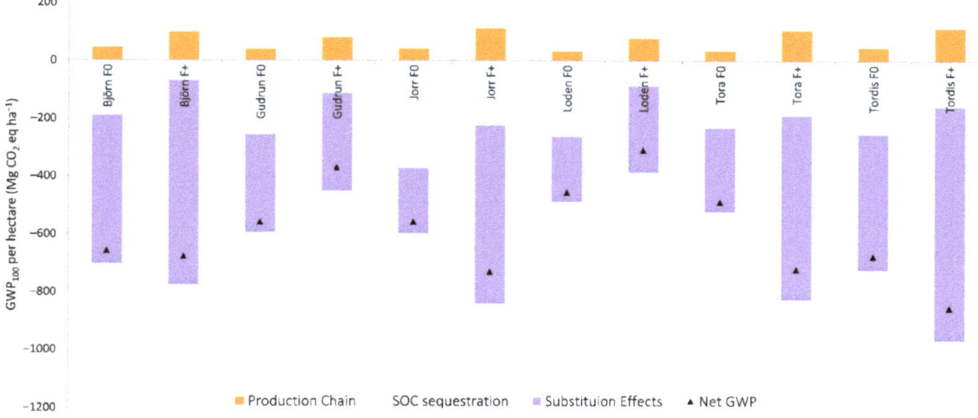

Figure 4. Contribution of the *Salix* production chain, SOC sequestration, substitution effects (from replacing green fallow and fossil energy) to the net GWP_{100} per hectare of each of the *Salix* bioenergy system.

3.5. Sensitivity Analysis

The sensitivity analysis results for net SOC increase and climate impacts (GWP_{100}) from considering a soil depth of 25 cm, compared with the base case of 20 cm, are shown in Table 8. Generally, a deeper soil layer gave a greater net SOC increase within the system boundary, leading to a lower climate impact. Fertilized 'Björn', 'Gudrun' and 'Loden' were exceptions to this, as the net SOC increase in the 0–25 cm layer was smaller than in the 0–20 cm layer. Consequently, the climate impacts for these three cases were also greater.

Table 8. Sensitivity analysis of soil organic carbon (SOC) sequestration and global warming potential (GWP_{100}) for the six *Salix* varieties in the fertilized and unfertilized treatments, when soil depth considered was increased from 20 to 25 cm. F0 and F+ refer to the unfertilized and fertilized treatments, respectively.

Variety and Treatment	0–20 cm Soil Layer				0–25 cm Soil Layer			
	Net SOC Increase	Annual SOC Uptake	GWP_{100}	GWP_{100}	Net SOC Increase	Annual SOC Uptake	GWP_{100}	GWP_{100}
	(Mg ha^{-1})	(Mg ha^{-1} yr^{-1})	(Mg ha^{-1})	(g MJ^{-1})	(Mg ha^{-1})	(Mg ha^{-1} yr^{-1})	(Mg ha^{-1})	(g MJ^{-1})
Björn F0	37.2	0.74	−143	−19	40.1	0.80	−100	−13
Björn F+	13.3	0.27	30	3	10.8	0.22	58	6
Gudrun F0	50.5	1.01	−220	−44	64.1	1.28	−193	−38
Gudrun F+	20.9	0.42	−31	−6	19.7	0.39	9	2
Jorr F0	73.6	1.47	−333	−97	82.0	1.64	−258	−75
Jorr F+	44.9	0.90	−114	−13	45.7	0.91	−55	−6
Loden F0	51.5	1.03	−231	−68	57.1	1.14	−175	−51
Loden F+	15.8	0.32	−10	−2	13.9	0.28	26	6
Tora F0	45.5	0.91	−198	−45	50.1	1.00	−147	−34
Tora F+	37.7	0.75	−84	−9	50.1	1.00	−34	−4
Tordis F0	50.7	1.01	−210	−31	58.2	1.16	−165	−24
Tordis F+	32.1	0.64	−47	−4	33.6	0.67	−9	−1

Fertilized 'Björn', 'Gudrun' and 'Loden' showed the lowest SOC increase in field measurements from 2001–2018, which led to lower SOC sequestration rates. On considering a deeper soil layer, the starting SOC level prior to *Salix* establishment was also higher. In absolute terms, the final SOC stock was greater with a deeper soil layer, but the net increase was lower for these three cases when compared with a shallower (20 cm) layer. Thus, a lower sequestration rate combined with a greater initial SOC level led to a smaller SOC increase for these fertilized varieties with increased soil depth. Overall, the changes in SOC stock and climate impacts were not highly influenced by considering a deeper soil layer of 25 cm.

4. Discussion

The analysis revealed that cultivation of the selected *Salix* varieties for bioenergy to substitute equivalent fossil fuels (under the given environmental and site conditions) can potentially mitigate climate change as it has a net cooling effect on global mean surface temperature over a 50-year time horizon. *Salix* variety had a major influence on the climate change mitigation potential. The *Salix* varieties in this study varied in some key factors derived from measured field data (SOC sequestration, biomass yield and response to fertilization) and these factors affected the overall climate impact between the different varieties. The major contribution to the climate mitigation effect comes from substitution of fossil fuels and SOC sequestration. While fossil fuel replacement is relatively easy to estimate using harvest yields, estimation of SOC change over time is complicated as it is subject to various environmental conditions and uncertainties.

The flue gas condensation technology assumed in the incineration plant with heat recovery gives high energy efficiency, leading to a greater output of energy delivered which puts the energy ratio in the higher range. This is a common technology in Swedish power plants [64], although it might not be common in other countries. The conversion efficiency of the thermochemical processes selected in a study determines the amount of useful energy production from the system and its subsequent ER. The ER in this study was within the range 16.1–28.2 for fertilized *Salix* varieties and 43.2–48.2 for non-fertilized varieties. Values of ER reported in the literature range from 16 to 79 in energy performance analysis studies [69–76], which are indicative of different methods and assumptions considered in individual studies.

The results in the present study indicate that *Salix* variety and fertilization regime strongly affect the NPP distribution between aboveground and belowground biomass. The ratio of NPP of annual aboveground biomass (AGB) to belowground biomass (BGB) in our study was estimated at 0.4–1.8 for unfertilized treatments and 1.9–8.0 for fertilized treatments (Table 3). The estimation of these values is based on the well-established conception that variety and fertilization influence the production of BGB relative to AGB, which leads to variation in SOC change.

Data on AGB and BGB production and allocation for *Salix* from some studies are presented in Table 9. Heinsoo et al. [77] reported large differences in the magnitude of the ratio between AGB and fine root production for fertilized and control plots in an Estonian *Salix* plantation with two species (*S. viminalis* and *S. dasyclados*). This study reported a significant reduction in annual production of fine root biomass under fertilization, while AGB production was greatly improved. The AGB to BGB production ratio for *S. viminalis* was 1.04–2.07 for lysimeter-grown *Salix* in sandy and clayey soils [61]. Rytter [78] found significant differences in biomass allocation to fine-roots between N-limited and unlimited growing conditions (*for S. viminalis*) but no change in annual turnover rates of fine roots. These studies support the idea that fertilization can lead to lower BGB production that leads to very different AGB to BGB ratios between unfertilized and fertilized treatments.

Table 9. Aboveground to Belowground biomass production and allocation ratios of *Salix* varieties reported under different environmental conditions.

Study	Description	Value
Heinsoo et al., 2009 [77]	Ratio of aboveground to fine root annual production	
	S. viminalis control	1.16–1.09
	S. viminalis fertilized	14.28–12.5
	S. dasyclados control	2.85–1.51
	S. dasyclados fertilized	20–16.67
Rytter, 2001 [61]	Ratio of total aboveground to belowground production of *S. viminalis L*	
	Year 1	1.04–0.73
	Year 2	1.73–2.07
	Year 3	1.63–1.5
Rytter, 2013 [78]	Ratio of annual production of stem to fine root of *S. viminalis L*	
	N limited	0.65
	Unlimited	1.84
Pacaldo et al., 2013 [79]	Ratio of biomass allocation of Aboveground biomass (Stem + Leaf) to Belowground biomass (FR + CR + stool) of *S. dasyclados*	0.32–0.61

Pacaldo et al. [79] reported biomass allocation for a single *Salix* variety (*S. dasyclados*) from two locations with different plantation ages and soil conditions; based on their data, the AGB to BGB allocation ratio was 0.32–0.61. The ratio of annual production of AGB to BGB in our study falls within the range of values reported for *Salix* in different studies, but these figures need to be validated by further studies on belowground biomass to increase accuracy in soil carbon modelling estimates.

Salix roots are characterized by high growth and mortality rates [80] and are not bound by seasonal patterns, with some growth and decay observed even during winter [61]. This indicates that root production is relatively higher under a non-fertilized regime, which combined with unchanged turnover rates would lead to higher belowground biomass input to the soil compared with a fertilized treatment, which can lead to greater SOC stocks.

The few previous studies on how biomass growth and allocation differ between *Salix* varieties [43,81,82] have shown that variety and growing environment can have significant impacts on biomass allocation and growth patterns. Cunniff et al. [43] found that belowground allocation differed up to 10% between *Salix* varieties and up to 94% between locations. Furthermore, a study by Gregory et al. [81] found significant differences in root density between *Salix* varieties, especially in the upper layers.

There is a scarcity of data on belowground biomass allocation and its variation between *Salix* varieties and environmental conditions. Only a few studies measured the production and turnover of roots (especially fine roots) as these analyses are time-consuming, labor-intensive and expensive [83]. Furthermore, the estimation of root growth and number can greatly vary due to the measurement method used [84]. A study including two *Salix* varieties [85] has also shown differences in decomposition rates of fine root litters, which further stresses the need for variety focused studies. This makes it difficult and complicated to compare data on aboveground to belowground biomass accumulation from different sources, as variations can occur owing to multiple factors. This is a source of variability in determining especially belowground biomass growth and its contribution to SOC sequestration. There is need for further research and standardization of methods to enable comparisons and calibration of soil carbon models to make more reliable long-term predictions.

In spite of the uncertainties regarding the variety-related input variables for soil carbon modelling, this investigation provides useful insights into the expected variety-related SOC changes over a longer period of time and based on measured data of above ground biomass and soil SOC over an 18-year period. While these uncertainties might affect all investigated varieties in a similar way, they are likely to result mostly in an uncertain absolute magnitude of SOC after a certain period of time, whereas the variety-specific pattern of SOC change is expected to be more robust. Thus, we believe that the use of *Salix* variety-specific data from the field study in this analysis is a clear improvement over previous studies dealing with SOC modelling in *Salix*. The scaling and extrapolation of soil carbon models is a challenge due to lack of long-term data and the complexity of SOC sequestration mechanisms. Despite the challenges, such approaches with assumed data are a necessary part of making sustainable management decisions. The accuracy of the models and their predictions can be constantly adjusted by feedback of new measured data and advancing knowledge of SOC.

The carbon modelling based on measured SOC levels from the measured field trial data, showed that non-fertilization led to a greater increase in SOC compared with fertilization of the same variety under the same soil conditions. A relationship between shoot biomass yield and increase in SOC was expected from other studies, but was not seen in our study, as greater yield did not correlate with more CO_2 being sequestered in the soil. For example, unfertilized 'Jorr' had one of the lowest shoot biomass yields among all varieties investigated here, but showed the highest increase in SOC in the top 20 cm soil layer; while fertilized 'Björn', with high biomass output, had one of the lowest increases in SOC stocks. This result questions the common assumption of higher shoot biomass yield leading to a greater increase in SOC due to higher production of leaf and root litter. While greater shoot biomass may lead to increased leaf litter production, root litter production might show a differential pattern. Interestingly, Pappas et al. [86] found that in boreal forests, aboveground biomass growth is decoupled from the carbon input to the ecosystem, highlighting the significance of belowground carbon inputs independent from aboveground growth. Also, Khan et al. [87] conclude that N-fertilization increases harvests for crops but can have a negative effect on SOC sequestration.

The SOC accumulation rate in our study was 0.24–1.29 Mg ha^{-1} yr^{-1} for the 0–20 cm soil layer over 50 years. Direct comparisons of SOC changes reported in different studies are difficult, because of variations in initial soil conditions, study period, growing conditions, methodology and depth of soil profile considered in the study. The test site had a clay content of 18%. This clay content promotes long-term carbon sequestration by stabilization of SOC against decomposition [88]. SOC sequestration rates of 1.44–2.27 Mg ha^{-1} yr^{-1} for the top 30 cm soil layer have been reported for two *Salix* varieties during a 6-year study period in the UK [81]. Other recent studies have recorded SOC sequestration rates of 1 Mg ha^{-1} yr^{-1} in the upper 10 cm in Italy [89] and high levels of 6.7–10.2 Mg ha^{-1} yr^{-1} in the upper 60 cm in Belgium [90]. In a meta-analysis by Agostini et al. [18], SOC accumulation rates in the range −0.06 to 3.57 Mg ha^{-1} yr^{-1} Mg ha^{-1} yr^{-1} were found for *Salix*. However, the studies in the meta-analysis varied greatly in methodology, soil conditions and length of study period, impeding comparisons. Greater accumulation rates have been reported for *Salix* grown on former arable land compared with grassland [91]. The amount and rate of SOC change are highly dependent on the previous land use, which consequently plays a major role in the climate impact. In any case, the annual SOC accumulation rates in our study clearly fall within the range reported from other sources. However, there is a need for further investigation of root production, turnover and decay based on soil types, plant variety, and nutrient regimes because the soil carbon change has an important effect in determining the climate impacts and should therefore be included in systems studies

From a land use perspective, the climate impact was governed by the *Salix* biomass yield. Higher biomass yields contributed to a greater replacement of fossil energy, thereby contributing to a greater cooling effect. Exceptions to this were the varieties 'Gudrun' and 'Loden', which showed almost no improvement in yield from fertilization. Thus, for optimum climate mitigation per unit land area, a high-yielding variety needs to be selected. However, on comparing the varieties from the functional unit of energy output (per MJ of energy output), the SOC sequestration potential played the major role in determining the climate impact. In this regard, the unfertilized varieties with good yields and SOC sequestration potential offered greater potential cooling effects. Hence the basis of comparison (land use or energy output) also plays an important role in the interpretation of climate impact results.

A literature review by Djomo et al. [54] reported that LCAs of short-rotation bioenergy crops often use very different system boundaries, impact indicators and conditions, which makes comparisons between studies difficult. All scenarios analyzed in the present study showed a GWP reduction potential of 95 to 237% compared with the fossil reference system (Table 7). This is much higher than the 90–99% reduction potential presented in the review by Djomo et al. [54], but only one study in that review had considered the effects of soil carbon sequestration. The high yield levels for unfertilized *Salix* varieties in the present study, combined with SOC sequestration, explain the much higher GWP reduction potential estimated in our study. However, the soil is not an endless C sink and increasing temperatures under climate change will accelerate the degradation of SOC, thereby reducing the size of the sink. Thus, the SOC sequestration potential is expected to decrease over time because of climate change. It is difficult to predict technological change during a long period, so in this study the systems were assumed to remain static during the 50-year period. Assuming a constant level of cultivation of *Salix* at the same location, the cooling effect from an increasing SOC pool will eventually decline, but the warming effect due to GHG emissions from the production system will continue to increase over time. The major sources of emissions are production of fertilizers and N_2O soil emissions. From a longer time perspective, these emissions will be of uppermost importance in improving the climate performance of *Salix* production systems.

Default IPCC values for calculation of nitrogen leaching from mineral fertilizers were used in this study, due to lack of site-specific data. *Salix* has been shown to have lower nitrogen leaching rates than other crops [40,92], and thus the default values used here might be on the higher side for *Salix* cultivation. In the field trials, all fertilized plots were enriched with the same quantity of mineral fertilizer, which might be higher or lower than the optimal fertilization level of the plant. Fertilization studies can help to determine the optimum fertilization by variety, which will greatly influence the emissions and energy input of the fertilization phase and the AGB to BGB production ratio.

The scarcity of complete data that are site- and variety-specific for all aspects of the *Salix* bioenergy production and decomposition poses some limitations. The SOC changes and climate impacts from one study should not be directly extrapolated to other cases as there are several factors (such as environmental conditions and previous land use) which can lead to different results. The results of this study stress the importance of accounting for variety and fertilization effects when estimating SOC changes and climate impacts of *Salix* bioenergy systems. As such, these effects should not be ignored in planning for bioenergy systems of the future. There is potential to develop varieties with high levels of both shoot and root biomass with efficient fertilizer utilization, which would give a greater climate mitigation benefit.

5. Conclusions

Soil carbon modelling based on Swedish field trial data showed that all *Salix* varieties tested can potentially increase the SOC stock in the soil over a period of 50 years under given soil conditions of vertic cambisols. *Salix* variety and fertilization treatment determined the magnitude of CO_2 sequestration. No clear relationship was found between biomass yield and SOC sequestration potential across the varieties and soil type used in this study, which indicates that belowground biomass accumulation and decomposition should not be directly estimated from shoot yield alone. High production and turnover rate of fine roots was estimated to be the major contributor to SOC inputs by *Salix*. Fertilization led to an increase in biomass yield (and therefore energy output), but a decrease in SOC sequestration potential, across all varieties.

The fertilized 'Björn' biomass systems showed a warming effect on the climate (positive GWP) without inclusion of substitution effects from replacing a natural gas-based reference case. However, all varieties and treatments showed the potential to mitigate climate change (negative GWP and ΔTs) on inclusion of substitution effects. High-yielding *Salix* varieties had the greatest potential to mitigate climate change when looking from a land-use perspective. When comparing per energy unit, the SOC sequestration effects become more prominent in determining the overall magnitude of the climate change mitigation potential of the different *Salix* varieties. System analysis approaches like LCA should incorporate SOC effects, which can significantly affect the climate impacts of biomass cultivation systems, as seen here for six *Salix* varieties.

Initial soil conditions are very important for biomass productivity because they influence the amount of leaf and root litter produced, which in turn influence the SOC accumulation rate. Hence, previous land use needs careful consideration when evaluating climate impacts. Results in previous studies, combined with our findings, show that there is some uncertainty about SOC sequestration rates, which makes it important to research belowground biomass production, including varietal and location effects.

The results from this study highlight the effects of variety on SOC sequestration, biomass yield, response to fertilization and, ultimately, climate impact. This shows the importance of selecting the appropriate variety of *Salix* and management practices based on the desired outcome from the bioenergy system.

Supplementary Materials: The following are available online at https://www.mdpi.com/article/10.3390/f12111529/s1, Table S1: Parameters used to model SOC changes in ICBM. Table S2: Initial values of aboveground (Y_a) and belowground (Y_b) young pool, and old pool (O) used in the ICBM calculation. Table S3: Values used to calculate the biomass allocation between the different pools (stems, leaves, fine roots and coarse roots) at stages of growth as a percentage of their 3-year net primary production. Table S4: The nitrogen content in leaf litter was calculated according to the abscission leaf N content by variety and fertilization as reported by Weih and Nordh, 2002. Table S5: The nitrogen (N) content of roots was calculated from the dataset by Manzoni et al., 2021. Table S6:

Energy input and emissions associated with production of pesticides, cutting, fertilizer and fossil fuels. Table S7: Data used to estimate emissions and energy usage for operations in the biomass procurement chain. Table S8: Data used to model emissions and energy for the reference case.

Author Contributions: Conceptualization, S.K., H.K.P., M.W., C.B., Å.N. and P.-A.H.; methodology, S.K.; formal analysis, S.K.; investigation, C.B.; data curation, M.W. and C.B.; writing—original draft preparation, S.K.; writing—review and editing, S.K., H.K.P., M.W., C.B., Å.N. and P.-A.H.; supervision, P.-A.H. and M.W. All authors have read and agreed to the published version of the manuscript.

Funding: This research received funding from Formas (The Swedish Research Council for Environment, Agricultural sciences and Spatial Planning) for the OPTUS and MixForChange projects under grants no 942-2016-31 and 2020-02339. Part of the soil analyses was funded by Deutsche Forschungsgesellschaft (DFG), project no. BA 1494/9-1.

Institutional Review Board Statement: Not applicable.

Informed Consent Statement: Not applicable.

Data Availability Statement: The data presented in this study are available the supplementary material and within the article. If required, further relevant details are available from the corresponding author on request.

Acknowledgments: We are grateful to Niclas Ericsson and Torun Hammar for their work on soil carbon modelling and climate impacts of *Salix*, and their help in adapting those methods for this study.

Conflicts of Interest: The authors declare no conflict of interest.

Appendix A

Table A1. Default parameters used in Equations (2) and (3) to calculate N_2O emissions as described in IPCC 2019 [56].

Parameter	Description	Value	Unit
EF_N	Direct emissions from applied N	0.01	kg N_2O-N kg^{-1} N
EF_D	N_2O emissions from volatilization and re-deposition	0.010	kg N_2O-N kg^{-1} NH_3-N
EF_L	N_2O emissions from N leaching	0.011	kg N_2O-N kg^{-1} N leached
F_A	Fraction of applied N lost as ammonia (for ammonia-N based fertilizer)	0.05	kg NH_3-N + NO_x-N kg^{-1} applied N
$N_{leached}$	Fraction of N lost by leaching	0.24	Kg N kg^{-1} applied N

References

1. Blunden, J.; Boyer, T. State of the Climate in 2020. *Bull. Am. Meteorol. Soc.* **2021**, *102*, S1–S475. [CrossRef]
2. Ciais, P.; Sabine, C.; Bala, G.; Bopp, L.; Brovkin, V.; Canadell, J.; Chhabra, A.; DeFries, R.; Galloway, J.; Heimann, M. Carbon and other biogeochemical cycles. In *Climate Change 2013: The Physical Science Basis*; Contribution of Working Group I to the Fifth Assessment Report of the Intergovernmental Panel on Climate Change; Cambridge University Press: Cambridge, UK, 2014; pp. 465–570.
3. European Comission. 2030 Climate & Energy Framework. YEAR. Available online: https://ec.europa.eu/clima/policies/strategies/2030_en (accessed on 10 December 2019).
4. European Comission. 2050 Long-Term Strategy. YEAR. Available online: https://ec.europa.eu/clima/policies/strategies/2050_en (accessed on 10 December 2019).
5. Allerup, J. Sweden's Climate Act and Climate Policy Framework. 2019. Available online: http://www.swedishepa.se/Environmental-objectives-and-cooperation/Swedish-environmental-work/Work-areas/Climate/Climate-Act-and-Climate-policy-framework-/ (accessed on 10 December 2019).
6. Farand, C. Sweden Just Committed to Having Zero Carbon Emissions, and Perfectly Trolled Trump at the Same Time. 2017. Available online: http://www.independent.co.uk/news/science/sweden-pledges-greenhouse-gas-emissions-zero-2045-paris-agreement-a7561111.html (accessed on 10 December 2019).
7. World Bioenergy Association. *WBA Global Bioenergy Statistics 2020*; World Bioenergy Association: Stockholm, Sweden, 2020; p. 64. Available online: https://www.worldbioenergy.org/uploads/201210%20WBA%20GBS%202020.pdf (accessed on 20 September 2021).

8. *Poplars and Willows: Trees for Society and the Environment*; Isebrands, J.G.; Richardson, J. (Eds.) CABI: Boston, MA, USA; FAO: Rome, Italy, 2013; ISBN 978-1-78064-108-9.
9. Karp, A.; Shield, I. Bioenergy from Plants and the Sustainable Yield Challenge. *New Phytol.* **2008**, *179*, 15–32. [CrossRef]
10. Ventura, M.; Panzacchi, P.; Muzzi, E.; Magnani, F.; Tonon, G. Carbon Balance and Soil Carbon Input in a Poplar Short Rotation Coppice Plantation as Affected by Nitrogen and Wood Ash Application. *New For.* **2019**, *50*, 969–990. [CrossRef]
11. Don, A.; Osborne, B.; Hastings, A.; Skiba, U.; Carter, M.S.; Drewer, J.; Flessa, H.; Freibauer, A.; Hyvönen, N.; Jones, M.B.; et al. Land-Use Change to Bioenergy Production in Europe: Implications for the Greenhouse Gas Balance and Soil Carbon. *Glob. Chang. Biol. Bioenergy* **2012**, *4*, 372–391. [CrossRef]
12. FAO. *Poplars and Other Fast-Growing Trees—Renewable Resources for Future Green Economies*. Synthesis of Country Progress Reports; 25th Session of the International Poplar Commision, Berlin, Germany; Working Paper IPC/15; Forestry Policy and Resources Division, FAO, Rome: Italy, Federal Republic of Germany, 2016. p. 120. Available online: http://www.fao.org/forestry/ipc2016/en (accessed on 20 September 2021).
13. Daoson , M.; Boström, B. *Vägen Till en Klimatpositiv Framtid. SOU 2020:4*; Betänkande Från Klimatpolitiska Vägvalsutredningen: Stockholm, Sweden, 2020; ISBN 978-91-38-25019-8. Available online: https://bioenergyinternational.com/app/uploads/2019/05/Monica_Daoson_Bjorn_Bostrom.pdf (accessed on 20 September 2021).
14. Lindegaard, K.N.; Adams, P.W.R.; Holley, M.; Lamley, A.; Henriksson, A.; Larsson, S.; von Engelbrechten, H.-G.; Lopez, G.E.; Pisarek, M. Short Rotation Plantations Policy History in Europe: Lessons from the Past and Recommendations for the Future. *Food Energy Secur.* **2016**, *5*, 125–152. [CrossRef]
15. Nicolescu, V.-N. *National Perspectives on Coppice from 35 EuroCoppice Member Countries*; Albert Ludwig University: Freiburg, Germany, 2017; ISBN 978-3-9817340-1-0.
16. Dimitriou, I.; Rosenqvist, H.; Berndes, G. Slow Expansion and Low Yields of Willow Short Rotation Coppice in Sweden; Implications for Future Strategies. *Biomass Bioenergy* **2011**, *35*, 4613–4618. [CrossRef]
17. Hammar, T.; Ericsson, N.; Sundberg, C.; Hansson, P.-A. Climate Impact of Willow Grown for Bioenergy in Sweden. *Bioenergy Res.* **2014**, *7*, 1529–1540. [CrossRef]
18. Agostini, F.; Gregory, A.S.; Richter, G.M. Carbon Sequestration by Perennial Energy Crops: Is the Jury Still Out? *Bioenergy Res.* **2015**, *8*, 1057–1080. [CrossRef] [PubMed]
19. Dimitriou, I.; Mola-Yudego, B.; Aronsson, P.; Eriksson, J. Changes in Organic Carbon and Trace Elements in the Soil of Willow Short-Rotation Coppice Plantations. *Bioenergy Res.* **2012**, *5*, 563–572. [CrossRef]
20. Gauder, M.; Billen, N.; Zikeli, S.; Laub, M.; Graeff-Hönninger, S.; Claupein, W. Soil Carbon Stocks in Different Bioenergy Cropping Systems Including Subsoil. *Soil Tillage Res.* **2016**, *155*, 308–317. [CrossRef]
21. Georgiadis, P.; Vesterdal, L.; Stupak, I.; Raulund-Rasmussen, K. Accumulation of Soil Organic Carbon after Cropland Conversion to Short-Rotation Willow and Poplar. *GCB Bioenergy* **2017**, *9*, 1390–1401. [CrossRef]
22. Rytter, R.-M. The Potential of Willow and Poplar Plantations as Carbon Sinks in Sweden. *Biomass Bioenergy* **2012**, *36*, 86–95. [CrossRef]
23. Harris, Z.M.; Alberti, G.; Viger, M.; Jenkins, J.R.; Rowe, R.; McNamara, N.P.; Taylor, G. Land-Use Change to Bioenergy: Grassland to Short Rotation Coppice Willow Has an Improved Carbon Balance. *GCB Bioenergy* **2017**, *9*, 469–484. [CrossRef]
24. Walter, K.; Don, A.; Flessa, H. No General Soil Carbon Sequestration under Central European Short Rotation Coppices. *GCB Bioenergy* **2015**, *7*, 727–740. [CrossRef]
25. Rytter, R.-M.; Rytter, L.; Högbom, L. Carbon Sequestration in Willow (*Salix* Spp.) Plantations on Former Arable Land Estimated by Repeated Field Sampling and C Budget Calculation. *Biomass Bioenergy* **2015**, *83*, 483–492. [CrossRef]
26. Garrigues, E.; Corson, M.S.; Angers, D.A.; van der Werf, H.M.G.; Walter, C. Soil Quality in Life Cycle Assessment: Towards Development of an Indicator. *Ecol. Indic.* **2012**, *18*, 434–442. [CrossRef]
27. Henryson, K.; Sundberg, C.; Kätterer, T.; Hansson, P.-A. Accounting for Long-Term Soil Fertility Effects When Assessing the Climate Impact of Crop Cultivation. *Agric. Syst.* **2018**, *164*, 185–192. [CrossRef]
28. Huijbregts, M.A.J. Application of Uncertainty and Variability in LCA. *Int. J. LCA* **1998**, *3*, 273. [CrossRef]
29. Huijbregts, M.A.J. Part II: Dealing with Parameter Uncertainty and Uncertainty Due to Choices in Life Cycle Assessment. *Int. J. LCA* **1998**, *3*, 343–351. [CrossRef]
30. Baumann, H.; Tillman, A.-M. *The Hitchhiker's Guide to LCA: An Orientation in Life Cycle Assessment Methodology and Application*; Studentlitteratur: Lund, Sweden, 2004; ISBN 978-91-44-02364-9.
31. Cherubini, F.; Strømman, A.H. Life Cycle Assessment of Bioenergy Systems: State of the Art and Future Challenges. *Bioresour. Technol.* **2011**, *102*, 437–451. [CrossRef]
32. Ericsson, N.; Sundberg, C.; Nordberg, Å.; Ahlgren, S.; Hansson, P.-A. Time-Dependent Climate Impact and Energy Efficiency of Combined Heat and Power Production from Short-Rotation Coppice Willow Using Pyrolysis or Direct Combustion. *GCB Bioenergy* **2017**, *9*, 876–890. [CrossRef]
33. Ahmadi Moghaddam, E.; Ericsson, N.; Hansson, P.-A.; Nordberg, Å. Exploring the Potential for Biomethane Production by Willow Pyrolysis Using Life Cycle Assessment Methodology. *Energy Sustain. Soc.* **2019**, *9*, 6. [CrossRef]
34. Hammar, T. Climate Impacts of Woody Biomass Use for Heat and Power Production in Sweden. Ph.D. Thesis, Swedish University of Agricultural Sciences, Uppsala, Sweden, 2017.

35. Ericsson, N.; Porsö, C.; Ahlgren, S.; Nordberg, Å.; Sundberg, C.; Hansson, P.-A. Time-Dependent Climate Impact of a Bioenergy System—Methodology Development and Application to Swedish Conditions. *GCB Bioenergy* **2013**, *5*, 580–590. [CrossRef]
36. Therasme, O.; Volk, T.A.; Eisenbies, M.H.; Amidon, T.E.; Fortier, M.-O. Life Cycle Greenhouse Gas Emissions of Ethanol Produced via Fermentation of Sugars Derived from Shrub Willow (*Salix* Ssp.) Hot Water Extraction in the Northeast United States. *Biotechnol. Biofuels* **2021**, *14*, 52. [CrossRef]
37. Yang, S.; Volk, T.A.; Fortier, M.-O.P. Willow Biomass Crops Are a Carbon Negative or Low-Carbon Feedstock Depending on Prior Land Use and Transportation Distances to End Users. *Energies* **2020**, *13*, 4251. [CrossRef]
38. Ericsson, N. Time-Dependent Climate Impact of Short Rotation Coppice Willow-Based Systems for Electricity and Heat Production. Ph.D. Thesis, Department of Energy and Technology, Swedish University of Agricultural Sciences, Uppsala, Sweden, 2015.
39. Murphy, F.; Devlin, G.; McDonnell, K. Energy Requirements and Environmental Impacts Associated with the Production of Short Rotation Willow (*Salix* Sp.) Chip in Ireland. *GCB Bioenergy* **2014**, *6*, 727–739. [CrossRef]
40. Heller, M.C.; Keoleian, G.A.; Volk, T.A. Life Cycle Assessment of a Willow Bioenergy Cropping System. *Biomass Bioenergy* **2003**, *25*, 147–165. [CrossRef]
41. Weih, M.; Nordh, N.-E. Characterising Willows for Biomass and Phytoremediation: Growth, Nitrogen and Water Use of 14 Willow Clones under Different Irrigation and Fertilisation Regimes. *Biomass Bioenergy* **2002**, *23*, 397–413. [CrossRef]
42. Adegbidi, H.G.; Volk, T.A.; White, E.H.; Abrahamson, L.P.; Briggs, R.D.; Bickelhaupt, D.H. Biomass and Nutrient Removal by Willow Clones in Experimental Bioenergy Plantations in New York State. *Biomass Bioenergy* **2001**, *20*, 399–411. [CrossRef]
43. Cunniff, J.; Purdy, S.J.; Barraclough, T.J.P.; Castle, M.; Maddison, A.L.; Jones, L.E.; Shield, I.F.; Gregory, A.S.; Karp, A. High Yielding Biomass Genotypes of Willow (*Salix* Spp.) Show Differences in below Ground Biomass Allocation. *Biomass Bioenergy* **2015**, *80*, 114–127. [CrossRef] [PubMed]
44. Baum, C.; Amm, T.; Kahle, P.; Weih, M. Fertilization Effects on Soil Ecology Strongly Depend on the Genotype in a Willow (*Salix* Spp.) Plantation. *For. Ecol. Manag.* **2020**, *466*, 118126. [CrossRef]
45. Weih, M.; Nordh, N.-E. Determinants of Biomass Production in Hybrid Willows and Prediction of Field Performance from Pot Studies. *Tree Physiol.* **2005**, *25*, 1197–1206. [CrossRef] [PubMed]
46. Andrén, O.; Kätterer, T. ICBM: The Introductory Carbon Balance Model for Exploration of Soil Carbon Balances. *Ecol. Appl.* **1997**, *7*, 1226–1236. [CrossRef]
47. Kahle, P.; Möller, J.; Baum, C.; Gurgel, A. Tillage-Induced Changes in the Distribution of Soil Organic Matter and the Soil Aggregate Stability under a Former Short Rotation Coppice. *Soil Tillage Res.* **2013**, *133*, 49–53. [CrossRef]
48. Rytter, R.-M.; Rytter, L. Growth, Decay, and Turnover Rates of Fine Roots of Basket Willows. *Can. J. For. Res.* **1998**, *28*, 893–902. [CrossRef]
49. Caslin, B.; Finnan, J.; Johnston, C.; McCracken, A.; Walsh, L. *Short Rotation Coppice Willow Best Practice Guidelines*; Teagasc, Crops Research Centre, Ireland and AFBI, Agri-Food and Bioscience Institute: Belfast, UK, 2015; ISBN 1-84170-610-8.
50. Dimitriou, I.; Rutz, D. *Sustainable Short Rotation Coppice: A Handbook*; WIP Renewable Energies: Munich, Germany, 2015; ISBN 978-3-936338-36-2.
51. Börjesson, P.; Tufvesson, L.; Lantz, M. *Life Cycle Assessment of Biofuels in Sweden*; (LUTFD2/TFEM–10/3061–SE + (1-88); Department of Technology and Society, Environmental and Energy Systems Studies, Lund University: Lund, Sweden, 2010; Volume 70, ISBN 91-88360-96-2.
52. Phyllis2—ECN Phyllis Classification. YEAR. #salix. Available online: https://phyllis.nl/Browse/Standard/ECN-Phyllis#salix (accessed on 20 August 2020).
53. Nilsson, D.; Bernesson, S. *Pelletering och Brikettering av Jordbruksråvaror: En Systemstudie (Processing Biofuels from Farm Raw Materials: A Systems Study)*; Rapport (Institutionen för energi och teknik, SLU), 001; Technical Report 001; Department of Energy and Technology, Swedish University of Agricultural Sciences: Uppsala, Sweden, 2008.
54. Djomo, S.N.; Kasmioui, O.E.; Ceulemans, R. Energy and Greenhouse Gas Balance of Bioenergy Production from Poplar and Willow: A Review. *GCB Bioenergy* **2011**, *3*, 181–197. [CrossRef]
55. Murphy, D.J.; Hall, C.A.S.; Dale, M.; Cleveland, C. Order from Chaos: A Preliminary Protocol for Determining the EROI of Fuels. *Sustainability* **2011**, *3*, 1888–1907. [CrossRef]
56. Buendia, E.E.C.; Tanabe, K.; Kranjc, A.; Jamsranjav, B.; Fukuda, M.; Ngarize, S.; Osako, A.; Pyrozhenko, Y.; Shermanau, P.; Federici, S. *2019 Refinement to the 2006 IPCC Guidelines for National Greenhouse Gas. Inventories*; Intergovernmental Panel on Climate Change Report; IPPC: Geneva, Switzerland, 2019.
57. Manzoni, S.; Lindh, M.; Hoeber, S.; Weih, M. *Salix Biomass and Nitrogen Content Measured in a Pot Experiment, Uppsala, Sweden, 2018–2019*; Bolin Centre Database: Stockholm, Sweden, 2021; Dataset Version 1.0.
58. Andrén, O.; Kätterer, T.; Karlsson, T. ICBM Regional Model for Estimations of Dynamics of Agricultural Soil Carbon Pools. *Nutr. Cycl. Agroecosyst.* **2004**, *70*, 231–239. [CrossRef]
59. Kirchmann, H.; Persson, J.; Carlgren, K. *The Ultuna Long-Term Soil Organic Matter Experiment, 1956–1991*; Department of Soil Sciences, *Reports and Dissertations—Swedish University of Agricultural Sciences*; Springer: Uppsala, Sweden, 1994.
60. Kätterer, T.; Bolinder, M.A.; Andrén, O.; Kirchmann, H.; Menichetti, L. Roots Contribute More to Refractory Soil Organic Matter than Above-Ground Crop Residues, as Revealed by a Long-Term Field Experiment. *Agric. Ecosyst. Environ.* **2011**, *141*, 184–192. [CrossRef]

61. Rytter, R.-M. Biomass Production and Allocation, Including Fine-Root Turnover, and Annual N Uptake in Lysimeter-Grown Basket Willows. *For. Ecol. Manag.* **2001**, *140*, 177–192. [CrossRef]
62. Rytter, R.-M.; Hansson, A.-C. Seasonal Amount, Growth and Depth Distribution of Fine Roots in an Irrigated and Fertilized Salix Viminalis L. Plantation. *Biomass Bioenergy* **1996**, *11*, 129–137. [CrossRef]
63. Han, C.; Young, S.L. Root Growth of Two Perennial Grass Types and Musk Thistle (Carduus Nutans) in Temperate Grasslands of North America. *Invasive Plant Sci. Manag.* **2014**, *7*, 387–397. [CrossRef]
64. Myhre, G.; Shindell, D.; Bréon, F.-M.; Collins, W.; Fuglestvedt, J.; Huang, J.; Koch, D.; Lamarque, J.-F.; Lee, D.; Mendoza, B.; et al. 2013: Anthropogenic and Natural Radiative Forcing. In *Climate Change 2013: The Physical Science Basis*; Contribution of Working Group I to the Fifth Assessment Report of the Intergovernmental Panel on Climate Change; Cambridge University Press: Cambridge, UK; New York, NY, USA, 2013; p. 82.
65. Yuan, C.; Wang, E.; Zhai, Q.; Yang, F. Temporal Discounting in Life Cycle Assessment: A Critical Review and Theoretical Framework. *Environ. Impact Assess. Rev.* **2015**, *51*, 23–31. [CrossRef]
66. Joos, F.; Prentice, I.C.; Sitch, S.; Meyer, R.; Hooss, G.; Plattner, G.-K.; Gerber, S.; Hasselmann, K. Global Warming Feedbacks on Terrestrial Carbon Uptake under the Intergovernmental Panel on Climate Change (IPCC) Emission Scenarios. *Glob. Biogeochem. Cycles* **2001**, *15*, 891–907. [CrossRef]
67. Röös, E.; Sundberg, C.; Hansson, P.-A. Uncertainties in the Carbon Footprint of Food Products: A Case Study on Table Potatoes. *Int. J. Life Cycle Assess.* **2010**, *15*, 478–488. [CrossRef]
68. Mueller, K.E.; Tilman, D.; Fornara, D.A.; Hobbie, S.E. Root Depth Distribution and the Diversity–Productivity Relationship in a Long-Term Grassland Experiment. *Ecology* **2013**, *94*, 787–793. [CrossRef]
69. Lettens, S.; Muys, B.; Ceulemans, R.; Moons, E.; Garcia, J.; Coppin, P. Energy Budget and Greenhouse Gas Balance Evaluation of Sustainable Coppice Systems for Electricity Production. *Biomass Bioenergy* **2003**, *24*, 179–197. [CrossRef]
70. Goglio, P.; Owende, P.M.O. A Screening LCA of Short Rotation Coppice Willow (Salix Sp.) Feedstock Production System for Small-Scale Electricity Generation. *Biosyst. Eng.* **2009**, *103*, 389–394. [CrossRef]
71. Heller, M.C.; Keoleian, G.A.; Mann, M.K.; Volk, T.A. Life Cycle Energy and Environmental Benefits of Generating Electricity from Willow Biomass. *Renew. Energy* **2004**, *29*, 1023–1042. [CrossRef]
72. Styles, D.; Jones, M.B. Energy Crops in Ireland: Quantifying the Potential Life-Cycle Greenhouse Gas Reductions of Energy-Crop Electricity. *Biomass Bioenergy* **2007**, *31*, 759–772. [CrossRef]
73. Boehmel, C.; Lewandowski, I.; Claupein, W. Comparing Annual and Perennial Energy Cropping Systems with Different Management Intensities. *Agric. Syst.* **2008**, *96*, 224–236. [CrossRef]
74. Börjesson, P.I.I. Energy Analysis of Biomass Production and Transportation. *Biomass Bioenergy* **1996**, *11*, 305–318. [CrossRef]
75. Gustavsson, L.; Börjesson, P.; Johansson, B.; Svenningsson, P. Reducing CO2 Emissions by Substituting Biomass for Fossil Fuels. *Energy* **1995**, *20*, 1097–1113. [CrossRef]
76. Matthews, R.W. Modelling of Energy and Carbon Budgets of Wood Fuel Coppice Systems. *Biomass Bioenergy* **2001**, *21*, 1–19. [CrossRef]
77. Heinsoo, K.; Merilo, E.; Petrovits, M.; Koppel, A. Fine Root Biomass and Production in a Salix Viminalis and Salix Dasyclados Plantation. *Est. J. Ecol.* **2009**, *58*, 27–37. [CrossRef]
78. Rytter, R.-M. The Effect of Limited Availability of N or Water on C Allocation to Fine Roots and Annual Fine Root Turnover in Alnus Incana and Salix Viminalis. *Tree Physiol.* **2013**, *33*, 924–939. [CrossRef]
79. Pacaldo, R.S.; Volk, T.A.; Briggs, R.D. Greenhouse Gas Potentials of Shrub Willow Biomass Crops Based on Below- and Aboveground Biomass Inventory Along a 19-Year Chronosequence. *Bioenergy Res.* **2013**, *6*, 252–262. [CrossRef]
80. Rytter, R.-M. Fine-Root Production and Turnover in a Willow Plantation Estimated by Different Calculation Methods. *Scand. J. For. Res.* **1999**, *14*, 526–537. [CrossRef]
81. Gregory, A.S.; Dungait, J.A.J.; Shield, I.F.; Macalpine, W.J.; Cunniff, J.; Durenkamp, M.; White, R.P.; Joynes, A.; Richter, G.M. Species and Genotype Effects of Bioenergy Crops on Root Production, Carbon and Nitrogen in Temperate Agricultural Soil. *Bioenergy Res.* **2018**, *11*, 382–397. [CrossRef]
82. Sevel, L.; Nord-Larsen, T.; Raulund-Rasmussen, K. Biomass Production of Four Willow Clones Grown as Short Rotation Coppice on Two Soil Types in Denmark. *Biomass Bioenergy* **2012**, *46*, 664–672. [CrossRef]
83. Pietrzykowski, M.; Woś, B.; Tylek, P.; Kwaśniewski, D.; Juliszewski, T.; Walczyk, J.; Likus-Cieślik, J.; Ochał, W.; Tabor, S. Carbon Sink Potential and Allocation in Above- and below-Ground Biomass in Willow Coppice. *J. For. Res.* **2021**, *32*, 349–354. [CrossRef]
84. Rytter, R.-M.; Rytter, L. Quantitative Estimates of Root Densities at Minirhizotrons Differ from Those in the Bulk Soil. *Plant Soil* **2012**, *350*, 205–220. [CrossRef]
85. Püttsepp, Ü.; Lõhmus, K.; Koppel, A. Decomposition of Fine Roots and α-Cellulose in a Short Rotation Willow (*Salix* Spp.) Plantation on Abandoned Agricultural Land. *Silva. Fenn.* **2007**, *41*, 247. [CrossRef]
86. Pappas, C.; Maillet, J.; Rakowski, S.; Baltzer, J.L.; Barr, A.G.; Black, T.A.; Fatichi, S.; Laroque, C.P.; Matheny, A.M.; Roy, A.; et al. Aboveground Tree Growth Is a Minor and Decoupled Fraction of Boreal Forest Carbon Input. *Agric. For. Meteorol.* **2020**, *290*, 108030. [CrossRef]
87. Khan, S.A.; Mulvaney, R.L.; Ellsworth, T.R.; Boast, C.W. The Myth of Nitrogen Fertilization for Soil Carbon Sequestration. *J. Environ. Qual.* **2007**, *36*, 1821–1832. [CrossRef]

88. Churchman, G.J.; Singh, M.; Schapel, A.; Sarkar, B.; Bolan, N. Clau Minerals as the Key to the Sequestration of Carbon in Soils. *Clays Clay Miner.* **2020**, *68*, 135–143. [CrossRef]
89. Chimento, C.; Almagro, M.; Amaducci, S. Carbon Sequestration Potential in Perennial Bioenergy Crops: The Importance of Organic Matter Inputs and Its Physical Protection. *GCB Bioenergy* **2016**, *8*, 111–121. [CrossRef]
90. Berhongaray, G.; Verlinden, M.S.; Broeckx, L.S.; Janssens, I.A.; Ceulemans, R. Soil Carbon and Belowground Carbon Balance of a Short-Rotation Coppice: Assessments from Three Different Approaches. *GCB Bioenergy* **2017**, *9*, 299–313. [CrossRef]
91. Qin, Z.; Dunn, J.B.; Kwon, H.; Mueller, S.; Wander, M.M. Soil Carbon Sequestration and Land Use Change Associated with Biofuel Production: Empirical Evidence. *GCB Bioenergy* **2016**, *8*, 66–80. [CrossRef]
92. Dimitriou, I.; Mola-Yudego, B.; Aronsson, P. Impact of Willow Short Rotation Coppice on Water Quality. *Bioenergy Res.* **2012**, *5*, 537–545. [CrossRef]

MDPI

Article

Potential Areas in Poland for Forestry Plantation

Piotr Boruszewski [1], Agnieszka Laskowska [1], Agnieszka Jankowska [1,*], Marcin Klisz [2] and Marcin Mionskowski [2]

1 The Institute of Wood Sciences and Furniture, 159 Nowoursynowska St., 02-776 Warsaw, Poland; piotr_boruszewski@sggw.edu.pl (P.B.); agnieszka_laskowska@sggw.edu.pl (A.L.)
2 Department of Silviculture and Genetics of Forest Trees, Forest Research Institute, Sękocin Stary, 3 Braci Leśnej St., 05-090 Raszyn, Poland; M.Klisz@ibles.waw.pl (M.K.); M.Mionskowski@ibles.waw.pl (M.M.)
* Correspondence: agnieszka_jankowska@sggw.edu.pl; Tel.: +48-225-938-634

Abstract: Plantations have many advantages when compared to natural or semi-natural forests, such as shortening production cycles, the production of wood with specific characteristics, and near-market production concentrations. The intensive development of this form of industrial wood production is practiced all over the world. The wood industry in Poland struggles in recent years, with a large shortage of wood. The deficit of wood has been accumulated for several years and is steadily increasing. One of the possibilities to change this trend can be development of fast-growing trees plantations. The main aim of this study was to determine the potential of land in Poland, which could be used for the cultivation of fast-growing trees plantations. The analyses took into account the area and marginal agricultural land. The potential plantation land areas were determined for poplar cultivar "Hybrid 275" and European larch (*Larix decidua* Mill.). The results show a possibility to generate a considerable area that can be developed into plantations of fast-growing trees in Poland. According to the analyses carried out for the purpose of this study, with only 5% use of the sown area and 5% use of forest lands, as well as the boscage (wooded land and bushy land), it is possible to obtain approximately 0.6 MM ha of land for fast-growing tree plantations. In the case of planting 50% of these lands with larch and 50% with poplar, and if a 50% capacity of the plantation is assumed, it will be possible to obtain nearly 6 MM m^3 of wood per year.

Keywords: capacity; European larch; fast-growing trees; plantations; plantation area; poplar cultivar "Hybrid 275"; sown area

Citation: Boruszewski, P.; Laskowska, A.; Jankowska, A.; Klisz, M.; Mionskowski, M. Potential Areas in Poland for Forestry Plantation. *Forests* **2021**, *12*, 1360. https://doi.org/10.3390/f12101360

Academic Editors: Dirk Landgraf and Timothy A. Martin

Received: 15 July 2021
Accepted: 3 October 2021
Published: 7 October 2021

Publisher's Note: MDPI stays neutral with regard to jurisdictional claims in published maps and institutional affiliations.

1. Introduction

According to the European Panel Federation (EPF), in recent years, there has been rapid growth in the development of European wood sectors. In 2019, in Poland, 11.7 MM m^3 of wood-based panels were produced, which made an 18% share in the UE market; therefore, Poland has become the second (after Germany)-largest manufacturer in Europe [1]. The strategic value of wood in Poland is confirmed by the fact that the industry based on the processing of this raw material is one of the pillars of the Polish economy. The market share of the forestry-based industry in GDP (gross domestic product) is ca. 1.7% and is higher than in the EU (about 1%) [2]. Despite that, the wood-based panels sector still faces serious challenges, especially a limited availability of raw wood. Since 2012, a shortage of wood has been observed for the wood-based panels industry in Poland, reaching 20%, and the trend is maintained [3]. Raw material shortages are now a widespread problem in the world. It is estimated that the wood deficiency would reach 200 MM m^3 and 300 MM m^3 in 2025 and 2030, respectively [4]. This is an apparent confirmation of the thesis about the strategic importance of wood as a raw material [5].

One of the possibilities to reduce the shortage of raw material can be creating an outer wood production ecosystem or a sub-ecosystem within forest ecosystems consisting of plantations of fast-growing trees. Already, over one-third of the global wood production comes from plantations [6]. There are different directions of using wood from plantations.

Initially, the leading direction was the establishment of plantations with the possibility of obtaining wood mainly for energy purposes. There are many local, national and international initiatives to support the development of renewable energy sources [7]. In the past decades, the most important tree species grown in European short-rotation coppices intended as a renewable energy source were willow, poplar and, to a lesser extent, aspen and robinia [8,9]. European commercial willow plantations were mainly located in Sweden, the UK, Poland and Germany [7,8,10,11]. In Poland, they covered a small area of approx. 6800 ha in 2008 [8] and approx. 5515 ha in 2011 [12]. A steady decrease of willow plantations in recent years in Sweden was observed too [13]. It should be noted that, in Poland, willow wood from plantation crops was grown mainly for energy purposes [12].

The growing deficit of wood, mainly in the wood-based panels sector, forced a change in the policy of establishing and running fast-growing tree plantations. Species intended for this type of purpose should be characterized by appropriate dimensional and qualitative characteristics [14]. Moreover, it was found that, in Poland, among the species of fast-growing trees, the best prognoses are: poplar (mainly the cultivar "Hybrid 275") and European larch (*Larix decidua* Mill.) [15,16]. When relatively dense spacing (3 × 3 m) is used, a very high annual increment is obtained—in the case of larch, up to 14 $m^3 \times ha^{-1}$, and poplars even up to 25 $m^3 \times ha^{-1}$—while, usually, it is 14–20 $m^3 \times ha^{-1}$ [16].

Warmbier et al., [17] pointed to the limited possibilities of using willow wood for the production of wood-based panels. Willow wood can be mainly dedicated for use only for core layers of particleboards, and replacing pine particles 25–30% with willow particles in the core layer of three-layer particleboards allows obtaining materials with properties that meet the requirements of the appropriate standards [18,19]. The wood of willow clones has an approx. 25% higher density than the wood of poplar clones. The use of wood of willow clones for the production of wood-based materials (especially with lower densities) is recommended to a lesser extent. This is due to the technological aspects and the anticipation of obtaining lower-strength parameters of boards made of higher-density wood, because, during pressing, particles from woods of higher density are less susceptible to compression [20–25]. In turn, the usefulness of plantation poplar and larch wood for the production of wood-based panels was confirmed by the published research results [26,27]. This was also evidenced by the implementation of technology and production of this type of material on an industrial scale in Europe (production based on plantation poplar wood).

Plantation development is a very important aspect for the Polish wood industry, especially for the manufacturers of wood-based panels. One of the most important facts justifying the establishment of plantations is the European context of strategic value of wood in the industry. According to the European Commission, wood is also considered an important source of raw materials for emerging bio-based industries [28]. Similar solutions to those already put in place by one of the leading manufacturers of wood pulp and paper can be successfully adopted by the manufacturers of wood-based panels who require large quantities of medium-size wood. Apart from the large corporations, State Forests National Forest Holding (SFN FH)—the main manager of Polish forests, as well as one of the leading manufacturers of wood—should undertake the production of commercial wood from plantations, which would be in compliance with the global tendency to ease the pressure on natural forests. Plantations of fast-growing trees managed only by SFN FH utilize a small area that currently amounts to 2547.94 ha (equivalent of 0.03% of total forest land) [29].

The formation of forests on former agricultural land can be the result of a natural secondary progression. Recently, particularly in Europe, that is the result of intentional afforestation. This was the effect of promotion by the European Parliament and of the Council the Regulation (EU) No 1305/2013 [28]. The Regulation established a community aid scheme for forestry measures in agriculture, with the general aim of transforming agricultural lands into forested areas. Parallel to this, due to the increase in the world's demand on wood, extensive research on harvesting wood from fast-growing trees plantations is being carried out [30]. The current trends concern mainly wood harvested from genetically modified trees (GMO) [31–34]. As a result of the insufficient recognition of genetically modified materials, ethical aspects and legislation shortages, the wide utilization of wood from genetically modified trees in Europe is presently ineffective [35–38]. Insufficient knowledge in the area of genetically modified species, especially in the utilitarian context, imposes the development of new research directions. Experienced foresters are exceptionally helpful in the breeding of fast-growing trees. The advantages of plantations are not only high productivity but, also, the availability and easy access to highly concentrated and deployed resource of raw materials for the wood industry [39–42].

Plantations of fast-growing trees on agricultural lands have a positive effect on the structure of the soil due to the lack of regular heavy agrotechnical treatment. In a longer perspective, the content of carbon and nitrogen in the soil will increase. This is an additional argument for locating plantations of fast-growing trees on low-class lands for a period of time and subsequently returning to cultivating agricultural crops after resources from the plantation are collected. The periodical usage of agricultural land for plantations of fast-growing trees facilitates the regeneration of overexploited land by improving the soil fertility and, consequently, boosts crop growth. This is due to the factors such as [16]: (I) by penetrating the ground, tree roots crumble the lower layer of soil, (II) the intensity of mineral fertilizing is decreased thanks to plants using nutrients that return into the soil with falling leaves, (III) the organic matter from the forest litter increases the humus layer, and (IV) plantations protect soil against excessive evaporation and erosion.

Agroforestry as a form of plantation of fast-growing trees is widely known and used in Western Europe, Asia and South America. Plantations as such are established on spaces larger than standard plantations, where agricultural and meadow plants are primarily cultivated within the first few years [43]. It is beneficial, as the crop residues left in the rows on the fields additionally fertilize the soil. Moreover, agroforestry may be a link between agricultural plant cultivation and tree plantations. It is estimated that the average annual gain from a poplar plantation in Poland is comparable to that from the production of oil-yielding rape or wheat. Moreover, the production cycles become shorter thanks to advancements in science and technology [44]. Fast-growing trees on agricultural land can achieve high biomass yields with a relatively low input of nitrogen fertilizer and are regarded as efficient nitrogen users [8,45]. Dimitrou and Rutz [7] indicated a number of other, nonproduction benefits resulting from the cultivation of short rotation woody crops (SRC or SRWC). The authors report that this type of coppices helps to improve the water quality; enhance biodiversity; provide ecosystem services, i.e., hunting, beekeeping, water supply and fire protection; mitigate animal diseases between farms; prevent erosion; reduce artificial input materials, i.e., fertilizers and pesticides, and mitigate climate change due to carbon storage.

Recently, in Poland, there has been a noticeable increase of the fallowing and set-aside process on agricultural land where the conditions for agricultural productions are unfavorable. Such land accounts for approximately 10% of the agricultural land [46]. This is caused mainly by a low availability of agricultural production areas, which results from unfavorable natural environment and soil quality [47]. Such land is located in areas where a combination of factors such as unfavorable conditions for agricultural production and unfavorable landscape features (eroded land, steppe formation, rocky ground and moorland) occur. The usefulness of agricultural land is also limited by the anthropogenic influence (land degraded and devastated due to careless human activity). The vicinity of factories, express roads and motorways is an additional factor that adds to the negative agricultural conditions. Agricultural land with unfavorable conditions for agricultural production as the above can be a potential area for plantations of fast-growing trees.

Collecting and processing data on forestry plantations is a common practice introduced in many countries around the world [48–50]. However, due to the lack of unambiguous criteria for classifying certain areas into the category of "forest plantation", it is difficult. The data should be derived from statistically designed inventories of forest plantations or statistics for planted areas reported by planting agencies or appearing in national reports. However, it often comes from many sources, e.g., nursery production, seedling distribution, estimates derived from agencies, industries and nongovernmental organizations participating in planting programs [51]. For this reason, each study in this area should be treated as a valuable source of information.

The main purpose of the study was to provide suggestions on the potential possibilities of meeting the demand for raw materials, which is considered to be the main problem of the wood industry. As part of the work, the areas available for the establishment of plantations of fast-growing trees were determined, taking into account the potential to generate wood resources as raw materials for the production of wood-based panels. The knowledge in this area is crucial to meet the needs of the wood industry— more, the wood-based panels industry in Poland is highly competitive. The potential plantation land areas were determined for poplar cultivar "Hybrid 275" and European larch. This work is an original study in terms of the definition of potential areas in Poland for forestry plantation.

2. Materials and Methods

In the first part of the work, the current data on the area and stock (taking into account age classes) of fast-growing trees: poplar cultivar "Hybrid 275" and European larch (*Larix decidua* Mill.) were presented. The data were prepared based on the State Forests Information System (SILP) [29]. The results of the study were presented in the form of maps prepared on the basis of Central Codification Information (COT) obtained from the Department of State Forestry Informatics (ZILP). The COT consists of SILP subsystem tables, excluding the planning system. The selections were made as follows:

1. Fast-growing tree plantations (in SILP, a specific separation is marked), with a poplar or larch in the main layer with a total share of at least 50%.
2. Area lists include the entire surface of the plantation areas that were not reduced by any fraction of a dominant species in the main layer.
3. The thickness of the selected species and not the entire plantation was specified in the thickness records.

The second stage of the research was carried out on the basis of data from the Central Statistical Office of Poland [46,52], and the Agency for Restructuring and Modernisation of Agriculture (ARMA) [53] consisted of an analysis of the structure of agricultural areas in Poland and the average area of agricultural holdings with a division into voivodeships. The adopted size of the assumptions resulted from the willingness to present the possibilities of developing agricultural land with a small and, at the same time, rational use of these areas without significantly limiting their area and importance. According to Zabielski [44], a single plantation should have a minimum area of 5 ha. A plantation area cannot be too small if all the cultivation and maintenance are mechanized with the use of special equipment to be cost-effective. Baum et al., [8] reported that the more diverse the surrounding landscape, the more species are able to be established in the plantation. Smaller plantations with longer-edged habitats facilitate species immigration from the surroundings better than larger plantations. Small plantations may increase the regional diversity.

For the purpose of this stage, the following were assumed:

1. The plantation-grown species for the wood industry are poplar cultivar "Hybrid 275" and European larch (*Larix decidua* Mill.).
2. An analysis of the division of the surveying area and land use in 2016 refers to the specification presented in the characteristics of the agricultural holdings [46].
3. The land shares to be dedicated to plantations:
    - not less than 5% and not more than 10% of the sown areas, permanent crop areas and forest land and
    - not less than 5% and not more than 30% of the areas of fallow land, wasteland and other land.

In the third stage, after indicating a potential area for plantation, the available amounts of alternative raw materials for the wood industry were determined based on the following assumptions:

1. Potential plantation land will be allotted completely for the cultivation of fast-growing trees; assuming a 100% production capacity of 14 $m^3 \times ha^{-1}$ (annually) for larch and 25 $m^3 \times ha^{-1}$ (annually) for the poplar cultivar "Hybrid 275" and 50% production capacity from the plantation [16], the production cycle should take 40 years in the case of European larch and 25 years in case of the poplar cultivar "Hybrid 275").
2. Total cover of the plantation with European larch, total cover of the plantation with poplar cultivar "Hybrid 275", 50% cover with larch and 50% cover with poplar.

Such extensive research dealing with the potential area for forest plantation has never been undertaken before. In the study, the potential areas in Poland for forestry plantations with the available amounts of raw materials were linked.

3. Results and Discussion

3.1. The Area and Stock of Raw Material from Plantation of Poplar Cultivar "Hybrid 275" and Larch (Larix decidua Mill.) in Poland—Current State

The data obtained from the State Forest Information System were imposed on the cartograms presented in Figure 1. The species dominating on the plantations of fast-growing trees in Poland are poplar and larch, which reach up to 80.29% of the overall plantation area [29]. In general, Poland is characterized by a low share of the poplar and larch plantations in the total forest areas (0.026%). The plantations with poplar as the dominant species or poplar single-species plantations (in the first stage of tree stands) are cultivated on 1156.05 ha, while plantations with larch as the dominant species or larch single-species plantations (in the first stage of tree stands) cover 889.7 ha. The highest share of the larch plantation is in Radom RDLP, which is reflected by the natural distribution of this species in Poland, excluding mountainous areas, while poplar plantations occupy the largest area in RDLP Gdańsk and Lublin, which means in the regions with the suitable soils conditions for this species. A nearly complete lack of fourth age class plantations seems to be understandable due to the purpose of the plantations of fast-growing trees characterized by short production cycles. In turn, the lack of plantations in mountain areas is associated with limiting climatic conditions. The total stock on plantations of fast-growing poplar trees (taking into account all age classes) was estimated at about 280,000 m^3, of which most are in the third age class, i.e., 190,000 m^3, whereas the total stock on plantations of fast-growing larch trees (taking into account all age classes) was estimated at about 160,000 m^3, of which most are in the second age class, i.e., 120,000 m^3. In 2015, the annual shortage of raw wood only for the wood-based composites industry in Poland, reached the level of around 7.7–11.4 MM m^3 [54]. On the basis of the presented data, it can be clearly stated that the acreage of fast-growing trees (poplar and larch) in Poland is insufficient to eliminate the deficit of wood to a large extent. Therefore, there is a need to increase plantation areas in Poland.

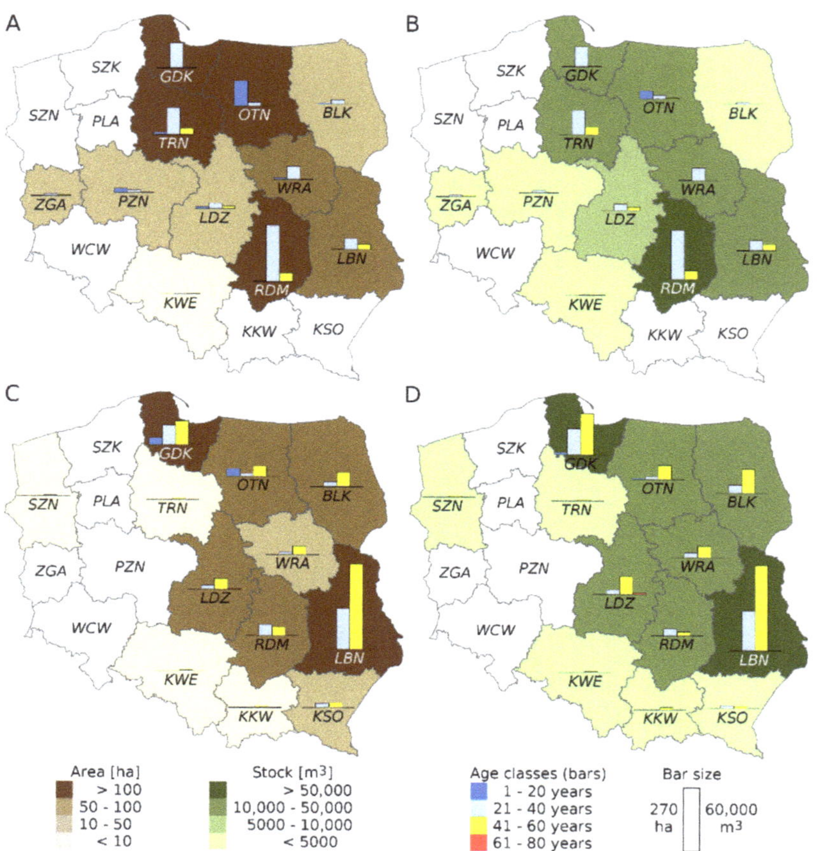

Figure 1. Arrangement of fast-growing tree plantations: (**A**,**B**) larch and (**C**,**D**) poplar in the regional LP Directorates. The carto diagram of the surface and stock (thickness) broken down into age classes (bars) and the cartogram of the total surface (bronze shades, **A**,**C**) and thicknesses (shades of green, **B**,**D**). In the RDLP without the colours and bars, there is a lack of fast-growing tree plantations. The sizes of the bars are linear. The symbols of the regional names of the LPs are explained in Abbreviations. In addition to the legend: On the left, cartographic explanations (colours RDLP: shades of brown-surface and green-stock). On the right-hand side, explanations for the carto diagrams (the colours of the bars and their sizes). Abbreviations: RDLP-Regional Directorates of the State Forests; BLK-RDLP in Białystok; GDK-RDLP in Gdańsk; KKW-RDLP in Kraków; KSO-RDLP in Krosno; KWE-RDLP in Katowice; LBN-RDLP in Lublin; LDZ-RDLP in Łódź; OTN-RDLP in Olsztyn; PLA-RDLP in Piła; PZN-RDLP in Poznań; RDM-RDLP in Radom; SZK-RDLP in Szczecinek; SZN-RDLP in Szczecin; TRN-RDLP in Toruń; WCW-RDLP in Wrocław; WRA-RDLP in Warszawa; ZGA-RDLP in Zielona Góra.

3.2. The Assessment of the Potential Area of Land for Plantations of Fast-Growing Trees in Poland

According to the Statistical Yearbook of Agriculture of the Central Statistical Office of Poland [46], in 2016, there were 14.5 MM ha of agricultural land, nearly 99.1% of which was in good agricultural condition. Of the above-mentioned land, plantations may be established on sown areas or fallow lands. Additionally areas used for permanent crops and permanent meadows can be partially prepared for plantations. A potential area for the plantations of fast-growing trees may also partially consist of forest land, as well as of woody and bushy land. The total area of this type of land was 944,031 ha in 2016 (Table 1). The data concerning the agricultural land and the number of agricultural farms did not include agricultural landowners who do not perform agricultural activity and owners of less than 1 ha of agricultural land who perform agricultural activities on a small scale [46]. Taking into consideration the geodesic area of the country, its utilization and the above assumptions, the potential area of land for plantations of fast-growing trees was shown in Table 1. Assuming only 5% use of the sown area, it is theoretically possible to obtain, respectively, 531,999 ha of land for plantations of fast-growing trees. If the utilized area constitutes 5% of the sown area and 5% of the fallow land, permanent meadows, forest land and woody and bushy land and wasteland and other land, a potential area for plantations of fast-growing trees may even reach almost 0.77 MM ha. If the target assumptions are reached, i.e., the land allotted to plantations of fast-growing trees constitutes 10% of the sown area, permanent meadows and forest land and 30% of the fallow land, uncultivated and other land and a potential area for plantations may reach approximately 1.52 MM ha. Xu and Mola-Yudego [13] analyzed the evolution and location of fast-growing plantations in Sweden for 30 years (for the period 1986–2017). The authors pointed out that willow tends to be planted on higher-productivity agricultural areas and poplar on less-productive lands. On this basis, it can be assumed that establishing poplar plantations on less fertile soils is an opportunity to change the management of the available land. According to the National Agricultural Census, since 2010, 300,000 farmers did not declare agricultural activity on their own farms, which theoretically constituted 447,000 hectares of uncultivated land potentially available for plantations [52]. However, these data do not fully reflect the situation in this kind of agricultural farm. Some of the agricultural areas (of the agricultural farms) were leased by the owners of other agricultural farms without an official contract. According to the data collected for the National Agricultural Census [52], the highest number of agricultural farms that do not carry out agricultural activities was in Śląskie Voivodeship (31.7%), Lubuskie Voivodeship (24.4%) and Małopolskie Voivodeship (20.4%), while the lowest number of such holdings was in Lubelskie Voivodeship (8.0%).

Table 1. The potential area of land for the plantation of fast-growing trees in Poland [46].

Land Type	The Area (ha)	The Use of Land for Plantations of Fast-Growing Trees in Poland (%)			
		5	10	20	30
		The Area (ha)			
I. Agricultural land area	14,543,282	-	-	-	-
1. In good agricultural condition	14,405,650	-	-	-	-
sown area	10,639,984	531,999	1,063,998	-	-
fallow land	165,627	8281	16,563	33,125	49,688
permanent crops	393,457	19,673	39,346	-	-
kitchen gardens	31,084	-	-	-	-
permanent meadows	2,698,018	134,901	269,802	-	-
permanent pastures	477,479	-	-	-	-
2. Others	137,632	6882	13,763	27,526	41,290
II. Forest land as well as woody and bushy land	944,031	47,202	94,403	-	-
III. Wasteland	748,884	37,444	74,888	149,777	224,665
Total	16,236,196	786,382	1,572,763	210,428	315,643

According to the adopted assumptions, the minimum area for plantation should be 5 ha [44]. An analysis of the data presented in Table 2 leads to a conclusion that agricultural farms meeting the expectations of an average area lower than 5 ha are located in Małopolskie and Podkarpackie Voivodeships. An observation can be made that Małopolskie Voivodeship characterizes the smallest average area of agricultural land per farm, i.e., 4.16 ha [53]. This is due to the fact that only one out of five agricultural farms in this voivodeship does not carry out any agricultural activities; it can be assumed that this is a region where small-sized areas are concentrated. Upon the merger of such lands, they will form an area large enough to be used for the plantation of fast-growing trees. This direction is in line with foreign trends. The need for the cultivation of different tree species in small-scale units is indicated [55]. In Sweden, there is a trend towards preferring smaller plantations (below 1 ha) versus large ones (above 10 ha) [13].

Table 2. Average area of land in an agricultural farm in 2020 in Poland [53].

Voivodships *	Average Area of Land in an Agricultural Farm (ha)
Małopolskie	4.16
Podkarpackie	4.94
Świętokrzyskie	5.88
Lubelskie	7.98
Łódzkie	7.98
Śląskie	8.14
Mazowieckie	8.77
Podlaskie	12.55
Wielkopolskie	14.09
Kujawsko-Pomorskie	16.58
Dolnośląskie	17.29
Opolskie	19.16
Pomorskie	19.62
Lubuskie	22.29
Warmińsko-Mazurskie	23.25
Zachodniopomorskie	31.75

* Poland is divided administratively into 16 voivodeships.

It is clear that plantations located on agricultural lands are the most efficient-speaking in terms of production [16]. However, as such locations are limited and utilized for more demanding crops, it is necessary in Poland to reach for low-quality, devastated and degraded soil. Such land is collectively referred to as marginal land. These areas at a low cost (cost of soil type and quality verification) can be adopted and serve as plantations of fast-growing trees. Marginal lands are lands currently in agricultural use or classified in the agricultural land records as not suitable for the production of healthy food, due to unfavorable environmental and anthropogenic conditions, are qualified for a different use form. In Poland, the area of marginal lands reaches 2.3 MM ha, which is equivalent to 16% of the agricultural lands. Approximately 1.7-MM ha (90%) of those lands are very light, dry and barren sandy areas. The marginal lands include the unfertile parts of agricultural lands where production is not cost-effective due to unfavorable environmental conditions and erosion. These lands are located in Małopolskie Voivodeship and Podkarpackie Voivodeship and amount to 370,000 hectares. The lands of various quality class but chemically polluted account for 140,000 hectares, and degraded or mechanically transformed lands that lack humus accumulate to 50,000 hectares. Furthermore, the marginal lands include lands with unfavorable environmental territorial conditions. This group includes cultivated areas difficult to access or difficult to cultivate mechanically [56].

As an illustration of a recultivation process of devastated soil in Poland, a forestation of a part of a dumping ground in an open pit sulfur mine in Piaseczno near Tarnobrzeg City can be set as an example [57]. Between 1967 and 1969, black locust was planted in a space of 1.2 m × 0.6 m on the slopes with unfavorable soil and untreatable agrotechnical (average slope of 60%) ravines. The tree stands were not fertilized or cut. In 2009, the 42-year-old tree stands were tested. It was calculated that the annual stand increment was 5.9 $m^3 \times ha^{-1}$ on the north slope, 4.8 $m^3 \times ha^{-1}$ on the southeast and 3.6 $m^3 \times ha^{-1}$ on the south slope. These numbers are not high; however, the single purpose of planting black locust was, first and foremost, soil reclamation. The wood produced was a byproduct of the process. On the other hand, in such extremely unfavorable soil, the average annual increment of 4.8 $m^3 \times ha^{-1}$ can be considered as a relatively high value, compared to the black locust average production capacity of 7.06 $m^3 \times ha^{-1}$ per year [57]. Hybrid poplar 275 is a cultivar with low requirements for soil quality conditions capable of adapting to soils of different types. In Poland, there is a long tradition of growing poplar cultivar "Hybrid 275" (as a clone obtained from the *S. Tacamahaca* species cross *P. maximowiczii* × *P. trichocarpa*) [42,58].

Therefore, it can be treated as the most appropriate mean in the process of the reclamation of lands offering annual increments higher than black locust. Moreover, poplar has the ability to accumulate substantial amounts of cadmium, zinc, lead and copper. Thus, in the areas polluted with heavy metals, poplar may be used for bioremediation of the environment [59].

The predicted values are vectors of probabilities of transitions to alternative land uses and the transition of land uses and forest management type conditional on the biophysical and socioeconomic factors. Regardless of the data presented, it may have some limitations. The potential of plantation crops of fast-growing trees in agriculture depends on the availability of the lands, climate conditions and water supply, as well as the quality of soils [12]. However, plantations can contribute to the enhancement of biodiversity in intensively used landscapes with low habitat heterogeneity [55]. In addition, some recommendations have been already developed to further improve the habitat function of fast-growing tree plantations and to increase their contributions to farmland biodiversity, especially to plant species diversity (phytodiversity). This includes the cultivation of different tree species in small-scale units; the sectional harvesting of trees in order to establish a mosaic of different growth stages side by side and the integration of accompanying structures such as headlands, clearings or rides to provide additional open habitat elements. The knowledge in this area can be used to promote phytodiversity in agricultural landscapes, as they contain relatively high species numbers (of mainly common and adaptable species) and support distinct plant communities that differ from other farmland habitats.

The presented results are related with Polish lands. Due to the fact that the Polish wood sector is important for all the European region, the information in that area is necessary. In the other regions of Europe (or the world), due to different forms of land and forest ownership, as well as regulations, there are no applications to obtain the results. The specificities of the regions require analyses at the regional level. For example, research determining potential areas for the establishment of commercial forest plantations were performed for Mexico for Tabebuia rosea (Bertol.) DC. by using geographic information systems [60]. A land suitability evaluation was evaluated for two forest plantations, including oak (*Quercus robur* L.) and pine (*Pinus sylvestris* L.) in the northeast of Iran [61]. The study area involved an area of about 394 km^2 with a total mainstream length of 35 km. The relatively small area allowed for the assessment of the potential area for trees plantation, taking into account the climate, soil and terrain data. Similar research was conducted for the southern part of the USA [62]. Land resources potentially available for pine plantations were determined using matrices of land and forest-type changes conditional on the biophysical and socioeconomic factors and applying them to the available land and forest resources to forecast the dynamics of pine plantations [62].

3.3. The Assessment of Potentially Available Quantities of Alternative Raw Material for the Wood Industry

The estimated increment of poplar and larch on sown and forest lands is shown in Table 3. The calculations assume 5% use of the sown lands (i.e., 531,999 ha) and 5% of the forest land (i.e., 47,202 ha) and 100% coverage of the land with the poplar plantation, with the average increase in the tree stand thickness for the poplar about 25 m^3 × ha^{-1}. In the case of 50% capacity of the poplar plantation, the annual wood production is estimated as 7 MM m^3. In the case of planting 50% of the land with larch and 50% with poplar, an assumed 50% capacity of the plantation should produce 5.63 MM m^3 of wood per year.

From the data presented in Table 3, it was concluded that, when plantations cover 5% of sown areas and 5% of forest areas, the annual increase in large-sized wood may be from 8.11 MM m^3 (100% coverage with larch) to 14.48 MM m^3 (100% coverage with poplar). Considering the fact that, in 2015, the annual shortage of wood only for the wood-based panels industry reached the level of around 7.7–11.4 MM m^3 [33] by using the indicated areas, it will be possible to significantly reduce the deficit of raw wood material in Poland.

Table 3. Data on the potential wood base from sown land and forest land.

Annual Increase in the Tree Stand Thickness	Cultivation Type/Plantation Coverage Level		
	100% of Larch	100% of Poplar	50% of Larch and 50% of Poplar
	from Sown Land (MM m^3 per year)		
14 (m^3 × ha^{-1})	7.45	-	3.72
25 (m^3 × ha^{-1})	-	13.30	6.65
For 50% capacity (MM m^3)	3.72	6.65	5.17
	from Forest Land (MM m^3 per year)		
14 (m^3 × ha^{-1})	0.66	-	0.33
25 (m^3 × ha^{-1})	-	1.18	0.59
For 50% capacity (MM m^3)	0.33	0.59	0.46

Lindegaard et al., [63] indicated that, with the intensification of the activities aimed at increasing the plantation areas of rapidly growing trees in Europe so far, each region has developed a number of recommendations for policymakers, public authorities and government agencies to support the development, production and use of biomass obtained from these types of crops for applications for energy and industrial purposes. It was pointed out that each region has many similarities regarding the restrictions on establishing and running fast-growing tree plantations. There is a need to educate farmers and policymakers about the multifunctional benefits of fast-growing tree plantations. In order to develop the market for planting fast-growing trees, more financial support is needed from regional and/or national authorities. Introducing targeted subsidies as an encouragement for growers could solve the problem of the lack of local supply chains. Haughton et al., [64] stated that promoting the plantation of fast-growing trees on agricultural land is important to increasing the landscape diversity and improving the ecosystem functions. Overall, fast-growing tree plantations are viewed positively by farmers to a limited extent, unless they receive the same benefits, subsidies and support that are offered to renewable energy providers. The relevant issue in this aspect also supports bureaucratic procedures at the regional and national levels. The role of scientists is also crucial in ensuring that there is clear and concrete evidence that planting fast-growing trees produces a range of environmental and socioeconomic benefits.

4. Conclusions

A constant supply of raw wood for the Polish market is a crucial matter for the Polish wood industry. Establishing fast-growing tree plantations would make it possible to obtain a supplementary source of lignocellulosic raw materials for wood-based panels production. Unlike traditional forestry based on natural or semi-natural forests, plantations offer a number of advantages, such as relatively short production cycles, as well as a possibility of producing raw material that meet specific requirements (for particular clients). Such wood may be produced in large quantities only thanks to plantations of vegetatively reproduced progeny from properly selected species. Moreover, the production of plantation woody raw materials may be located near end customers. Such and many other benefits have led to a noticeable worldwide development of this form of commercial wood production. In order to maintain a leading position in the Polish industry for the production of wood-based panels in the world, Poland must implement deep changes not only in the process of plantations creation but, also, in the legislation area, emphasizing how strategic creating a secondary source of raw materials is.

The analyzed data evidenced that the research conducted in Poland for several decades showing that the species with the highest productivity in plantations are poplar (mainly the cultivar "Hybrid 275") and, also, European larch (less popular in other European countries). Based on the presented research results, it can be concluded that, using only 5% of the sown area or 5% of the forest land and woody and bushy land, it is theoretically possible to obtain, respectively, 531,999 or 47,202 hectares of land for plantations of fast-growing trees. In the case of planting 50% of these lands with larch and 50% with poplar, at a 50% capacity of the plantation, it will be possible to obtain nearly 6 MM m^3 of wood per year. When the land allotted to plantations of fast-growing trees constitutes 10% of the sown area, permanent meadows and forest land and 30% fallow land, uncultivated and other land, the potential area for plantations may reach approximately 1.52 MM ha. It was concluded that, using an estimated plantation area covering only 5% of the sown areas and 5% of the forest areas, the annual increase in large-sized wood may be from 8.11 MM m^3 (100% coverage with larch) to 14.48 MM m^3 (100% coverage with poplar).

From a practical point of view, based on the obtained results, it can be said that, in Poland, plantations can be established on sown lands; forest lands and, partially, on areas under cultivation, permanent meadows and on fallow and idle lands. Plantations can be established by large companies through land leases, by farmers on their own lands and, also, by SFN FH on lands transformed by agricultural activity placed at their disposal under the provisions of the National Programme for the Augmentation of Forest Cover. Taking that into consideration, another conclusion can be made that such research plays a part in ensuring there is clear and concrete evidence in the field of the environmental and socioeconomic benefits of the cultivation of fast-growing tree plantations. Such a quantity of potential of the raw material makes it possible to eliminate the wood deficit existing in Poland as a crucial global producer of wood-based materials.

Author Contributions: Conceptualization, P.B.; methodology, P.B.; formal analysis, A.J. and A.L.; investigation, A.J., A.L., M.K. and M.M.; resources A.J., A.L., M.K. and M.M.; data curation, P.B., A.J., A.L., M.K. and M.M.; writing—original draft preparation, A.J. and A.L.; writing—review and editing, P.B., A.J. and A.L.; visualization, M.K. and M.M.; supervision, A.J. and A.L.; project administration, P.B. and funding acquisition, P.B. All authors have read and agreed to the published version of the manuscript.

Funding: This research was carried out under development project No. LIDER/002/406/L-4/NCBR/2013, financed by The National Centre for Research and Development.

Data Availability Statement: Not applicable.

Conflicts of Interest: The authors declare no conflict of interest.

References

1. EPF. Available online: www.europanels.org/facts--figures/market-information (accessed on 17 March 2021).
2. Respondek, T.; Adamowicz, M.; Wiktorski, T. Drewno—Strategiczny surowiec branży meblarskiej (Wood—A strategic raw material of furniture industry). In *Drewno—Surowiec Strategiczny? (Wood—A Strategic Raw Material?)*; Strykowski, W., Gałecka, A., Pawłowska, J., Eds.; Instytut Technologii Drewna, Centrum Informacyjne Lasów Państwowych: Warsaw, Poland, 2012; pp. 57–64.
3. Boruszewski, P. Raw materials for the wood based composites industry. In *Raw Materials and Particleboards—A Current Status and Perspectives*; Boruszewski, P., Mamiński, M., Ružinska, E., Eds.; Warsaw University of Life Sciences—SGGW Press: Warsaw, Poland, 2012; pp. 5–20.
4. European Economic and Social Committee. *Opinion of the European Economic and Social Committee on 'Opportunities and Challenges for a More Competitive European Woodworking and Furniture Sector' (Own-Initiative Opinion)*; European Parliament: Brussels, Belgium, 2011.
5. Strykowski, W. Czy drewno jest surowcem strategicznym? (Is wood a strategic raw material?). In *Drewno—Surowiec Strategiczny? (Wood—A Strategic Raw Material?)*; Strykowski, W., Gałecka, A., Pawłowska, J., Eds.; Instytut Technologii Drewna, Centrum Informacyjne Lasów Państwowych: Warsaw, Poland, 2012; pp. 11–25.
6. Kelty, M.J. The role of species mixtures in plantation forestry. *For. Ecol. Manag.* **2006**, *233*, 195–204. [CrossRef]
7. Dimitriou, I.; Rutz, D. *Sustainable Short Rotation Coppice—A Handbook*; WIP Renewable Energies: Munich, Germany, 2015.
8. Baum, S.; Weih, M.; Busch, G.; Kroiher, F.; Bolte, A. The impact of Short Rotation Coppice plantations on phytodiversity. *Landbauforsch. Vti Agric. For. Res.* **2009**, *3*, 163–170.
9. Baum, C.; Leinweber, P.; Weih, M.; Lamersdorf, N.; Dimitriou, I. Effects of short rotation coppice with willows and poplar on soil ecology. *Landbauforsch. Vti Agric. For. Res.* **2009**, *3*, 183–196.
10. Mola-Yudego, B. Regional Potential Yields of Short Rotation Willow Plantations on Agricultural Land in Northern Europe. *Silva Fenn.* **2010**, *44*, 63–76. [CrossRef]

11. Liesebach, M. *Poplars and Other Fast Growing Tree Species in Germany: Report of the National Poplar Commission. 2016–2019*; Thünen Working Paper, No. 141a; Johann Heinrich von Thünen-Institut: Braunschweig, Germany, 2020. [CrossRef]
12. Szostak, A.; Bidzińska, G.; Ratajczak, E.; Herbeć, M. Wood biomass from plantations of fast-growing trees as an alternative source of wood raw material in Poland. *Drewno* **2013**, *56*, 85–113. [CrossRef]
13. Xu, X.; Mola-Yudego, B. Where and when are plantations established? Land-use replacement patterns of fast-growing plantations on agricultural land. *Biomass Bioenergy* **2021**, *144*, 105921. [CrossRef]
14. Thoemen, H.; Irle, M.; Sernek, M. *Wood-Based Panels: An. Introduction for Specialists*, 1st ed.; Brunel University Press: London, UK, 2010.
15. Bielawska, K. Uprawy plantacyjne alternatywą? (Plantation crops as an alternative?). *Głos Lasu* **2009**, *12*, 6–8.
16. Zajączkowski, J.; Zajączkowski, K. *Hodowla Lasu (Silviculture)*; Państwowe Wydawnictwo Rolnicze i Leśne: Warsaw, Poland, 2013.
17. Warmbier, K.; Wilczyński, A.; Danecki, L. Properties of one-layer experimental particleboards from willow (*Salix viminalis*) and industrial wood particles. *Eur. J. Wood Wood Prod.* **2013**, *71*, 25–28. [CrossRef]
18. Sean, S.T.; Labrecque, M. Use of short-rotation coppice willow clones of *Salix viminalis* as furnish in panel production. *For. Prod. J.* **2006**, *56*, 47–52.
19. Warmbier, K.; Wilczyński, A.; Danecki, L. Particle size dependent properties of three-layer particleboards with the core layer made from willow (*Salix viminalis*). *Ann. Wars. Univ. Life Sci. SGGW For. Wood Technol.* **2010**, *71*, 405–409.
20. Suchsland, O.; Woodson, G.E. *Fiberboard Manufacturing Practices in the United States*; Forest Products Research Society USDA Forest Service: Madison, WI, USA, 1991.
21. Chen, T.Y.; Wu, J.P. Studies on the manufacturing of particleboard from fast-growing tree species (IV-1)—Effect of particleboard density on the properties of the PMDI and UF glue bonded particleboard. *For. Prod. Ind.* **1993**, *12*, 72–88.
22. Medved, S. Impact of wood species used in surface layer on density distribution of particleboard. *Drewno* **2007**, *50*, 17–26.
23. Irle, M.; Barbu, M.C. Chapter 1 Wood-based panels technology. In *Wood-Based Panels. An introduction for Specialists*; Thoemen, H., Irle, M., Sernek, M., Eds.; Brunel University Press: London, UK, 2010; pp. 1–94.
24. Habibi, M.R.; Hosseinkhani, H.; Mahdavi, S.; Sepidehdam, J. Effect of wood species on particleboard properties. *Iran. J. Wood Pap. Sci. Res.* **2011**, *26*, 58–71.
25. Boruszewski, P.; Borysiuk, P.; Mamiński, M.; Czechowska, J. Mat compression measurements during low-density particleboard manufacturing. *BioResources* **2016**, *11*, 6909–6919. [CrossRef]
26. Zaraziński, K.; Boruszewski, P. Analysis of the influence of particle and poplar fibres share on selected properties of particle-fibre bords. *Ann. Wars. Univ. Life Sci. SGGW For. Wood Technol.* **2020**, *112*, 22–31. [CrossRef]
27. Pazio, B.; Boruszewski, P. Analysis of the influence of larch fibers and particles on selected properties of fiber- and particleboards. *Ann. Wars. Univ. Life Sci. SGGW For. Wood Technol.* **2020**, *111*, 43–52. [CrossRef]
28. European Parliament. *Communication From the Commission to the European Parliament, the Council, the European Economic and Social Committee and the Committee of the Regions. A new EU Forest Strategy: For Forests and the Forest-Based Sector*; European Parliament: Brussels, Belgium, 2013.
29. State Forest Information System (SILP). (accessed on 17 March 2021).
30. Zajączkowski, K. Dobór odmian topól i wierzb do uprawy na plantacjach drzew szybko rosnących (Selection of varieties of poplar and larch for plantations of fast-growing trees). In *Elementy Genetyki i Hodowli Selekcyjnej Drzew Leśnych (Elements of Genetics and Selective Cultivation of Forest Trees)*; Sabor, J., Ed.; Centrum Informacyjne Lasów Państwowych: Warsaw, Poland, 2006; pp. 281–301.
31. Bradshaw, H.D., Jr.; Strauss, S.H. Breeding strategies for the 21st Century: Domestication of poplar. In *Poplar Culture in North America*; Dickmann, D.I., Isebrands, J.G., Eckenwalder, J.H., Richardson, J., Eds.; National Research Council Press: Ottawa, ON, Canada, 2001; pp. 383–394.
32. Strauss, S.H.; Brunner, A.M.; Busov, V.B.; Ma, C.; Meilan, R. Ten lessons from 15 years of transgenic Populus research. *Forestry* **2004**, *77*, 455–465. [CrossRef]
33. Brunner, A.M.; Li, J.; DiFazio, S.P.; Shevchenko, O.; Montgomery, B.E.; Mohamed, R.; Wei, H.; Ma, C.; Elias, A.A.; Van Wormer, K.; et al. Genetic containment of forest plantations. *Tree Genet. Genomes* **2007**, *3*, 75–100. [CrossRef]
34. Flachowsky, H.; Hanke, H.; Peil, M.V.; Strauss, S.H.; Fladung, M. A review on transgenic approaches to accelerate breeding of woody plants. *Plant Breed.* **2009**, *128*, 217–226. [CrossRef]
35. Strauss, S.H.; Raffa, K.F.; List, P.C. Ethics and Genetically Engineered Plantations. *J. For.* **2000**, *98*, 47–48.
36. Lang, C. *Genetically Modified Trees. The Ultimate Threat to Forests*; World Rainforest Movement and Friends of the Earth: Montevideo, Uruguay, 2004.
37. Gamborg, C.; Sandøe, P. Ethical considerations regarding genetically modified trees. In *Forests and Genetically Modified Trees*; El-Kassaby, Y.A., Prado, J.A., Eds.; The Food and Agriculture Organization of the United Nations (FAO): Rome, Italy, 2010; pp. 163–175.
38. Muhs, H.J. Regulation for genetically modified forest reproductive material moving in international trade. In *Forests and Genetically Modified Trees*; El-Kassaby, Y.A., Prado, J.A., Eds.; The Food and Agriculture Organization of the United Nations (FAO): Rome, Italy, 2010; pp. 227–235.

39. Rowe, R.L.; Hanley, M.E.; Goulson, D.; Clarke, D.J.; Doncaster, C.P.; Taylor, G. Potential benefits of commercial willow Short Rotation Coppice (SRC) for farm-scale plant and invertebrate communities in the agri-environment. *Biomass Bioenergy* **2011**, *35*, 325–336. [CrossRef]
40. Rodriguez-Pleguezuelo, C.R.; Duran-Zuazo, V.H.; Bielders, C.; Jimenez-Bocanegra, J.A.; Perea-Torres, F.; Francia Martinez, J.R. Bioenergy farming using woody crops. A review. *Agron. Sustain. Dev.* **2015**, *35*, 95–119. [CrossRef]
41. Parmar, K.; Keith, A.M.; Rowe, R.L.; Sohi, S.P.; Moeckel, C.; Pereira, M.G.; McNamara, N.P. Bioenergy driven land use change impacts on soil greenhouse gas regulation under Short Rotation Forestry. *Biomass Bioenergy* **2015**, *82*, 40–48. [CrossRef]
42. Niemczyk, M.; Wojda, T.; Kaliszewski, A. Biomass productivity of selected poplar (*Populus* spp.) cultivars in short rotations in northern Poland. *N. Z. J. For. Sci.* **2016**, *46*, 22. [CrossRef]
43. Zajączkowski, K.; Wojda, T. Plantacje topolowe w przyrodniczych warunkach Polski (Poplar plantations in the environmental conditions in Poland). *Studia Mater. CEPL Rogowie* **2012**, *33*, 136–142.
44. Zabielski, S. *Plantacyjna Uprawa Drzew i Krzewów Szybko Rosnących (Plantations of Fast-Growing Trees and Bushes)*; Akademia Rolnicza: Poznań, Poland, 1998.
45. Karp, A.; Shield, I. Bioenergy from plants and the sustainable yield challenge. *New Phytol.* **2008**, *179*, 15–32. [CrossRef]
46. GUS. Available online: https://stat.gov.pl/obszary-tematyczne/rolnictwo-lesnictwo/rolnictwo/charakterystyka-gospodarstw-rolnych-w-2016-r-,5,5.html (accessed on 31 March 2021).
47. Sajnóg, N.; Wójcik, J. Możliwości zagospodarowania gruntów marginalnych i nieużytków gruntowych w scalaniu gruntów (Possibilities of developing degraded and unculitivated lands in land consolidation). *Infrastrukt. Iekologia Teren. Wiej.* **2013**, *2*, 155–166.
48. Cubbage, F.; Koesbandana, S.; Mac Donagh, P.; Rubilar, R.; Balmelli, G.; Olmos, V.M.; De La Torre, R.; Murara, M.; Hoeflich, V.A.; Kotze, H.; et al. Global timber investments, wood costs, regulation, and risk. *Biomass Bioenergy* **2010**, *34*, 1667–1678. [CrossRef]
49. Walker, J. Wood Quality: A Perspective from New Zealand. *Forests* **2013**, *4*, 234–250. [CrossRef]
50. Korhonen, J.; Toppinen, A.; Cubbage, F.; Kuuluvainen, J. Factors Driving Investment in Planted Forests: A Comparison between OECD and Non-OECD Countries. *Int. For. Rev.* **2014**, *16*, 67–77. [CrossRef]
51. Carle, J.; Vuorinen, P.; Del Lungo, A. Status and Trends in Global Forest Plantation Development. *For. Prod. J.* **2002**, *52*, 1–13.
52. GUS. Available online: http://stat.gov.pl/obszary-tematyczne/rolnictwo-lesnictwo/psr-2010/powszechny-spis-rolny-2010-obszary-wiejskie,1,1.html (accessed on 16 March 2021).
53. ARMA. Available online: https://www.gov.pl/web/arimr/srednia-powierzchnia-w-2020r (accessed on 12 July 2021).
54. Grzegorzewska, E.; Burawska-Kupniewska, I.; Boruszewski, P. Economic profitability of particleboards production with a diversified raw material structure. *Maderas. Cienc. Tecnol.* **2020**, *22*, 537–548. [CrossRef]
55. Zitzmann, F.; Rode, M. Short-Rotation Coppice Managed According to Ecological Guidelines—What Are the Benefits for Phytodiversity? *Forests* **2021**, *12*, 646. [CrossRef]
56. Wójcik, J.; Balawejder, M.; Leń, P. Grunty marginalne, propozycje sposobów ich zagospodarowania w pracach scaleniowych w powiecie brzozowskim (Marginal lands, proposals for their management in the consolidating works in the District of Brzozów). *Infrastrukt. Ekol. Teren. Wiej.* **2014**, *2*, 399–410.
57. Węgorek, T. Warunki produkcji leśnej na skarpach zwałowiska zewnętrznego po kopalni siarki w Piasecznie w aspekcie pozyskania drewna opałowego (The conditions of forest production on scarps of the externalwaste bank after sulphur mine in piaseczno in the aspect of firewood logging). In *Tereny Zdegradowane i Rekultywowane—Możliwości ich Zagospodarowania (Reclaimed and Degraded Areas—Possibilities of Their Development)*; Stankowski, S., Pacewicz, K., Eds.; P.P.H. Zapol Dmochowski, Sobczyk Sp. j.: Szczecin, Poland, 2009; pp. 219–229.
58. Niemczyk, M.; Wojda, T.; Kantorowicz, W. Przydatność hodowlana wybranych odmian topoli w plantacjach energetycznych o krótkim cyklu produkcji. *Sylwan* **2016**, *160*, 292–298.
59. Kabała, C.; Karczewska, A.; Kozak, M. Przydatność roślin energetycznych do rekultywacji i zagospodarowania gleb zdegradowanych (Energetic plants in reclamation and management of degraded soils). *Zesz. Nauk. Uniw. Przyr. Wrocławiu Rol.* **2010**, *576*, 97–118.
60. Muñoz-Flores, H.J.; Castillo-Quiroz, D.; Castillo-Reyes, F.; Sáenz-Reyes, J.T.; Avila-Flores, D.; Rueda-Sánchez, A. Potential Areas for Commercial Timber Plantations of *Tabebuia rosea* (Bertol.) DC. in Michoacan, Mexico. *Open J. For.* **2017**, *7*, 48–57. [CrossRef]
61. Gholizadeh, A.; Bagherzadeh, A.; Keshavarzi, A. Model application in evaluating land suitability for OAK and PINE forest plantations in Northeast of Iran. *Geol. Ecol. Landsc.* **2020**, *4*, 236–250. [CrossRef]
62. Zhang, D.; Polyakov, M. The geographical distribution of plantation forests and land resources potentially available for pine plantations in the U.S. South. *Biomass Bioenergy* **2010**, *34*, 1643–1654. [CrossRef]
63. Lindegaard, K.N.; Adams, P.W.R.; Holley, M.; Lamley, A.; Henriksson, A.; Larsson, S.; Von Engelbrechten, H.G.; Lopez, G.E.; Pisarek, M. Short Rotation Plantations Policy History in Europe: Lessons from the Past and Recommendations for the Future. *Food Energy Secur.* **2016**, *5*, 125–152. [CrossRef]
64. Haughton, A.J.; Bohan, D.A.; Clark, S.J.; Mallott, M.D.; Mallott, V.; Sage, R.; Karp, A. Dedicated biomass crops can enhance biodiversity in the arable landscape. *GCB Bioenergy* **2015**, *8*, 1071–1081. [CrossRef] [PubMed]

MDPI

Article

Short-Rotation Coppice Managed According to Ecological Guidelines—What Are the Benefits for Phytodiversity?

Felix Zitzmann * and Michael Rode

Institute of Environmental Planning, Leibniz Universität Hannover, Herrenhäuser Str. 2, 30419 Hannover, Germany; rode@umwelt.uni-hannover.de

* Correspondence: zitzmann@umwelt.uni-hannover.de; Tel.: +49-511-762-19506

Abstract: In recent years, the impact of short-rotation coppice (SRC) on biodiversity has been a regular subject of research and ecological guidelines have been developed to make biomass cultivation on SRC more compatible with biodiversity concerns. However, since these guidelines are only implemented voluntarily by farmers, there are barely any SRC that are managed according to ecological guidelines. Consequently, knowledge about their importance for farmland biodiversity and about the impact of different measures for increasing biodiversity remains scarce. Therefore, three experimental SRC, which are managed according to ecological guidelines and thus include stands of different tree species (varieties of poplar (*Populus*) and willow (*Salix*), rowan (*Sorbus aucuparia*), silver birch (*Betula pendula*)) and different growth-stages within the same site, were investigated with regard to their importance as habitat for vascular plants. Species numbers and species composition were compared with the following habitat types: afforestations (AFO), young (HE-Y) and old hedges (HE-O), field margins (FM) and arable land (AL). Furthermore, different stand types (i.e., stands with different tree species and growth-stages, headlands, clearings) within these SRC were surveyed and compared. Species numbers of SRC were similar to HE-Y, AFO and FM and significantly higher than in AL and HE-O. The composition of plant communities in SRC differed considerably from the other farmland habitats, especially from AL, HE-O and FM. Within the SRC, most stand types had similar species numbers. Only the non-harvested poplar stands were particularly species-poor. Harvesting led to increased species numbers. This increase was significant for the poplar stands but only moderate for the willow stands. With regard to their species composition, the different stand types differed considerably in many cases. We conclude that SRC, which are managed according to ecological guidelines, can be an additional measure to promote phytodiversity in agricultural landscapes as they contain relatively high species numbers (of mainly common and adaptable species) and support distinct plant communities that differ from other farmland habitats. Therefore, measures such as the cultivation of different tree species or sectional harvesting could be offered as agri-environmental schemes to further increase the ecological sustainability of biomass production on SRC.

Keywords: woody biomass crops; bioenergy; biodiversity; species richness; flora; vascular plants

Citation: Zitzmann, F.; Rode, M. Short-Rotation Coppice Managed According to Ecological Guidelines—What Are the Benefits for Phytodiversity?. *Forests* **2021**, *12*, 646. https://doi.org/10.3390/f12050646

Academic Editor: Dirk Landgraf

Received: 28 April 2021
Accepted: 18 May 2021
Published: 19 May 2021

Publisher's Note: MDPI stays neutral with regard to jurisdictional claims in published maps and institutional affiliations.

1. Introduction

The decline of biodiversity is progressing rapidly. Thereby, this negative development is particularly evident in agricultural landscapes and even species that were formerly common and widespread are now affected [1–6]. A major reason for the decline in biodiversity is an increase in intensive agricultural use [7], for example, through the increasing cultivation of biomass crops [8–14]. However, biomass cultivation also offers opportunities to promote farmland biodiversity, since extensively managed perennial biomass crops, such as short-rotation coppice (SRC), can provide new habitats for wildlife and plants in agricultural landscapes [15–18]. SRC are biomass crops that consist of fast-growing trees (mostly cultivated varieties of poplar (*Populus*) or willow (*Salix*)), which are harvested in short cycles in order to use their wood for energy purposes [19].

In previous studies on the biodiversity of these woody crops, their importance as habitat for vascular plants was a frequent subject of investigation [20,21]. Thereby, SRC were found to be more species-rich than conventional arable crops [22,23] and clearly differed from arable land and forests with regard to their plant species composition [24–26]. Studies on the development of plant communities within SRC have shown dynamic changes with increasing age of the plantation and within rotation cycles [27–30] as well as differences in species numbers and species composition between plantations with different tree species or varieties [23,31,32].

From these results, it was concluded that SRC can contribute to the enhancement of biodiversity in intensively used landscapes with low habitat heterogeneity [33]. In addition, recommendations were developed to further improve the habitat function of SRC and to increase their contribution to farmland biodiversity, especially to plant species diversity ("phytodiversity"). The proposed measures aim to increase the structural diversity of these woody biomass crops. This includes the cultivation of different tree species in small-scale units, sectional harvesting of trees in order to establish a mosaic of different growth-stages side by side and the integration of accompanying structures such as headlands, clearings or rides to provide additional open habitat elements within SRC [cf. [15,27,34]]. Currently, these measures can only be implemented by farmers on a voluntary basis and there is no regular financial compensation (e.g., agri-environmental schemes) for the associated management effort or yield losses [cf. [35]]. Therefore, SRC managed according to ecological guidelines are quite rare. Consequently, there are no studies available that evaluate their importance as habitats for vascular plants and that examine and directly compare the effects of the different measures on one and the same plantation (i.e., on the same site with similar conditions with regard to plantation age, land use history or adjacent habitats, which are all factors that can have a strong influence on the composition of the recent vegetation of SRC [27,30,36]).

Therefore, in this study, we want to determine the importance of appropriately managed SRC as habitat for vascular plants on three experimental SRC, which are managed according to the ecological guidelines mentioned above. Our aim is to evaluate how these SRC perform in comparison to other farmland habitats with regard to their plant species diversity and whether they can provide an additional value for phytodiversity in an intensively used agricultural landscape. Furthermore, we want to evaluate the effects of the different measures implemented within the investigated SRC. For this purpose, we want to compare different stand types (i.e., stands with different tree species and growth-stages, headlands, clearings) within our SRC study sites with regard to their species numbers and species composition.

2. Materials and Methods

2.1. Study Area and Study Sites

Investigations were carried out in the municipality of Schapen (Emsland district, Lower Saxony) in north-western Germany. The landscape in this rural region is dominated by intensive agriculture (approx. 70%, predominantly arable land) and contains a relatively low (14%) proportion of woodland [37]. The study area is located 30–40 m above sea level. Mean annual precipitation accounts for 800 mm and mean annual temperature for 10 °C (long-term recordings from 1981–2010, Climate station Lingen, [38]). The region is dominated by sandy soils such as Podzols, Gleyic Podzols and, in areas closer to the groundwater, Gleysols, which developed from glacial sand deposits [39].

In addition to SRC managed according to ecological guidelines, five other farmland habitat types were selected for the study. In addition to arable land (previous land use), this includes a range of typical farmland habitats (afforestations, young and old hedges, field margins), which are regularly implemented as measures to increase biodiversity in agricultural landscapes (e.g., as agri-environmental schemes or greening measures in the context of the Common Agricultural Policy of the European Union). Therefore, the study focuses on the six following habitat types: short-rotation coppice (SRC), afforestations (AFO), young (HE-Y) and old hedges (HE-O), field margins (FM) and arable land (AL) (Table 1).

Table 1. Surveyed habitat types and number of surveyed sites, stands and plots per type.

Habitat Type	Area (ha) or Length (m)	No. of Sites (No. of Stands)	No. of Plots Per Site	No. of Plots Per Habitat Type
Short-rotation coppice (SRC)	2.0–2.2 ha	3 (38)	135, 195, 240	570
Afforestation (AFO)	0.9–3.4 ha	3 (13)	45, 105, 45	195
Hedge 'young' (HE-Y)	180–430 m	3	15	45
Hedge 'old' (HE-O)	110–230 m	6	15	90
Field margin (FM)	70–200 m	6	15	90
Arable land (maize) (AL)	1.6–6.0 ha	6	15	90

The three SRC sites, each about 2 ha in size, were established in spring 2011 and 2012, respectively, and were thus in their 8th or 9th growing season after establishment at the time of the study (June 2019). These small-scale and structurally diverse plantations are experimental sites which were established and are managed according to ecological guidelines [cf. [15,27,34]] in order to increase their contribution to farmland biodiversity. Therefore, different tree species and varieties were cultivated within the same site (see Figure 1) and are managed without any use of fertilisers or pesticides. In addition to varieties of poplar (Max 3, Hybride 275) and willow (Inger, Tordis), some native tree species (rowan (*Sorbus aucuparia*) and silver birch (*Betula pendula*)) were also planted. The trees were each planted in species-specific stands of 20 m width. Within the poplar and willow stands, harvesting was carried out in sections, so that three different growth-stages existed side by side at the time of the study: stands that were not harvested so far and stands that had been harvested in February 2018 or in February 2019, respectively, i.e., regrowth in the first or second growing season after harvesting (Table 2, Figure 1).

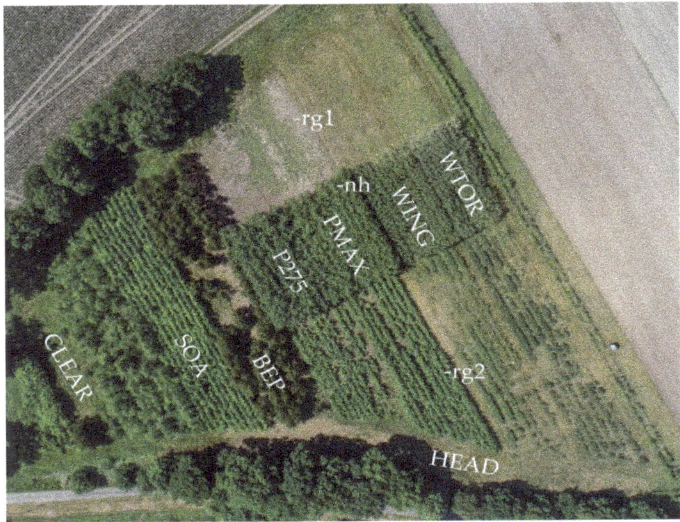

Figure 1. Aerial view on one of the three surveyed experimental SRC sites (site no. 3). The stands with different varieties of poplar and willow (in different growth-stages) and with different native tree species as well as the accompanying structures headland and clearing are clearly visible (Recording date: June 2019). See Table 2 for abbreviations of the SRC stand types.

Table 2. Characteristics of the surveyed SRC stand types.

SRC Stand Type		Abbrev.	No. of Stands	Planting Scheme (stools/ha)	Stools/ha (Incl. Failures)	Tree Height (m)
Poplar 'Max 3'	non-harvested	PMAX-nh	3	2 × 0.5 (10,000)	6000–7500	12–15
	2nd-year regrowth	PMAX-rg2	3			3–5
	1st-year regrowth	PMAX-rg1	3			<0.5
Poplar 'Hybride 275'	non-harvested	P275-nh	2	2 × 0.5 (10,000)	5000–7500	12–15
	2nd-year regrowth	P275-rg2	2			2–3
	1st-year regrowth	P275-rg1	2			<0.5
Willow 'Inger'	non-harvested	WING-nh	2	2 × 0.75 × 0.6 (12,000)	8000–10,000	8–10
	2nd-year regrowth	WING-rg2	2			2–3
	1st-year regrowth	WING-rg1	2			<0.5
Willow 'Tordis'	non-harvested	WTOR-nh	2	2 × 0.75 × 0.6 (12,000)	8000–10,000	8–10
	2nd-year regrowth	WTOR-rg2	2			2–3
	1st-year regrowth	WTOR-rg1	2			<0.5
Sorbus aucuparia	non-harvested	SOA	3	2 × 0.5 (10,000)	8000–9000	2–5
Betula pendula	non-harvested	BEP	3	2 × 1 (5000)	1000–3000	8–12
Clearings		CLEAR	3	–	–	–
Headlands		HEAD	2	–	–	–

The three afforestations (AFO) were established at the end of 2012. At the time of the study, they were in their 7th growing season since establishment. Hence, they were similar in age to the SRC. Within the afforestations, the deciduous tree species *Quercus robur*, *Q. petraea*, *Betula pendula*, *Fagus sylvatica*, *Carpinus betulus* and *Acer pseudoplatanus* were planted in varying proportions per site. The habitat type young hedge (HE-Y) included three hedge plantings of about 8 m width, also planted at the end of 2012, which directly bordered on the three AFO. Old hedges (HE-O) were mature hedgerows that had already been established decades or centuries ago in order to separate different fields from each other. The old hedges studied were characterised by shrubs and trees that had last been coppiced at least 10 to max. 50 years ago. The width of the hedges ranged from 4 to 10 m. Due to the intensive agricultural use of the adjacent fields, the hedges did not have any fringes, but bordered directly on intensively used arable land. Field margins (FM) were 1–3 m wide strips of herbaceous vegetation between two adjacent arable fields or at the edge of a field between the field and a track. Arable land (AL) comprised arable fields cultivated with maize (*Zea mays*), since maize represents the most commonly cultivated crop in the region.

We selected three sites of each habitat type SRC, AFO and HE-Y for the study, as only few (AFO) or no (SRC, HE-Y) other sites were available in the region and a random selection was therefore not possible. For the habitat types AL, HE-O and FM, the six study sites per type were randomly selected. Therefore, one site each for AL, HE-O and FM was randomly selected within a radius of 500 m around each of the three SRC and around each of the three AFO sites.

In addition to the six different habitat types, various stand types within the SRC were investigated (Table 2, Figure 1). The SRC stand types included stands with different native tree species (BEP, SOA) and different tree varieties (PMAX, P275, WING, WTOR) typically grown on SRC in various growth-stages (-nh, -rg1, -rg2) as well as accompanying structures such as headlands (HEAD) and clearings (CLEAR, areas > 500 m^2 where trees failed to establish). The average size of the individual stands within the SRC was 1510 ± 720 m^2 (Range: 400–3850 m^2). Each stand type was present once per site. Certain stand types were absent on individual SRC sites. Therefore, the number of replications per stand type was two or three (Table 2). Since the afforestations also contained different structures within the sites (besides tree stands there were also rides, clearings and margins), different stands were also investigated within the individual sites (Table 1).

The selection of the study sites resulted in six spatial clusters, each containing several sites (located within a radius of 500 m around each of the three SRC and AFO sites) or stands (located within each SRC and AFO site) that were spatially aggregated. For the statistical analysis, these clusters were considered as a fixed effect in the model (see data preparation and statistical analysis). Sites of SRC and AFO were located between 1 and 5.5 km apart.

2.2. *Survey of the Flora*

Surveys were carried out in June 2019. For each site (AL, HE-O, HE-Y and FM) and for each stand within the SRC and AFO sites, the flora was recorded using 15 randomly distributed quadratic sample plots (hereafter "plots"), each with a size of 1 m^2 (1 × 1 m). In order to minimize edge effects, the edge-zones were excluded from the surveys as far as possible and the 15 plots were distributed within the sites/stands at a minimum distance of 2 m from the edge. For HE-O and FM, it was not possible to exclude the edge-zones from the surveys, since these linear habitats were too narrow. Since the sites of arable land (AL) were fairly large, a 20 m wide area (corresponding to the width of the individual tree stands on the SRC) on a randomly selected side of the field was demarcated for the surveys.

All vascular plant species growing within each sample plot (of 1 m^2) were identified according to Jäger [40] and noted (presence/absence; no recording of the cover of individual species). The scientific nomenclature also followed Jäger [40]. Cultivated non-native plant species (e.g., varieties of poplar or willow, *Zea mays*) were excluded from sampling. Due to the high number of stands studied, especially on the SRC, but also on the AFO, there were major differences with regard to the total number of plots sampled per habitat type (Table 1).

2.3. *Data Preparation and Statistical Analysis*

The different habitat types as well as the different stand types within the SRC were compared with regard to their species numbers at (i) plot-level and at (ii) stand-level. Therefore, the species numbers were log-transformed before analysis (loge(y + 1), y = species number) to account for samples with no species detections. For the log-transformed data, a linear mixed effect model (LMM) was fitted to account for the unbalanced hierarchical sampling design. Spatial clusters (i.e., sites of different habitat types clustered within a 500 m radius around each SRC and AFO site, see section study area and study sites), habitat types and stand types were included as fixed effects. Two random effects were included: the variance between sites, and the variance between stands nested within sites. The residuals variance was thus the variance between plots nested within stands. For the number of species per site/stand (i.e., per 15 m^2) a simplified mixed model was fitted with the spatial clusters (see above), habitat types and stand types as fixed effects, and variance between sites as random effect. Assessment of normality assumption of the residuals was performed using R package 'hnp' [41]: the empirical distribution of residuals was symmetric, with no indication of skewedness or heterogeneity of variance and no indication of extreme values. Based on the fitted LMM, all pairwise comparisons of means between habitat types and between stand types were computed (analogously to the Tukey test). Significant differences at the 5%-level ($p < 0.05$) are shown in graphs by compact letter display. The statistical analysis was performed in R [42], using packages 'lme4' [43] for fitting mixed models, package 'lmerTest' [44] for ANOVA tables of the fixed effects, package 'emmeans' [45] for pairwise comparisons and compact letter display and 'ggplot' [46] for graphs.

To compare species numbers of SRC and AFO at site-level, despite different numbers of stands and plots sampled per site (Table 1), the species numbers for 15 and 45 plots were estimated for each of the three sites of these two habitat types by a sample-based rarefaction [cf. [47]] using the software BioDiversity Pro [48]. Therefore, samples of the dataset of each site (from plots of all stand types occurring there) were pooled in random order and the number of random sorts to perform on each pass was set to 50. For the three SRC sites, this estimation was also performed for 135 plots, since this was the minimum number of sampled plots per site (see Table 1).

Species composition of the habitat types and stand types was compared with regard to the habitat preferences of the occurring plant species. Therefore, the detected species were assigned to one of the following categories according to Oberdorfer [49]: (a) species of arable land or short-lived ruderal habitats, (r) species of persistent ruderal habitats, (g) grassland species, (h) species of heathland or nutrient-poor/dry grasslands, (w) species of woodlands incl. herbaceous vegetation of woodland margins, shrubs or hedges and (x) indifferent, not stated or species of other habitats. For the comparison of habitat types and stand types, all sites, stands and plots of the respective type were combined. For the comparison of the stand types, different varieties of the same genus (poplar or willow) were combined. For all comparisons, each species was weighted with its frequency, i.e., the number of plots in which the species was detected in the respective habitat type or stand type.

To compare the species composition of the different habitat types a detrended correspondence analysis (DCA) was applied using R package 'vegan' [50] for analysis and 'gplots' [51] for graphs. All stands of the different habitat types, respectively, stand types were included in this analysis. The species within the individual stands/sites were weighted with their frequency (i.e., occurrence in $n/15$ plots per stand/site). The default options were used, including detrending by 26 segments, non-linear re-scaling of axes with 4 iterations and no downweighting of rare species.

3. Results

3.1. Comparison of Habitat Types

3.1.1. Species Numbers

A total of 182 species were found across all surveyed habitat types. The majority of these species are widespread and common, no species classified as threatened in Germany were detected [52]. At the regional level (Lower Saxony), two of the species found are classified as threatened [53]: *Malus sylvestris* and *Ulmus minor*, the former was recorded on the AFO and the latter on some HE-O sites.

The highest number of species was recorded on SRC, with a total of 123 and the lowest on AL, with a total of 18. AFO ranked between SRC and AL, with a total of 108 species. HE-Y and FM both had 68 and HE-O had 41 species (Table 3). However, these species numbers are not directly comparable due to the different numbers of sites and stands surveyed per habitat type and thus different numbers of sampled plots (Table 1). Therefore, species numbers per stand (=15 m^2) and species numbers per plot (=1 m^2) were used for comparisons.

HE-Y was the most species-rich habitat type in terms of the number of species recorded per plot and per stand, followed by AFO, SRC and FM (Figure 2, Table 3). There were no differences between these four habitat types in terms of species numbers at either plot- or stand-level. AL had the lowest number of species at both levels and differed significantly from all other habitat types. HE-O took an intermediate position between the four species-rich habitat types and AL and differed significantly from them in terms of species number per plot. With regard to the species numbers per stand, HE-O differed significantly from AFO, HE-Y and AL. Like AFO and HE-Y, SRC and FM also had higher species numbers per stand than HE-O, but the differences were not significant.

Table 3. Number of species (overall and separated by species of different habitat preferences) per habitat type (total number and mean ± SD per site/stand and per plot). Values with no consistent letter indicate significant differences ($p < 0.05$).

		SRC	AFO	FM	HE-Y	HE-O	AL
Total no. of species (all sites included per type)		123	108	68	68	41	18
Mean no. of species [1] per site/stand (=15 m^2)	overall	25.9 ± 6.6 [ab]	30.2 ± 7.9 [a]	24.2 ± 7.7 [ab]	35 ± 4 [a]	16.2 ± 2.8 [b]	6 ± 2 [c]
	a	10.7 ± 4.9 [a]	5.9 ± 4.5 [ab]	6.8 ± 3.4 [a]	4.3 ± 4.5 [ab]	1.5 ± 1.6 [b]	4.2 ± 1.9 [ab]
	r	8 ± 2.9 [ab]	11.7 ± 2.1 [a]	7 ± 2.5 [b]	11.3 ± 1.5 [ab]	3.8 ± 1.3 [c]	1 ± 0 [d]
	g	2.8 ± 1.9 [a]	4.5 ± 2.5 [ab]	7 ± 2.7 [b]	6 ± 1 [ab]	1.8 ± 0.8 [a]	0 ± 0 [c]
	w	2.8 ± 2 [ab]	5 ± 1.7 [ac]	1.5 ± 1.1 [b]	10.7 ± 2.1 [c]	8.5 ± 1.6 [c]	0.8 ± 0.8 [b]
Mean no. of species [1] per plot (=1 m^2)	overall	7.5 ± 3.5 [a]	8.2 ± 2.7 [a]	8 ± 3 [a]	8.4 ± 2.7 [a]	4.1 ± 1.7 [b]	1.5 ± 1.3 [c]
	a	3.4 ± 2.8 [a]	1 ± 1.3 [ab]	1.7 ± 1.3 [a]	1 ± 1.4 [ab]	0.3 ± 0.6 [b]	1.2 ± 1.3 [ab]
	r	2.6 ± 1.6 [a]	4.6 ± 1.6 [a]	3.2 ± 1.1 [a]	3.7 ± 1.4 [a]	0.8 ± 1 [b]	0.2 ± 0.4 [b]
	g	0.7 ± 0.8 [a]	0.9 ± 1.1 [ab]	2.3 ± 1.5 [c]	1.2 ± 1 [ac]	0.2 ± 0.6 [ab]	0 ± 0 [b]
	w	0.5 ± 0.8 [ab]	1.1 ± 0.9 [ac]	0.4 ± 0.6 [ab]	1.8 ± 1 [cd]	2.6 ± 1.2 [d]	0.1 ± 0.3 [b]

[1] Abbreviations for habitat preferences: a = species of arable land or short-lived ruderal habitats, r = species of persistent ruderal habitats, g = grassland species, w = species of woodland, shrubs or hedges (incl. vegetation of woodland margins).

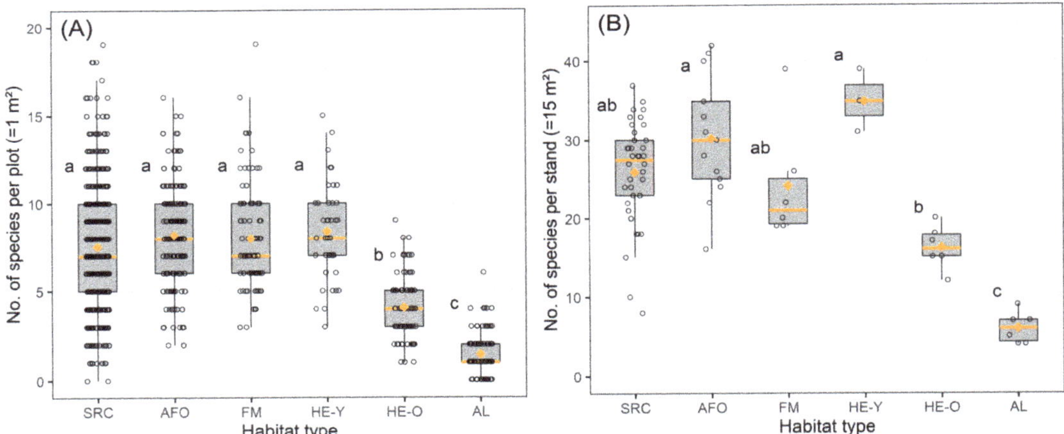

Figure 2. Species numbers of the different habitat types at plot-level (**A**) and at stand-level (**B**). Types with no consistent letter indicate significant differences ($p < 0.05$). Median values are presented as horizontal orange lines, mean values as orange diamonds. Unfilled circles show the data of the single plots or stands sampled per type. In each boxplot, the boundaries of the box are the 25th and 75th percentiles and the whiskers represent the lowest and largest values no further than 1.5 times away from the 25th and 75th percentiles.

Since several stands per site were investigated for SRC and AFO, a comparison of species numbers for a larger number of plots was possible. Therefore, a sample-based rarefaction was used to estimate the species number per site for a given number of sampled plots. Estimated species numbers for the sites of both habitat types were consistent, ranging from 35–43 species for 15 plots and 52–59 species for 45 plots (Figure 3). The SRC sites also had similar estimated species numbers (74–80 species) per site for 135 plots.

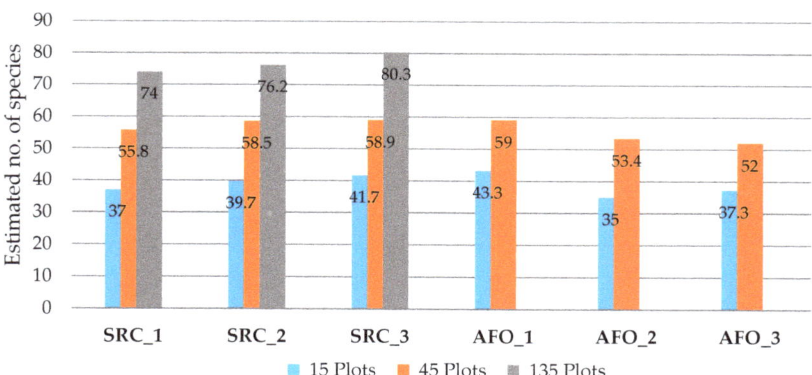

Figure 3. Estimated number of species for the different short-rotation coppice (SRC) and afforestation (AFO) sites for a given number of sampled plots (based on *sample-based rarefaction*, see data preparation and statistical analysis).

3.1.2. Species Composition and Species Numbers with Regard to Habitat Preferences

Figure 4 shows the species composition within the habitat types with regard to the habitat preferences of the detected species weighted with their frequencies. AL was dominated by species of arable land or short-lived ruderal habitats, while HE-O was dominated by species of woody habitats. The highest proportion of grassland species (almost 30%) was found in FM. SRC had high proportions of species from arable land or short-lived ruderal habitats (45%) and from persistent ruderal habitats (35%). With regard to the arable species, although the frequency proportion was lower for SRC than for AL, considerably more arable species were detected on the SRC than on AL due to the significantly higher total species number (Table 3). Compared to FM, HE-Y and AFO, the proportion of arable species in SRC was higher but the proportion of species from persistent ruderal habitats was lower.

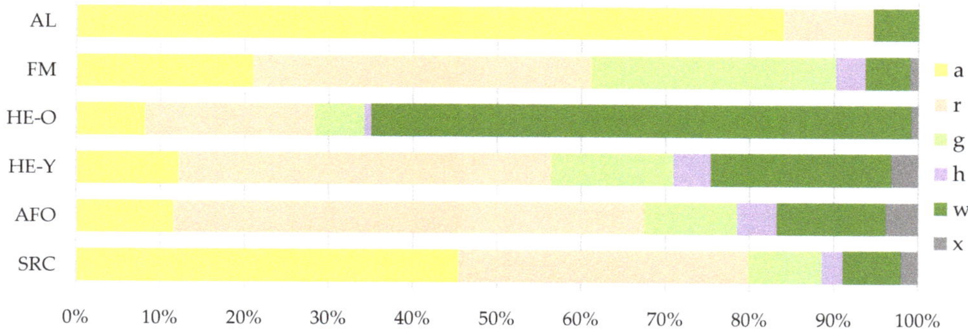

Figure 4. Proportions of plant species of different habitat preferences in the different habitat types (see Table 1 for abbreviations) weighted with their frequencies. Data of all sites, stands and plots included per type. Abbreviations for habitat preferences: a = species of arable land or short-lived ruderal habitats, r = species of persistent ruderal habitats, g = grassland species, h = species of heathland or nutrient-poor/dry grasslands, w = species of woodland, shrubs or hedges (incl. vegetation of woodland margins), x = indifferent, not stated or other habitat.

In addition to Figure 4, Table 3 shows the mean numbers of species of different habitat preferences at site/stand-level and at plot-level. In comparison to the other habitat types, SRC had the highest number of species of arable land and short-lived ruderal habitats, both at site/stand-level and at plot-level. The number of species of persistent ruderal habitats was particularly high in AFO and HE-Y. In SRC, the number of species of this category was slightly lower, but the differences with AFO and HE-Y were not significant. In contrast, HE-O and AL had significantly lower numbers of species of persistent ruderal habitats than SRC, AFO, FM and HE-Y. The number of grassland species was highest in FM at both site- and plot-level. Their number was significantly higher there than in SRC. Species of woodland, shrubs or hedges were particularly frequent in HE-O and HE-Y. Both hedge types differed significantly from SRC with regard to the number of species of this category.

3.1.3. Similarity of Plant Communities

On the DCA graph, the stands are separated into four different groups (Figure 5), which are, however, very heterogeneous and widely scattered. AL, HE-O and FM are each separated from a large group consisting of SRC, AFO and HE-Y. Within the latter group, a certain separation into two groups, SRC and AFO/HE-Y, is visible, but there are also smooth overlaps between both groups, respectively, the three habitat types. In addition, one FM site and single AFO and SRC stands were fairly similar. These stands were a margin on a AFO site and a clearing on a SRC site.

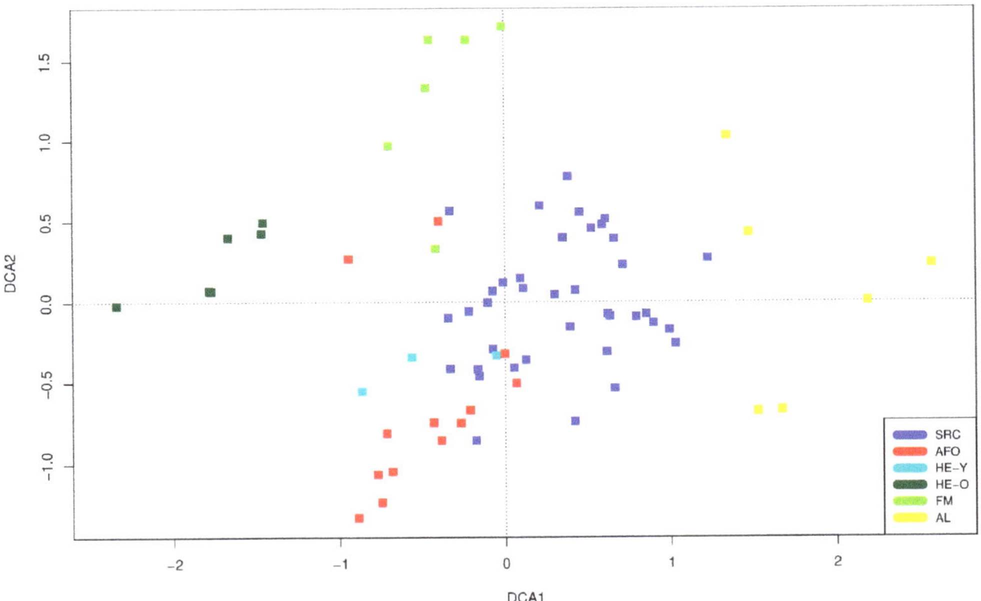

Figure 5. DCA ordination of the individual stands/sites of the different habitat types (see Table 1 for abbreviations). Eigenvalue axis 1: 0.4266, axis 2: 0.3283, length of axis 1: 4.9152, axis 2: 3.0447.

The considerable dispersion within all groups illustrates that the individual sites or stands of the respective habitat types did not show homogeneous plant communities and sometimes differed significantly with regard to their qualitative (occurring species) and quantitative (frequency of the species) species composition. For SRC and AFO/HE-Y, there were somewhat higher similarities between individual stands, but also considerable differences between others (in detail see Section 3.2.3).

3.2. Comparison of SRC Stand Types

3.2.1. Species Numbers

When comparing the different stand types within the SRC, the number of species at plot-level did not differ significantly in most cases (Figure 6). The lowest species numbers were found in the non-harvested poplar stands. Stands with the poplar variety Hybride 275 (P275-nh) were particularly species-poor and differed significantly from all other stand types. For all poplar and willow stands, an increase in species numbers was observed after harvesting. This was particularly evident for the poplar stands. Here, non-harvested stands (nh) mostly differed significantly from stands in the first and second growing season after harvesting (rg-1, rg-2). Notably, PMAX-rg1 showed the highest number of species of all investigated stand types. For the willows, however, differences between the harvested and the non-harvested stands were less pronounced and not significant. Stand types with native tree species (SOA, BEP) as well as accompanying structures (HEAD, CLEAR) showed average species numbers. Species numbers at stand-level (Figure A1 in Appendix A) followed the trend of species numbers at plot-level. Since only two or three individual stands were surveyed per stand type, a statistical comparison at this level was not applicable.

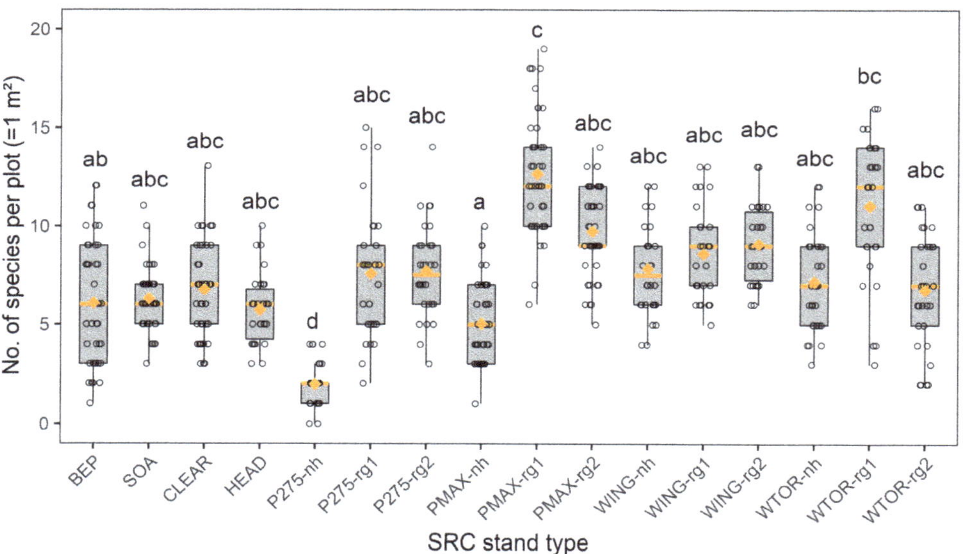

Figure 6. Species numbers of the different SRC stand types at plot-level. Types with no consistent letter indicate significant differences ($p < 0.05$). Median values are presented as horizontal lines, mean values as orange diamonds. Unfilled circles show the data of the single plots sampled per type. In each boxplot, the boundaries of the box are the 25th and 75th percentiles and the whiskers represent the lowest and largest values no further than 1.5 times away from the 25th and 75th percentiles. See Table 2 for abbreviations of stand types.

3.2.2. Species Composition and Species Numbers with Regard to Habitat Preferences

Figure 7 shows the species composition within the different SRC stand types with regard to the habitat preferences of the detected species weighted with their frequencies. Willow and poplar stands (varieties of the same genus were combined for this analysis) had particularly high proportions (45–70%) of species from arable land or short-lived ruderal habitats, both when harvested (-rg1, -rg2) and when not harvested (-nh). The highest proportions (almost 70%) were found in the poplar stands in the first growing season after harvest (P-rg1). Species of persistent ruderal habitats had particularly high proportions (50–60%) in SOA, BEP and CLEAR, while grassland species (almost 35%) were most frequently in HEAD. Species of woody habitats had low proportions overall, but were slightly more frequent in stands that had not been harvested since the establishment of the SRC (SOA, BEP, W-nh, P-nh).

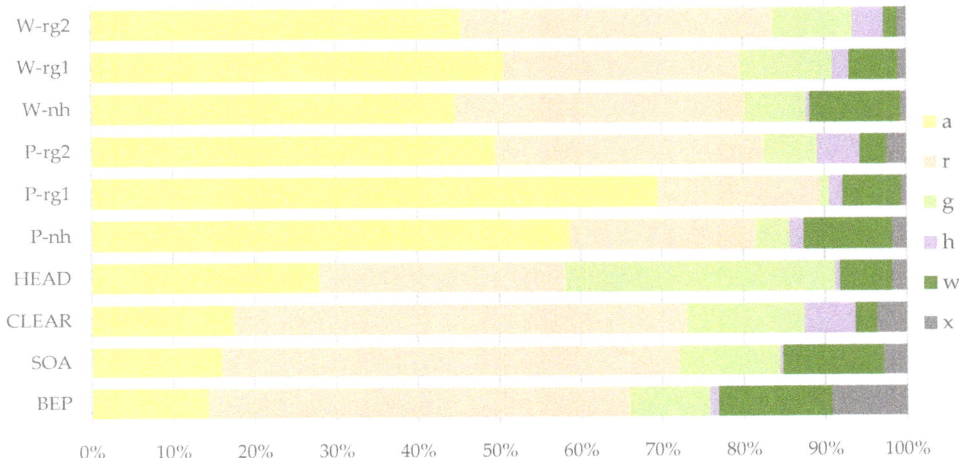

Figure 7. Proportions of plant species of different habitat preferences within the SRC stand types weighted with their frequencies (see Table 2 for abbreviations of stand types; different poplar (P) and willow (W) varieties are combined). Abbreviations for habitat preferences: a = species of arable land or short-lived ruderal habitats, r = species of persistent ruderal habitats, g = grassland species, h = species of heathland or nutrient-poor/dry grasslands, w = species of woodland, shrubs or hedges (incl. vegetation of woodland margins), x = indifferent, not stated or other habitat.

In addition to Figure 7, Table A1 shows the mean numbers of species of different habitat preferences within the different SRC stand types at plot-level. Particularly high numbers of arable species were found within the different willow (WING, WTOR) and poplar (P275, PMAX) stands in the first (-rg1) and second (-rg2) growing season after harvesting. In contrast, BEP, SOA and CLEAR had particularly low numbers of arable species. Species of persistent ruderal habitats were found with similar numbers per plot in most of the surveyed stand types. Only the non-harvested poplar stands (P275-nh, PMAX-nh) had comparatively low numbers of species belonging to this category. Grassland species were also recorded with similar numbers per plot in most of the stand types. HEAD had the highest number of grassland species per plot while P275-nh and PMAX-nh had the lowest. The number of woodland species was generally low in all stand types and there were no significant differences between the different types.

3.2.3. Similarity of Plant Communities

The DCA graph (Figure 8) shows a wide dispersion of the different SRC stands both within and between the sites. Stands within the same site were therefore not clearly separated into distinct groups, instead there were overlaps between the three sites and their individual stands. These overlaps indicate some similarities with regard to the development of the flora and the species composition on the three surveyed SRC sites (as complexes consisting of different stand types).

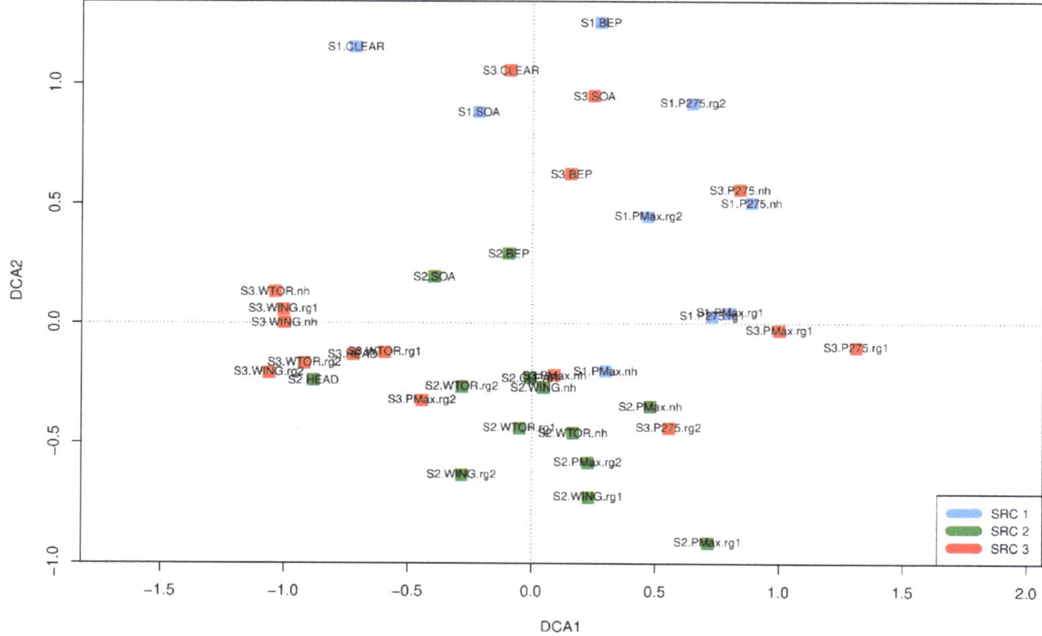

Figure 8. DCA ordination of the individual stands of the different SRC stand types. See Table 2 for abbreviations of SRC stand types. Eigenvalue axis 1: 0.3243, axis 2: 0.2540, length of axis 1: 2.3706, axis 2: 2.1698.

In terms of qualitative (occurring species) and quantitative (frequency of the species) species composition there were, in some cases, somewhat greater similarities between stands from different sites, but in other cases there were also clear differences. Overall, the graph of the DCA shows a heterogeneous result in which no clear patterns of similarity between certain stands or stand types are discernable. On the one hand, it is not clearly discernible that stands of the same stand type (on different sites) were regularly very similar. At the same time, stands (of different stand types) within the same sites were not necessarily very similar. Some stands differed considerably from other stands within the same site with regard to their species composition. This was most evident for SOA and BEP on all three sites. In other cases, however, there were also greater similarities between different stands within the same site. This was particularly noticeable at sites 2 and 3. For example, on site 3 there was a fairly high similarity between the willow stands of different varieties and growth-stages.

4. Discussion

In accordance with previous studies on phytodiversity of SRC [22,23,26,31], we also found significantly higher species numbers compared to conventional arable fields with annual crops. Species of arable land and short-lived ruderal habitats still had high proportions on the surveyed SRC in the 8th and 9th year after establishment and were particularly frequent in comparison to species of other habitat preferences. A reason for this is the landscape context (high proportion of arable land) and the previous agricultural use of the SRC sites. Both factors can influence the species composition of SRC for a long time after its establishment [27,30,36]. On the other hand, the high proportion of arable species can also be explained by the sectional harvesting, since it repeatedly creates suitable conditions for these light-demanding species in certain areas of the plantations [cf. [31,34]]. Especially within the harvested stands, high numbers and frequency proportions of arable species (such as *Aphanes australis, Myosotis arvensis, Spergula arvensis* or *Veronica arvensis*) were found. Many of these species were absent or less frequent in the surveyed arable fields with maize cultivation. Therefore, recently harvested (as well as recently established) SRC can provide a substitute habitat for common arable species that do not find suitable conditions on intensively used conventional arable fields anymore. However, SRC are not expected to have any potential to promote rare or threatened arable species, since the shading from trees and competition with the accompanying perennial herbaceous vegetation does not provide favorable conditions for highly specialized and low-competitive arable species [23,31,54,55]. In order to promote rare and threatened arable species, specific measures within arable fields are required [56,57], and SRC do not offer an alternative to these.

In comparison to AFO, HE-Y and FM, the surveyed SRC had similar species numbers. In comparison to HE-O, the species numbers of SRC were even higher. At the same time, species composition of SRC was quite different from the other habitat types, in particular from HE-O, FM and AL. Even though many species found on the SRC also occurred in the other habitat types, the combination of these species on the SRC was different from the other habitat types in qualitative (species inventory) and quantitative (species frequencies) terms. Furthermore, the SRC also contained some species that were not found in the other surveyed habitats. Therefore, the SRC formed distinct plant communities that differed from the other farmland habitat types (cf. similar findings of Baum et al. [26]). Similarity of the surveyed SRC with AFO and HE-Y was highest in comparison to the other habitat types. This is not a surprising result, as these habitats had a similar age and vegetation structure as the SRC and were also established on arable land. However, during succession, the similarity of AFO and HE-Y with SRC will further decrease. This is already indicated by the higher numbers and proportions of perennial species and the lower numbers and proportions of short-lived species in AFO and HE-Y (Table 3, Figure 4). Despite their similar species numbers, SRC are not an immediate substitute for other habitats of agricultural landscapes, since they have their own distinct plant communities. Instead, SRC are novel habitats that can increase the habitat diversity of a landscape [24,26,33] and complement the range of existing agricultural crops and the range of regularly implemented measures to promote farmland biodiversity such as afforestations, hedges or field margins. Thus, SRC can contribute to phytodiversity at the landscape-level, especially in intensively used agricultural landscapes with low habitat heterogeneity [cf. [25,33]].

A major result of our surveys of different stands within the SRC is that different stand types, either with different tree species or in different growth-stages, can significantly increase plant species diversity within a plantation. This confirms recommendations [cf. [15,27,34]] for implementing appropriate measures to promote phytodiversity on SRC. The positive effect of such a mosaic is shown by the fact that the species composition of different stand types within the same site differed considerably in several cases. Different stand types create heterogeneous conditions (e.g., light availability, intensity of disturbance), providing habitats for species with different habitat requirements within the same site [34,58]. Regular disturbance by harvesting was particularly beneficial for phytodiversity at stand-level. This was particularly evident for the poplar stands, where species numbers strongly increased after harvesting. For the willow stands, species numbers also increased after harvesting, but more moderately. In contrast, stands with native tree species (SOA, BEP) as well as headlands (HEAD) and clearings (CLEAR) contained fairly average species numbers. SOA and BEP had not been disturbed by harvesting so far. In addition, these stand types were either characterized by numerous tree failures (BEP) or tree growth was significantly lower (SOA) than for the poplar and willow clones. Due to the high light availability and the lack of any disturbance, these stand types, as well as HEAD and CLEAR, were often dominated by highly competitive species (e.g., *Elymus repens, Festuca rubra*) of persistent ruderal habitats or grasslands (similar observations were made by Glaser and Schmidt [23] within gaps of SRC, where trees failed to grow), while species of arable land or short-lived ruderal habitats had lower numbers and proportions

(Figure 7, Table A1). However, despite their more average species numbers, these stand types were also important for the phytodiversity of the entire plantation, as they differed considerably from the poplar and willow stands in terms of their species composition (see Figures 7 and 8). In addition, stands with native tree species and/or accompanying structures are important components of SRC for other species groups such as ground beetles [59], large and medium-sized mammals [60], breeding birds [61,62] or birds in the winter season [63] and are therefore important features of SRC with regard to overall biodiversity. Furthermore, plant species richness in non-wooded accompanying structures of SRC such as headlands, clearings/gaps and rides can be further increased by additional measures, e.g., by establishing species-rich fringe vegetation by seeding of native seed mixtures (see Kiehl et al. [64] for general recommendations).

In addition to the partly considerable differences between harvested and non-harvested stands within the SRC (especially within the poplar stands), the non-harvested poplar and willow stands also differed considerably with regard to their species numbers and species composition. These differences can essentially be explained by the specific local conditions that exist in SRC stands with different tree species. Specific characteristics (e.g., leaf shape and size) of different tree species and varieties can influence the environmental conditions within the stands; large-leafed poplar clones lead to much greater shading than narrow-leafed willow clones at similar planting densities [31]. The resulting low species numbers in the non-harvested poplar stands, however, lead to a significant increase in species numbers directly after harvesting, as many plant species are able to establish spontaneously in these almost vegetation-free stands [cf. [31,54]]. In the willow stands, on the other hand, the increase in species numbers after harvest was much more moderate. This was due to the fact, that there still existed a comparatively species-rich and dense herb layer before harvest which meant that there was less space for additional species to establish spontaneously. Based on these findings, it can be concluded that stand types that differ significantly in terms of their structural characteristics (e.g., poplar and willow stands) are particularly beneficial for the phytodiversity of the entire plantation (as a mosaic of different stands) [cf. 34]. In contrast, if only different clones of the same tree genus are cultivated within a SRC, the increase in phytodiversity is likely to be less pronounced since the habitat conditions in stands of the same genera are more similar [cf. [31]].

A finding that seems unusual at first sight is the high frequency of species from arable land and short-lived ruderal habitats within the non-harvested willow and poplar stands (Figure 7). However, it must be taken into account that these were, especially in the case of the poplar stands, only a few frequently occurring species (in P-nh especially *Chenopodium album* and *Stellaria media*), which colonized the stands at the beginning of the vegetation period (April–June), but which had already disappeared in summer when the canopy closed. Similar findings of seasonal changes on flora are reported by Gustafsson [28] for SRC in Sweden and by Heilmann et al. [31] for SRC in Germany.

In addition to the positive contribution of a mosaic of different stand types to the phytodiversity of the SRC at the site-level, the small size of the individual stands is likely to have had a beneficial effect on phytodiversity at stand-level, since edge-zones of SRC are usually more species-rich than central areas [22,25,32,54,65]. To promote phytodiversity, SRC should either be small-scale and established in an elongated and rectangular rather than a square shape, or larger plantations should be managed in a small-scale manner (like on the investigated SRC sites) in order to increase the proportion of edge-zones and improve immigration opportunities for plants [34].

Basically, the plant communities of the surveyed SRC and all other investigated habitat types were dominated by widespread, common and non-threatened plant species. In our study region, the presence of species of conservation concern was not expected, since the establishment potential for these species is limited due to many decades of intensive agricultural use [66–68]. However, in many other studies on SRC in different regions of Germany and in other European countries [22–26,29,33,54,55,69], these woody crops did not prove to be a habitat of major importance for rare or threatened plant species. Instead, they supported adaptable, mostly competitive, widespread and common plant species. The few threatened or rare species found on SRC so far were mostly light-demanding pioneer species that were present in the first few years after establishment and disappeared as the SRC matured [34,70]. Therefore, SRC cannot be expected to be a suitable measure to promote rare, threatened or specialized plant species, even when managed according to ecological guidelines.

5. Conclusions

Small-scale and structurally diverse SRC which are managed according to ecological guidelines provide suitable habitats for a variety of different plant species due to their diverse habitat conditions within the same site. They have similar species numbers as hedges and afforestations of the same age or narrow field margins. Due to their specific habitat characteristics, they form distinct plant communities that differ considerable from other farmland habitats. Therefore, they can increase habitat diversity in intensively used agricultural landscapes and thus be an additional tool to promote farmland phytodiversity [cf. [33]]. However, the species that benefit from their establishment are mostly adaptable, widespread and common species of no conservation concern [cf. [34]]. On the investigated SRC, it was confirmed that measures such as harvesting in sections or cultivation of different tree species in small-scale units within the same plantation, are particularly effective in promoting phytodiversity of these woody biomass crops. Therefore, these measures could be offered as agri-environmental schemes in order to compensate for the associated effort or yield reduction and to further increase the ecological sustainability of biomass production on SRC [cf. [71,72]].

Author Contributions: Conceptualization, F.Z.; methodology, F.Z. and M.R.; formal analysis, F.Z.; investigation, F.Z.; data curation, F.Z.; writing original draft preparation, F.Z.; writing review and editing, F.Z. and M.R.; visualization, F.Z.; supervision, M.R.; project administration, M.R.; funding acquisition, M.R. All authors have read and agreed to the published version of the manuscript.

Funding: This research was funded by the Lower Saxony Ministry of Food, Agriculture and Consumer Protection [Niedersächsisches Ministerium für Ernährung, Landwirtschaft und Verbraucherschutz], grant number 105.2-3234/1-13-4. The publication of this article was funded by the Open Access Fund of Leibniz University Hannover.

Institutional Review Board Statement: Not applicable.

Informed Consent Statement: Not applicable.

Data Availability Statement: The raw-datasets generated during and/or analysed during the current study are available from the corresponding author on reasonable request.

Acknowledgments: We are grateful to the Lower Saxony Ministry of Food, Agriculture and Consumer Protection for the funding of our research. Furthermore, we thank K. Grobe, C. Meiser, J. Nöhren and G. Wienrich for their assistance in field work and L. von Falkenhayn for proofreading the English manuscript. We are also grateful to S. Budig, F. Schaarschmidt and A. Grobe for advice on the statistical analysis and for their help with some statistical analyses using the programme R. The publication of this article was funded by the Open Access Fund of Leibniz University Hannover.

Conflicts of Interest: The authors declare no conflict of interest. The funding organisation had no role in the design, execution, interpretation or writing of the study.

Appendix A

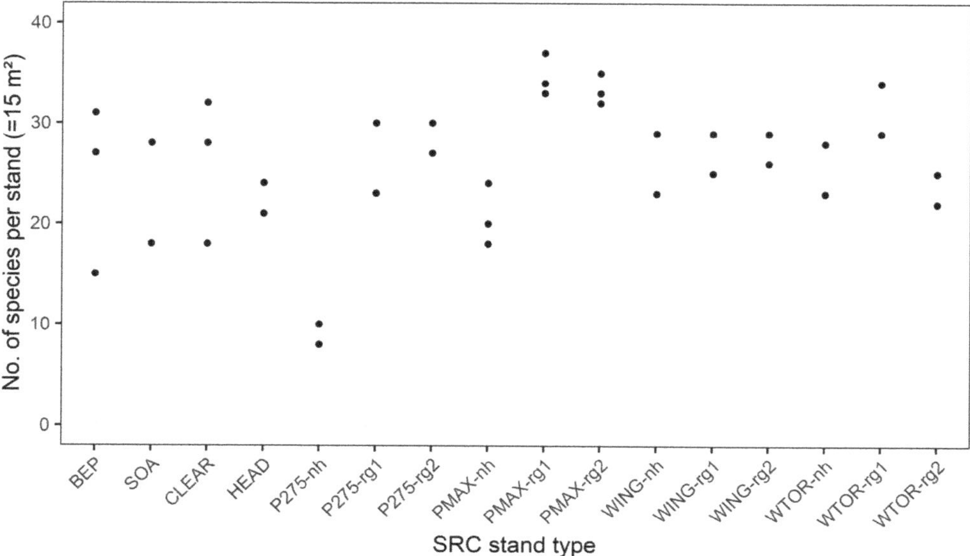

Figure A1. Species numbers of the different stand types at stand-level. Filled circles show the data of the single stands sampled per stand type (with n = 2 or 3 replications per type). See Table 2 for abbreviations of stand types.

Table A1. Number of species (overall and separated by species of different habitat preferences) of the different SRC stand types at plot-level. Values with no consistent letter indicate significant differences ($p < 0.05$). For abbreviations of stand types see Table 2.

No. of Species [1]	BEP	SOA	CLEAR	HEAD	P275			PMAX			WING			WTOR		
					-nh	-rg1	-rg2	-nh	-rg1	-rg2	-nh	-rg1	-rg2	-nh	-rg1	-rg2
overall	6.1 ± 3.1^{ab}	6.3 ± 1.6^{abc}	6.8 ± 2.4^{abc}	5.7 ± 1.7^{abc}	2 ± 1.1^{d}	7.6 ± 3.3^{abc}	7.7 ± 2.4^{abc}	5 ± 2^{a}	12.6 ± 3.1^{c}	9.8 ± 2.2^{abc}	7.8 ± 2.3^{abc}	8.6 ± 2.3^{abc}	9.1 ± 1.9^{abc}	7.2 ± 2.5^{abc}	11 ± 3.6^{bc}	6.8 ± 2.9^{abc}
a	0.9 ± 1^{a}	1 ± 0.9^{ab}	1.2 ± 1.4^{ab}	1.6 ± 1^{abcd}	1 ± 0.9^{abc}	5.4 ± 2^{de}	3.4 ± 1.7 abcde	3.1 ± 1.5 abcde	8.6 ± 2.4^{e}	5.1 ± 2^{de}	3.6 ± 1.9 abcde	4.3 ± 2.2^{bcde}	4.5 ± 1.4^{cde}	3.1 ± 1.9 abcde	5.6 ± 2.5^{de}	2.7 ± 1.7^{abcd}
r	3.1 ± 1.7^{a}	3.5 ± 1.3^{a}	3.8 ± 1.9^{a}	1.7 ± 0.9^{abc}	0.7 ± 0.7^{b}	1.6 ± 1.5^{abc}	2.8 ± 1.9^{abc}	1 ± 0.8^{bc}	2.5 ± 1.3^{abc}	3 ± 1.7^{a}	2.8 ± 1.2^{ac}	2.4 ± 0.7^{ac}	3.1 ± 0.8^{a}	2.5 ± 1.2^{ac}	3.3 ± 0.9^{a}	3 ± 0.9^{a}
g	0.6 ± 0.7^{ab}	0.8 ± 0.8^{ab}	1 ± 1^{ab}	1.9 ± 0.6^{a}	0 ± 0^{b}	0.1 ± 0.3^{b}	0.6 ± 0.6^{ab}	0.3 ± 0.5^{b}	0.2 ± 0.4^{b}	0.6 ± 0.8^{ab}	0.5 ± 0.5^{ab}	1.2 ± 1.1^{ab}	0.8 ± 0.9^{ab}	0.6 ± 0.8^{ab}	1.1 ± 0.9^{ab}	0.8 ± 0.8^{ab}
w	0.8 ± 0.9	0.8 ± 0.6	0.2 ± 0.4	0.4 ± 0.7	0.2 ± 0.5	0.4 ± 0.7	0.3 ± 0.6	0.6 ± 0.8	1 ± 0.9	0.3 ± 0.5	0.8 ± 1	0.3 ± 0.6	0.2 ± 0.4	0.8 ± 1	0.8 ± 1.5	0.1 ± 0.3

[1] Abbreviations for habitat preferences: a = species of arable land or short-lived ruderal habitats, r = species of persistent ruderal habitats, g = grassland species, w = species of woodland, shrubs or hedges (incl. vegetation of woodland margins).

References

1. Donald, P.F.; Sanderson, F.J.; Burfield, I.J.; van Bommel, F.P.J. Further evidence of continent-wide impacts of agricultural intensification on European farmland birds, 1990–2000. *Agric. Ecosyst. Environ.* **2006**, *116*, 189–196. [CrossRef]
2. Eichenberg, D.; Bowler, D.E.; Bonn, A.; Bruelheide, H.; Grescho, V.; Harter, D.; Jandt, U.; May, R.; Winter, M.; Jansen, F. Widespread decline in Central European plant diversity across six decades. *Glob. Chang. Biol.* **2020**, 1–14. [CrossRef]
3. Cardoso, P.; Barton, P.S.; Birkhofer, K.; Chichorro, F.; Deacon, C.; Fartmann, T.; Fukushima, C.S.; Gaigher, R.; Habel, J.C.; Hallmann, C.A. Scientists' warning to humanity on insect extinctions. *Biol. Conserv.* **2020**, *242*, 108426. [CrossRef]
4. Hallmann, C.A.; Ssymank, A.; Sorg, M.; de Kroon, H.; Jongejans, E. Insect biomass decline scaled to species diversity: General patterns derived from a hoverfly community. *Proc. Natl. Acad. Sci. USA* **2021**, *118*, e2002554117. [CrossRef] [PubMed]
5. Kamp, J.; Frank, C.; Trautmann, S.; Busch, M.; Dröschmeister, R.; Flade, M.; Gerlach, B.; Karthäuser, J.; Kunz, F.; Mitschke, A. Population trends of common breeding birds in Germany 1990–2018. *J. Ornithol.* **2021**, *162*, 1–15. [CrossRef]
6. Warren, M.S.; Maes, D.; van Swaay, C.; Goffart, P.; van Dyck, H.; Bourn, N.A.D.; Wynhoff, I.; Hoare, D.; Ellis, S. The decline of butterflies in Europe: Problems, significance, and possible solutions. *Proc. Natl. Acad. Sci. USA* **2021**, *118*, e2002551117. [CrossRef]
7. Stoate, C.; Boatman, N.D.; Borralho, R.J.; Carvalho, C.R.; de Snoo, G.R.; Eden, P. Ecological impacts of arable intensification in Europe. *J. Environ. Manag.* **2001**, *63*, 337–365. [CrossRef]
8. Huston, M.A.; Marland, G. Carbon management and biodiversity. *J. Environ. Manag.* **2003**, *67*, 77–86. [CrossRef]
9. Robertson, G.P.; Dale, V.H.; Doering, O.C.; Hamburg, S.P.; Melillo, J.M.; Wander, M.M.; Parton, W.J.; Adler, P.R.; Barney, J.N.; Cruse, R.M.; et al. Sustainable Biofuels Redux. *Science* **2008**, *322*, 49–50. [CrossRef]
10. Eggers, J.; Tröltzscht, K.; Falcucci, A.; Maiorano, L.; Verburg, P.H.; Framstad, E.; Louette, G.; Maes, D.; Nagy, S.; Ozinga, W.I.M. Is biofuel policy harming biodiversity in Europe? *Glob. Chang. Biol. Bioenergy* **2009**, *1*, 18–34. [CrossRef]
11. Fletcher, R.J.; Robertson, B.A.; Evans, J.; Doran, P.J.; Alavalapati, J.R.R.; Schemske, D.W. Biodiversity conservation in the era of biofuels: Risks and opportunities. *Front. Ecol. Environ.* **2011**, *9*, 161–168. [CrossRef]
12. Everaars, J.; Frank, K.; Huth, A. Species ecology and the impacts of bioenergy crops: An assessment approach with four example farmland bird species. *Glob. Chang. Biol. Bioenergy* **2014**, *6*, 252–264. [CrossRef]
13. Immerzeel, D.J.; Verweij, P.A.; van der Hilst, F.; Faaij, A.P.C. Biodiversity impacts of bioenergy crop production: A state-of-the-art review. *Glob. Chang. Biol. Bioenergy* **2014**, *6*, 183–209. [CrossRef]
14. Sauerbrei, R.; Ekschmitt, K.; Wolters, V.; Gottschalk, T.K. Increased energy maize production reduces farmland bird diversity. *Glob. Chang. Biol. Bioenergy* **2014**, *6*, 265–274. [CrossRef]
15. Sage, R.B. Short rotation coppice for energy: Towards ecological guidelines. *Biomass Bioenergy* **1998**, *15*, 39–47. [CrossRef]
16. Dauber, J.; Jones, M.B.; Stout, J.C. The impact of biomass crop cultivation on temperate biodiversity. *Glob. Chang. Biol. Bioenergy* **2010**, *2*, 289–309. [CrossRef]
17. Englund, O.; Dimitriou, I.; Dale, V.H.; Kline, K.L.; Mola-Yudego, B.; Murphy, F.; English, B.; McGrath, J.; Busch, G.; Negri, M.C.; et al. Multifunctional perennial production systems for bioenergy: Performance and progress. *WIREs Energy Environ.* **2020**. [CrossRef]
18. Pedroli, B.; Elbersen, B.; Frederiksen, P.; Grandin, U.; Heikkilä, R.; Krogh, P.H.; Izakovičová, Z.; Johansen, A.; Meiresonne, L.; Spijker, J. Is energy cropping in Europe compatible with biodiversity?—Opportunities and threats to biodiversity from land-based production of biomass for bioenergy purposes. *Biomass Bioenergy* **2013**, *55*, 73–86. [CrossRef]
19. Dimitriou, I.; Rutz, D. (Eds.) *Sustainable Short Rotation Coppice: A Handbook*; WIP Renewable Energies: München, Germany, 2015.
20. Vanbeveren, S.P.P.; Ceulemans, R. Biodiversity in short-rotation coppice. *Renew. Sustain. Energy Rev.* **2019**, *111*, 34–43. [CrossRef]
21. Rowe, R.L.; Street, N.R.; Taylor, G. Identifying potential environmental impacts of large-scale deployment of dedicated bioenergy crops in the UK. *Renew. Sustain. Energy Rev.* **2009**, *13*, 271–290. [CrossRef]
22. Cunningham, M.D.; Bishop, J.D.; McKay, H.V.; Sage, R.B. *ARBRE Monitoring—Ecology of Short Rotation Coppice: Four Year Study Involving Wildlife Monitoring of Commercial SRC Plantations Planted on Arable Land and Arable Control Plots*; Department of Trade and Industry: London, UK, 2004.
23. Glaser, T.; Schmidt, P.A. Auswirkungen von Kurzumtriebsplantagen auf die Phytodiversität. In *AGROWOOD: Kurzumtriebsplantagen in Deutschland und Europäische Perspektiven*; Bemmann, A., Knust, C., Eds.; Weißensee-Verlag: Berlin, Germany, 2010; pp. 153–161.
24. Britt, C.P.; Fowbert, J.; McMillan, S.D. The ground flora and invertebrate fauna of hybrid poplar plantations: Results of ecological monitoring in the PAMUCEAF project. *Asp. Appl. Biol.* **2007**, *82*, 83–90.
25. Weih, M.; Karacic, A.; Munkert, H.; Verwijst, T.; Diekmann, M. Influence of young poplar stands on floristic diversity in agricultural landscapes (Sweden). *Basic Appl. Ecol.* **2003**, *4*, 149–156. [CrossRef]
26. Baum, S.; Bolte, A.; Weih, M. High value of short rotation coppice plantations for phytodiversity in rural landscapes. *GCB Bioenergy* **2012**, *4*, 728–738. [CrossRef]
27. Gustafsson, L. Plant conservation aspects of energy forestry-a new type of land use in Sweden. *For. Ecol. Manag.* **1987**, *21*, 141–161. [CrossRef]
28. Gustafsson, L. Vegetation dynamics during the establishment phase of an energy forest on a riverside in south-western Sweden. *Studia For. Suec.* **1988**, *178*, 1–16.
29. Baum, S.; Weih, M.; Bolte, A. Stand age characteristics and soil properties affect species composition of vascular plants in short rotation coppice plantations. *BioRisk* **2012**, *7*, 51–71. [CrossRef]

30. Sage, R.B. Factors affecting wild plant communities occupying short rotation coppice crops on farmland in the UK and Eire. In *Brighton Crop Protection Conference: Weeds. Proceedings of an International Conference, Brighton, UK, 20–23 November 1995*; British Crop Protection Council: Farnham, UK, 1995; pp. 985–990.
31. Heilmann, B.; Makeschin, F.; Rehfuess, K.E. Vegetationskundliche Untersuchungen auf einer Schnellwuchsplantage mit Pappeln und Weiden nach Ackernutzung. *Forstwiss. Cent.* **1995**, *114*, 16–29. [CrossRef]
32. Birmele, J.; Kopp, G.; Brodbeck, F.; Konold, W.; Sauter, U.H. Successional changes of phytodiversity on a short rotation coppice plantation in Oberschwaben, Germany. *Front. Plant Sci.* **2015**, *6*, 1–8. [CrossRef] [PubMed]
33. Baum, S.; Bolte, A.; Weih, M. Short rotation coppice (SRC) plantations provide additional habitats for vascular plant species in agricultural mosaic landscapes. *BioEnergy Res.* **2012**, *5*, 573–583. [CrossRef]
34. Baum, S.; Weih, M.; Busch, G.; Kroiher, F.; Bolte, A. The impact of short rotation coppice plantations on phytodiversity. *Appl. Agric. Forestry Res.* **2009**, *59*, 163–170.
35. Hennemann-Kreikenbohm, I.; Jennemann, L.; Peters, W.; Wilhelm, E.-G. Nature Conservation Requirements of Short Rotation Coppice Management. In *Bioenergy from Dendromass for the Sustainable Development of Rural Areas*; Butler Manning, D., Bemmann, A., Bredemeier, M., Lamersdorf, N., Ammer, C., Eds.; Wiley: Weinheim, Germany, 2015; pp. 97–104. ISBN 3527337644.
36. Soo, T.; Tullus, A.; Tullus, H.; Roosaluste, E. Floristic diversity responses in young hybrid aspen plantations to land-use history and site preparation treatments. *For. Ecol. Manag.* **2009**, *257*, 858–867. [CrossRef]
37. Landkreis Emsland. *Die Gemeinden im Landkreis Emsland: Strukturdaten im Vergleich 2016, Meppen*. 2016. Available online: https://www.emsland.de/das-emsland/zahlen-und-daten/strukturdaten/strukturdaten.html (accessed on 4 February 2021).
38. DWD. Deutscher Wetterdienst (Ed.): Temperatur und Niederschlag, vieljährige Mittelwerte 1981–2010. Available online: https://www.dwd.de/DE/leistungen/klimadatendeutschland/vielj_mittelwerte.html (accessed on 4 February 2021).
39. LBEG. *Bodenkundliche Übersichtskarte von Niedersachsen und Bremen 1:500 000*; Landesamt für Bergbau, Energie und Geologie: Hannover, Germany, 1999.
40. Rothmaler—Exkursionsflora von Deutschland. *Gefäßpflanzen: Grundband*; Jäger, E.J., Ed.; 21 Durchgesehene Auflage; Springer: Berlin/Heidelberg, Germany, 2017; ISBN 978-3-662-49707-4.
41. Moral, R.d.A.; Hinde, J.; Demetrio, C. *R Package 'hnp'*; R Foundation for Statistical Computing: Vienna, Austria, 2018.
42. R Core Team. *R Version 4.0.4*; R Core Team: Vienna, Austria, 2021.
43. Bates, D.; Maechler, M.; Bolker, B.; Walker, S.; Christensen, R.H.B.; Singmann, H.; Dai, B.; Scheipl, F.; Grothendieck, G.; Green, P.; et al. *R Package 'lme4'*; R Core Team: Vienna, Austria, 2019.
44. Kuznetsova, A.; Brockhoff, P.B.; Christensen, R.H.B.; Jensen, S.P. *R Package 'lmerTest'*; R Foundation for Statistical Computing: Vienna, Austria, 2020.
45. Lenth, R.; Singmann, H.; Love, J.; Buerkner, P.; Herve, M. *R Package 'emmeans'*; R Foundation for Statistical Computing: Vienna, Austria, 2019.
46. Wickham, H.; Chang, W.; Henry, L.; Pedersen, T.L.; Takahashi, K.; Wilke, C.; Woo, K.; Yutani, H. *R Package 'ggplot2'*; R Foundation for Statistical Computing: Vienna, Austria, 2019.
47. Gotelli, N.J.; Colwell, R.K. Quantifying biodiversity: Procedures and pitfalls in the measurement and comparison of species richness. *Ecol Lett* **2001**, *4*, 379–391. [CrossRef]
48. McAleece, N.; Gage, J.D.G.; Lambshead, P.J.D.; Paterson, G.L.J. *BioDiversity Professional Statistics Analysis Software*; Scottish Association for Marine Science: Oban, UK, 1997.
49. Oberdorfer, E. *Pflanzensoziologische Exkursionsflora*, 8th ed.; Verlag Eugen Ulmer: Stuttgart, Germany, 2001.
50. Oksanen, J.; Blanchet, F.G.; Kindt, R.; Legendre, P.; McGlinn, D.; Minchin, P.R.; O'hara, R.B.; Simpson, G.L.; Solymos, P.; Stevens, M.H.H.; et al. *R Package 'vegan'*; R Foundation for Statistical Computing: Vienna, Austria, 2020.
51. Warnes, G.R.; Bolker, B.; Bonebakker, L.; Gentleman, R.; Huber, W.; Liaw, A.; Lumley, T.; Maechler, M.; Magnussen, A.; Moeller, S.; et al. *R Package 'gplots'*; R Foundation for Statistical Computing: Vienna, Austria, 2020.
52. Metzing, D.; Garve, E.; Matzke-Hajek, G.; Adler, J.; Bleeker, W.; Breunig, T.; Caspari, S.; Dunkel, F.G.; Fritsch, R.; Gottschlich, G.; et al. Rote Liste und Gesamtartenliste der Farn- und Blütenpflanzen (Trachaeophyta) Deutschlands. *Nat. Biol. Vielfalt* **2018**, *70*, 13–358.
53. Garve, E. Rote Liste und Florenliste der Farn- und Blütenpflanzen in Niedersachsen und Bremen. 5. Fassung, Stand 1.3.2004. *Inform. Nat. Niedersachs.* **2004**, *24*, 1–76.
54. Seifert, C.; Leuschner, C.; Culmsee, H. Short rotation coppice as habitat for vascular plants. In *Bioenergy from Dendromass for the Sustainable Development of Rural Areas*; Butler Manning, D., Bemmann, A., Bredemeier, M., Lamersdorf, N., Ammer, C., Eds.; Wiley: Weinheim, Germany, 2015; pp. 63–78, ISBN 3527337644.
55. Pöltl, M.; Berg, C. Vegetationskundliche Untersuchung von Kurzumtriebsplantagen (KUP) bestockt mit Populus × canadensis im südoststeirischen Alpenvorland. *Mitt. Nat. Ver. Steiermark* **2016**, *146*, 5–18.
56. Meyer, S.; Wesche, K.; Krause, B.; Leuschner, C. Dramatic losses of specialist arable plants in Central Germany since the 1950s/60s–a cross-regional analysis. *Divers. Distrib.* **2013**, *19*, 1175–1187. [CrossRef]
57. Lang, M.; Kollmann, J.; Prestele, J.; Wiesinger, K.; Albrecht, H. Reintroduction of rare arable plants in extensively managed fields: Effects of crop type, sowing density and soil tillage. *Agric. Ecosyst. Environ.* **2021**, *306*, 107187. [CrossRef]
58. Baum, S.; Weih, M.; Bolte, A. Floristic diversity in Short Rotation Coppice (SRC) plantations: Comparison between soil seed bank and recent vegetation. *Landbauforsch. Appl. Agric. For. Res.* **2013**, *63*, 221–228.

59. Kriegel, P.; Fritze, M.-A.; Thorn, S. Surface temperature and shrub cover drive ground beetle (Coleoptera: Carabidae) assemblages in short-rotation coppices. *Agric. For. Entomol.* **2021**. [CrossRef]
60. Zitzmann, F.; Reich, M.; Schaarschmidt, F. Potential of small-scale and structurally diverse short-rotation coppice as habitat for large and medium-sized mammals. *Biologia* **2021**. [CrossRef]
61. Gruß, H.; Schulz, U. Brutvogelfauna auf Kurzumtriebsplantagen: Besiedlung und Habitateignung verschiedener Strukturtypen. *Nat. Landsch.* **2011**, *43*, 197–204.
62. Hanowski, J.M.; Niemi, G.J.; Christian, D.C. Influence of Within-Plantation Heterogeneity and Surrounding Landscape Composition on Avian Communities in Hybrid Poplar Plantations. *Conserv. Biol.* **1997**, *11*, 936–944. [CrossRef]
63. Zitzmann, F.; Reich, M. Bedeutung von Kurzumtriebsbeständen mit unterschiedlichen Gehölzarten als Lebensraum für Vögel im Winterhalbjahr. *Vogelwelt* **2019**, *139*, 261–272.
64. Kiehl, K.; Kirmer, A.; Jeschke, D.; Tischew, S. Restoration of speciesrich field margins and fringe communities by seeding of native seed mixtures. In *Guidelines for Native Seed Production and Grassland Restoration*; Kiehl, K., Kirmer, A., Shaw, N., Tischew, S., Eds.; Cambridge Scholars Publishing: Newcastle upon Tyne, UK, 2014; pp. 244–273.
65. Rowe, R.L.; Hanley, M.E.; Goulson, D.; Clarke, D.J.; Doncaster, C.P.; Taylor, G. Potential benefits of commercial willow Short Rotation Coppice (SRC) for farm-scale plant and invertebrate communities in the agri-environment. *Biomass Bioenergy* **2011**, *35*, 325–336. [CrossRef]
66. Meyer, S.; Bergmeier, E.; Becker, T.; Wesche, K.; Krause, B.; Leuschner, C. Detecting long-term losses at the plant community level–arable fields in Germany revisited. *Appl. Veg. Sci.* **2015**, *18*, 432–442. [CrossRef]
67. Wietzke, A.; van Waveren, C.-S.; Bergmeier, E.; Meyer, S.; Leuschner, C. Current State and Drivers of Arable Plant Diversity in Conventionally Managed Farmland in Northwest Germany. *Diversity* **2020**, *12*, 469. [CrossRef]
68. Wietzke, A.; Albert, K.; Bergmeier, E.; Sutcliffe, L.M.E.; van Waveren, C.-S.; Leuschner, C. Flower strips, conservation field margins and fallows promote the arable flora in intensively farmed landscapes: Results of a 4-year study. *Agric. Ecosyst. Environ.* **2020**. [CrossRef]
69. Cunningham, M.D.; Bishop, J.D.; Watola, G.; McKay, H.V.; Sage, R.B. *The Effects on Flora and Fauna of Converting Grassland to Short Rotation Coppice: Four Year Study Involving Wildlife Monitoring of Commercial SRC Plantations Planted on Grassland and Grassland Control Plots*; The Game Conservancy Trust: Fordingbridge, UK, 2006.
70. Delarze, R.; Ciardo, F. Rote Liste Arten in Pappelplantagen. *Inf. Forsch. Wald* **2002**, *9*, 3–4.
71. Sage, R.; Cunningham, M.; Boatman, N. Birds in willow short-rotation coppice compared to other arable crops in central England and a review of bird census data from energy crops in the UK. *IBIS* **2006**, *148*, 184–197. [CrossRef]
72. Fry, D.A.; Slater, F.M. Early rotation short rotation willow coppice as a winter food resource for birds. *Biomass Bioenergy* **2011**, *35*, 2545–2553. [CrossRef]

MDPI

Article

Effects of Leaf Loss by Artificial Defoliation on the Growth of Different Poplar and Willow Varieties

Christiane E. Helbig [1,*], Michael G. Müller [1] and Dirk Landgraf [2]

[1] Chair of Forest Protection, Institute of Silviculture and Forest Protection, Technische Universität Dresden, Pienner Straße 8, 01737 Tharandt, Germany; michael.mueller@tu-dresden.de

[2] Forestry and Ecosystem Management, Erfurt University of Applied Sciences, Leipziger Straße 77, 99085 Erfurt, Germany; dirk.landgraf@fh-erfurt.de

* Correspondence: christiane.helbig@tu-dresden.de; Tel.: +49-351-463-31285

Abstract: The cultivation of fast-growing tree species in short rotation coppices has gained popularity in Germany in recent years. The resilience of these coppices to phyllophagous pest organisms is crucial for their profitable management, since the loss of a single annual increment can lead to uncompensable economic losses. To study the effects of leaf loss on the growth of poplar and willow varieties that are frequently cultivated under local conditions, three sample short rotation coppices including five poplar (*Populus* spp.) and three willow (*Salix* spp.) varieties were established in a randomized block design with four artificial defoliation variants and, on one site, with three different variants regarding the number of defoliation treatments. After up to three defoliation treatments within two growing seasons, the results show negative effects of leaf loss on the height growth and the fresh weight of the aboveground biomass of plants. Our data also suggests a lasting effect of defoliation on plant growth and re-growth after the end of the treatment. In general, defoliation had a greater impact on the growth of poplars than on willows. We conclude that even minor leaf loss can have an impact on plant growth but that the actual effects of defoliation clearly depend on the site, tree species, and variety as well as the extent and number of defoliations, which determine the ability of plants for compensatory growth.

Keywords: short rotation coppices; poplars; willows; feeding simulation; defoliation; herbivory

Citation: Helbig, C.E.; Müller, M.G.; Landgraf, D. Effects of Leaf Loss by Artificial Defoliation on the Growth of Different Poplar and Willow Varieties. *Forests* **2021**, *12*, 1224. https://doi.org/10.3390/f12091224

Academic Editor: Daniele Castagneri

Received: 28 July 2021
Accepted: 3 September 2021
Published: 8 September 2021

Publisher's Note: MDPI stays neutral with regard to jurisdictional claims in published maps and institutional affiliations.

1. Introduction

Several global developments, such as the depletion of fossil fuels, the increasing demand for wood products, and the striving for climate protection, have been the reason for an increasing importance of the cultivation of fast-growing tree species in short rotation coppices on agricultural land in Germany. Short rotation coppices are defined as high-density plantations with rotation times between 2 and 20 years [1,2]. Poplars (*Populus* spp.) and willows (*Salix* spp.), which are characterized by a very fast juvenile growth, a great resprouting ability, and an easy propagation, have proven to be particularly suitable and are widely used for this kind of land use [3–7].

As typical monocultures with a high plant density, a low genetic diversity, and a distinctive spatial homogeneity, short rotation coppices generally hold a high risk for the occurrence of plant diseases and the outbreaks of pest organisms [8–13]. In addition, poplars and willows are naturally associated with an exceptionally high number of insect species in comparison to other tree species [14–17]. Accordingly, many studies have reported a large number of pest insects in poplar and willow short rotation coppices with a particular emphasis on phyllophagous species, which find ideal living conditions in these plantations [18–22]. The feeding activities of their larvae and/or adults cause a loss of leaf area but only in rare cases lead to the death of plants [23]. That is, in most cases, no lasting impact of leaf feeding can be directly seen. Several studies have shown, however, that the natural or artificial reduction of the leaf area of plants can already lead to a reduction of biomass yield after a relatively short period of time, and can still be detected after several months or years without this kind of damage [24,25]. Negative consequences of a loss of leaf area by insect feeding have also been documented on several other parameters of plant fitness, such as seed production [26–28]. For these reasons, the resilience of plants to biotic pest organisms is one of the most crucial preconditions for a large-scale, reliable, and profitable cultivation of fast-growing tree species on agricultural land, in particular because the loss of a single annual increment can lead to economic losses that may not be compensated within the short rotation times [23]. With regard to the strong economic focus of short rotation coppices, the actual effects of plant damage by insects on the yield of the coppice is a crucial aspect for their cultivation and management, for example when deciding for or against the use of insecticides.

A common procedure to study the effects of leaf area loss on the growth and yield of plants is the simulation of leaf feeding of phyllophagous pest insects by artificial defoliation [24,29–32]. The main advantage of this procedure is the ability to precisely control and modify the extent, number, and timing of defoliation, whereas the difference in the duration between natural and artificial defoliation as well as the potential lack of herbivore-induced plant volatiles due to a merely mechanical damage are disadvantageous [30,33,34]. Since the reaction of plants to herbivory is a very complex process that is not only determined by the actual leaf loss, studies comparing artificial and natural defoliation often show differences in the reaction of plants to these procedures [30]. Chen et al. (2002), for example, documented a greater reduction in plant height, height increment, and root to shoot ratio by an artificial defoliation of three-year-old Douglas fir seedlings in comparison to a natural defoliation, whereas the natural process had a greater impact on the diameter growth [29]. In contrast, Coyle et al. (2002) reported greater effects of natural feeding by *Chrysomela scripta* (Coleoptera: Chrysomelidae) on poplars than by artificial defoliation [35]. Nevertheless, many studies came to the conclusion that artificial defoliations are generally suitable to demonstrate the effects of natural defoliations [25,31,36–42].

The aims of this study were to transfer the approaches of existing studies, which were mainly carried out on potted plants [29,31,32,38–40], into the field, where the competition between plants is not excluded, and to examine the short-term and long-term effects of different extents and frequencies of leaf loss under local conditions on those poplar and willow varieties that are mainly planted in Germany.

2. Materials and Methods

2.1. Sites, Plant Material, and Experimental Design

For the leaf feeding simulation experiment sample short rotation coppices with a size of about 0.2 ha were established on one site in the federal state of Saxony [Obercarsdorf (50°51′35.0″ N 13°39′10.5″ E), 400 m a. s. l., 8.2 °C mean annual temperature, 786 mm average annual precipitation, typical cambisol, former pasture] and two sites in the south of the federal state of Brandenburg [Großthiemig (51°23′34.5″ N 13°40′35.8″ E), 94 m a. s. l., 8.6 °C mean annual temperature, 561 mm average annual precipitation, sandy gleyic cambisol, tree nursery land and Schönheide (51°34′39.4″ N 14°30′17.6″ E), 120 m a. s. l., 9.6 °C mean annual temperature, 568 mm average annual precipitation, slightly-loamy sandy cambisol, former grassland] [43].

Each of the three coppices consisted of a poplar area planted in double rows with a distance of 0.75 m within the double rows, 1.50 m between double rows, and 1.00 m between plants within the rows (8888 plants ha^{-1}), and a willow area planted with the same row spacing and a distance of 0.70 m between plants within rows (12,698 plants ha^{-1}) (Figure 1).

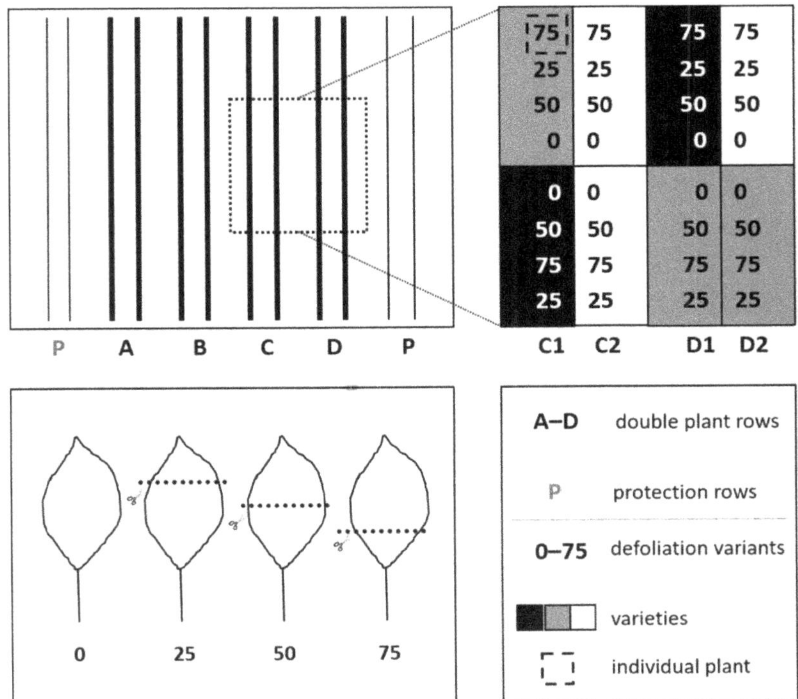

Figure 1. Planting design of the sample short rotation coppices (**above**) and design of leaf clipping of the different defoliation treatments (**below**).

The experiment was set up as randomized block design according to Powers et al. (2006) and Peacock et al. (2002) with five poplar varieties (Androscoggin, Max 1, Max 3, Max 4, Muhle Larsen) and three willow varieties (Sven, Tora, Tordis), respectively [32,44], and four different defoliation treatment variants (0%, 25%, 50%, 75% leaf loss) that were based on Reichenbacker et al. (1996) [41]. The cuttings were obtained from the Research Institute for Post-Mining Landscapes (FIB) in Finsterwalde (Max 1–4), Lantmännen Agroenergi AB (Sven, Tora, Tordis), and P&P Tree Nursery in Großthiemig (Androscoggin, Muhle Larsen) and planted manually. In order to avoid micro-spatial differences among treatments, varieties were planted with four consecutive plants representing the four treatments (Figure 1). Furthermore, the poplar and the willow area of each sample short rotation coppice were bordered with one double row of the other species on each side and two plants of the other species on each beginning and end of rows to reduce potential edge effects. On each of the three locations, the four defoliation treatments were represented with 32 plants per variety, that is, there were 32 plots with four plants each of every variety.

In accordance with Powers et al. (2006), the simulation of leaf feeding was carried out as a reduction of leaf area on every single leaf instead of relating the intended proportion of defoliation to the total plant leaf mass [32]. Leaves were cut with paper scissors across the midvein, similar to Peacock et al. (2002) [44], to simulate the feeding of phyllophagous insects as accurately as possible (Figure 1).

2.2. Experimental Process and Data Recording

The three sample short rotation coppices were established at the end of March (Großthiemig, Schönheide) and beginning of April 2007 (Obercarsdorf) after a standard soil preparation including ploughing, tilling, and, except for the organic farming site in

Obercarsdorf, the application of pre-emergent herbicides was carried out by the site owners. The first defoliation treatment took place four months after planting in July (Großthiemig, Schönheide) and August 2007 (Obercarsdorf). The willow varieties in Schönheide were excluded from the first defoliation due to their weak growth. Extensive browsing made it necessary to fence the willow area in Großthiemig in July 2007 and the whole coppice in Schönheide in May 2008. The second defoliation was carried out in June 2008 and the third defoliation in August 2008. During these two treatments, only a part of the plants in Großthiemig were defoliated so that at the end of the experiment the site held plants treated once, twice, or three times.

Data recordings on all three study sites took place directly prior to the first defoliation treatments in July/August 2007 and in December 2007. Recorded parameters were plant height, number of shoots, and plant damage (verbal description of damage and its potential abiotic or biotic causal factor). After a final data recording in Großthiemig in February 2009, the short rotation coppice was harvested manually in March 2009 to determine the fresh and dry weight of the aboveground biomass. Due to organizational reasons, it was not possible to determine the weight of the individual plants. Instead, the total weight of all plants per variety, defoliation variant, and number of defoliations was recorded. Determination of dry weight was only carried out on samples: four plots without plant losses were chosen for all poplars and those willows that were defoliated once, whereas for the willows that were defoliated twice or three times, dry weight was determined for all plants that were defoliated 0% or 75%. To study a potential long-term effect of defoliation on the re-growth of plants, additional data recordings in Großthiemig took place in June 2009, after which the resprouting shoots were reduced to the highest shoot per stool, and in September 2009. The short rotation coppices in Obercarsdorf and Schönheide were not harvested after the end of the defoliation treatments. Their final data recording took place in early April 2009 prior to bud burst. An additional data recording to study the potential long-term effect of defoliation on plant growth was carried out in early April 2010.

2.3. Data Analysis

All analyses were carried out using IBM SPSS Statistics 27.0 [45]. The significance level for all statistical tests was set at $\alpha = 0.05$. Data were analyzed for normality using the Shapiro–Wilk test and for homogeneity of variances using the Levene test. Based on the results, data was either further analyzed using parametric or non-parametric tests. For parametric tests, the *t* test (TT) or Welch test was used to compare the mean values of two independent samples or an analysis of variance (ANOVA, AN) with Tukey or Tukey-Kramer post hoc tests in case of multiple samples. For non-parametric tests, the Mann–Whitney U test (MU) or the Kruskall–Wallis test with Dunn-Bonferroni post hoc tests (KW) were used to analyze differences in the central tendencies of bivariate or multivariate datasets. Relevant *p* values of statistical analyses are either included in the text or in tables. In addition, statistical tests are denoted using the abbreviations stated above. Statistical analyses regarding plant height are based on the main shoot of each plant, which is the highest one. Plants that showed significant damage by browsing, insects, or other biotic and abiotic factors were excluded from analyses.

3. Results

3.1. Plant Growth Directly Prior and after the First Defoliation Treatment

In summer 2007, four months after the establishment of the short rotation coppices and directly prior to the start of the first defoliation treatment, plant losses on the study site in Großthiemig (Ø 5.2%) were considerably lower than on the sites in Obercarsdorf (Ø 17.0%) and Schönheide (Ø 17.3%). Differences in site conditions and management are assumed to be the reason for this difference in plant survival. Optimal mechanical and chemical soil preparation, low ground vegetation cover, and a potentially better nutrient supply on the tree nursery site in Großthiemig facilitated a high plant survival rate and fast growth. In contrast, on the sites in Obercarsdorf and Schönheide, which had previously been used as pasture and grassland, a full ground vegetation cover quickly re-developed despite the mechanical (Obercarsdorf) or mechanical and chemical (Schönheide) soil preparation, leading to greater plant losses in the newly established short rotation coppices on these two sites. In spite of these site-related differences, the trends regarding plant survival on tree genus and variety level were the same on all three study sites (Table A1 in Appendix A). Willows showed a slightly higher number of surviving plants than poplars. Within poplars, an average of 14.2% more plants of the three Max varieties survived in comparison to Androscoggin and Muhle Larsen, whereas within willows, an average of 6.7% more plants of the Tordis variety survived in comparison to Sven and Tora.

Clear differences among the three study sites were also visible with regard to plant heights and reflected the different site conditions similarly to the data on plant survival (Figure A1 in Appendix A). All eight varieties reached a significantly greater height in Großthiemig than in Obercarsdorf and Schönheide (KW: $p = 0.000$ for all pairwise comparisons). Despite the differences in site conditions, the height growth trends of the individual varieties are very similar on all three study sites and indicate a certain genetic fixation of height growth among varieties.

After the first defoliation treatment in July/August 2007, from which all willows in Schönheide were excluded due to their weak growth, plant heights were recorded again in December 2007 (Figure A2 in Appendix A). In general, data do not show a statistically significant effect of the four defoliation variants, except for Muhle Larsen in Großthiemig (AN: $p = 0.032$) and Obercarsdorf ($p = 0.010$). In Großthiemig, post hoc tests reveal significantly greater heights of undefoliated in comparison to 75% defoliated plants ($p = 0.046$) and in Obercarsdorf significantly greater heights of 25% defoliated in comparison to 50% defoliated plants ($p = 0.017$).

Although only very few statistically significant differences were detected among the four defoliation variants after the first treatment, the direct comparison of heights between 75% defoliated and undefoliated plants shows height losses for all poplar varieties on all three study sites with the only exception of Max 3 in Obercarsdorf (Table 1). In general, leaf loss had a greater impact on the growth of Androscoggin, Max 3 and Muhle Larsen than on Max 1 and Max 4. In contrast, data on all willow varieties in Großthiemig and on Sven and Tordis in Obercarsdorf indicate a positive effect of defoliation on their height growth. Height reductions due to defoliation were only visible for all willow varieties in Schönheide and for Tora in Obercarsdorf. However, statistical analyses again resulted in very few significant differences in the height between 75% defoliated and undefoliated plants: on the variety level for Muhle Larsen in Großthiemig (TT: $p = 0.010$) and on the tree genus level for poplars in Großthiemig (TT: $p = 0.005$) and Schönheide ($p = 0.007$).

Table 1. Height differences of 75% defoliated and undefoliated plants (Δ 75-0) in December 2007 after the first defoliation treatment (bold values indicate mean values of all varieties of a tree genus, * statistically significant difference according to *t* test).

	Study Site					
	Großthiemig		**Obercarsdorf**		**Schönheide**	
Genus/Variety	**Δ 75-0 [cm]**	**Δ 75-0 [%]**	**Δ 75-0 [cm]**	**Δ 75-0 [%]**	**Δ 75-0 [cm]**	**Δ 75-0 [%]**
Poplars	**−13.4 ***	**−10.7**	**−5.8**	**−8.3**	**−13.0 ***	**−16.0**
Androscoggin	−14.6	−10.3	−20.7	−24.8	−34.3	−38.2
Max 1	−1.2	−1.1	−7.0	−10.8	−0.8	−1.2
Max 3	−15.9	−12.3	10.3	13.9	−17.1	−19.1
Max 4	−11.2	−9.1	−6.4	−9.4	−9.8	−12.5
Muhle Larsen	−30.8 *	−24.2	−15.2	−23.6	−23.3	−28.9
Willows	**16.5**	**11.7**	**−1.4**	**−1.7**	**−20.8**	**−22.4**
Sven	6.2	4.2	5.0	7.5	−3.5	−12.1
Tora	16.7	11.8	−26.5	−30.1	−18.8	−18.8
Tordis	26.5	19.8	14.3	15.2	−7.0	−5.2

3.2. Plant Growth after the Last Defoliation

After the first defoliation treatment in July/August 2007, two more treatments were carried out in June and in August 2008, and data was recorded again in early spring 2009, prior to the start of the growing season. In Großthiemig, only a part of the plants were defoliated during the treatments in 2008, so that data from this site cannot only be grouped by variety and defoliation variant but also by number of defoliation treatments. With the mere regard to the number of defoliation treatments on the genus level, a different reaction of poplars and willows to the increasing number of defoliation treatments was recorded when considering the average of all defoliation variants. While there is no statistically significant difference in the plant height of poplars defoliated once (Ø 176.6 cm) and twice (Ø 177.9 cm), the poplars treated three times (Ø 164.9 cm) had a significantly reduced plant height in comparison to both, with a mean height reduction of 7%. Willows, in contrast, showed again a promotion of plant growth by defoliation. Plants defoliated twice had a significantly greater height (Ø 277.3 cm) than plants defoliated once (Ø 259.6 cm). However, defoliation carried out three times led to a significant height loss (Ø 239.3 cm) in comparison to plants defoliated once and twice, with a mean height reduction of 11%.

Taking into account not only the number of defoliation treatments but also the variety and defoliation variant, trends show a decreasing height with increasing leaf loss in several cases, in particular for poplar varieties, even though statistically significant differences on the group level only exist in the five cases marked with an asterisk (Figure 2). The p values for the pairwise comparisons of defoliation variants on tree genus level and those comparisons with at least one significant value on the variety level show an increasing number of statistically significant differences with an increasing number of defoliation treatments (Table 2). These especially occur when comparing 75% defoliated and undefoliated plants but also in parts among the three variants that included leaf loss. It can be noted that defoliation particularly led to significant differences in plant height among defoliation variants for the Muhle Larsen and Tora varieties.

Table 2. p values of pairwise comparisons of plant heights among defoliation variants (DVs) for both tree genera (green) as well as for all varieties with at least one significant value (orange) on the study site in Großthiemig with regard to the number of defoliation treatments prior to the start of the growing season in 2009 via Tukey HSD test (bold values indicate statistically significant differences).

	Genus Level						Variety Level				
	Großthiemig (1 defoliation treatment)										
			Willows								
	DV	**0%**	**25%**	**50%**	**75%**		**DV**	**0%**	**25%**	**50%**	**75%**
Poplars	0%	—	0.675	0.997	0.604	Max 3	0%	—			
	25%	0.507	—	0.800	0.999		25%	0.290	—		
	50%	1.000	0.513	—	0.736		50%	0.574	**0.024**	—	
	75%	0.305	0.987	0.315	—		75%	0.876	0.774	0.229	—
	Großthiemig (2 defoliation treatments)										
			Willows						Tora		
	DV	**0%**	**25%**	**50%**	**75%**		**DV**	**0%**	**25%**	**50%**	**75%**
Poplars	0%	—	0.964	0.998	0.547	Muhle Larsen	0%	—	0.440	0.064	**0.005**
	25%	0.917	—	0.919	0.268		25%	0.430	—	0.666	0.134
	50%	0.399	0.795	—	0.651		50%	0.123	0.883	—	0.666
	75%	**0.024**	0.115	0.515	—		75%	**0.020**	0.378	0.773	—
	Großthiemig (3 defoliation treatments)										
			Willows						Tora		
	DV	**0%**	**25%**	**50%**	**75%**		**DV**	**0%**	**25%**	**50%**	**75%**
Poplars	0%	—	0.123	0.290	**0.027**	Muhle Larsen	0%	—	**0.013**	**0.035**	**0.004**
	25%	0.592	—	0.969	0.943		25%	0.107	—	0.981	0.978
	50%	**0.012**	0.276	—	0.733		50%	0.195	0.982	—	0.857
	75%	**0.001**	0.062	0.893	—		75%	**0.018**	0.892	0.683	—

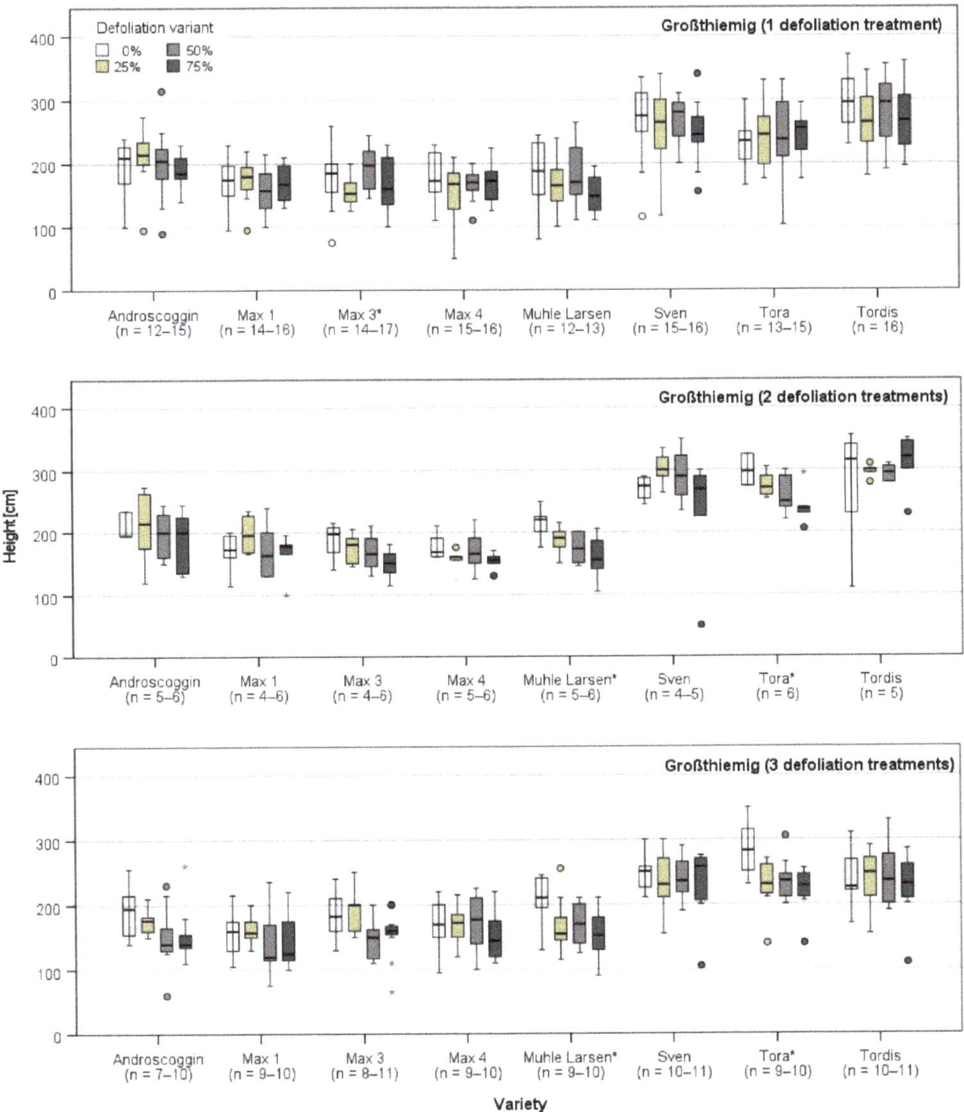

Figure 2. Height of poplar and willow varieties on the study site in Großthiemig with regard to the number of defoliation treatments prior to the start of the growing season in 2009 (n = range of number of plants per defoliation variant, * statistically significant difference according to ANOVA).

Despite the three defoliation treatments, analyses resulted in no statistically significant differences in plant heights among defoliation variants in Obercarsdorf, whereas in Schönheide all poplar varieties did show significant differences in plant heights with decreasing heights at increasing leaf loss (Figure 3). Data on the pairwise comparisons bet-ween defoliation variants generally show a p value decrease with an increasing difference in leaf loss (Table 3).

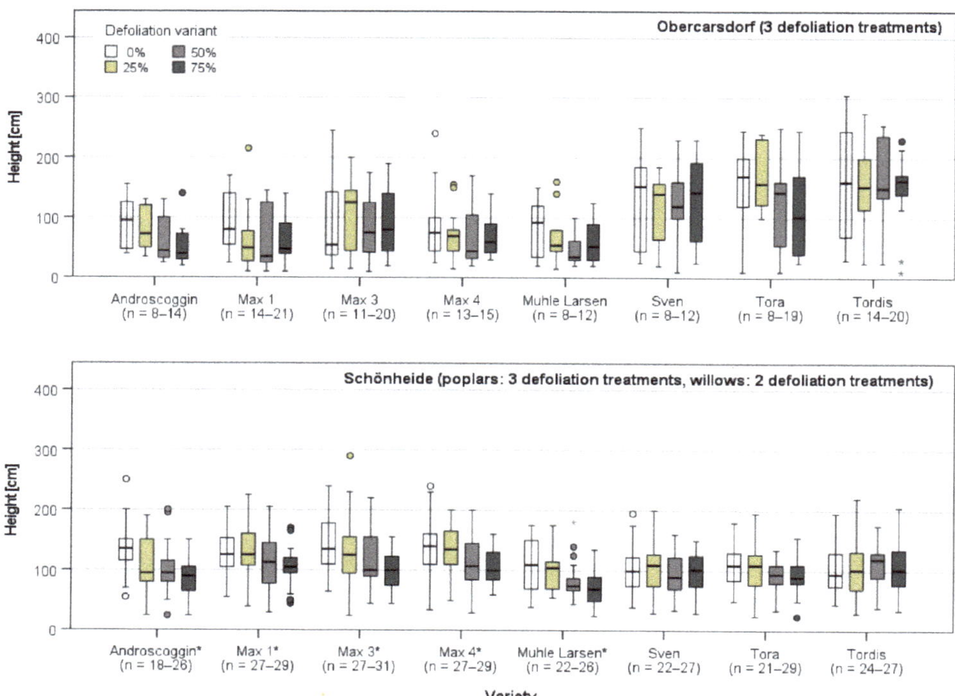

Figure 3. Height of poplar and willow varieties on the study sites in Obercarsdorf and Schönheide after two and/or three defoliation treatments prior to the start of the growing season in 2009 (n = range of number of plants per defoliation variant, * statistically significant difference according to ANOVA).

When only comparing the heights of those plants with the greatest leaf loss, that is 75%, with undefoliated plants, height reductions of up to 42% are visible, with only a few exceptions for willow varieties (Table 4). The data shows again that the effects of defoliation on plant height depend on the site, tree genus, variety, and frequency of defoliation. Generally, height reduction increased with increasing defoliation frequency, and defoliation had a greater impact on poplar than on willow varieties. On average for all three study sites, for poplars, Max 1 had the least height reduction with 11% and Muhle Larsen the greatest with 25%. For willows, Tora was most impacted by defoliation with a height reduction of 18%, whereas Tordis and Sven showed a reduction of 5% on average. Statistically significant differences were detected for all poplar varieties (TT: p = 0.001–0.005) as well as the Tora variety (p = 0.017). On the tree genus level, a statistically significant effect was only existent for poplars (p = 0.000), although the p value for willows (0.051) came very close to a significance.

Table 3. *p* values of pairwise comparisons of plant heights among defoliation variants (DVs) for both tree genera (green) as well as for all varieties with at least one significant value (orange) on the study sites in Obercarsdorf and Schönheide with regard to the number of defoliation treatments prior to the start of the growing season in 2009 via Tukey HSD test (bold values indicate statistically significant differences).

Genus/Variety Level						Variety Level					
Obercarsdorf (3 defoliation treatments)											
		Willows									
	DV	**0%**	**25%**	**50%**	**75%**		no case of pairwise comparisons among defoliation variants with at least one significant value				
Poplars	**0%**	—	0.995	0.905	0.752						
	25%	0.827	—	0.981	0.914						
	50%	**0.043**	0.261	—	0.992						
	75%	0.057	0.317	0.999	—						
Schönheide (poplars: 3 defoliation treatments, willows: 2 defoliation treatments)											
		Willows									
	DV	**0%**	**25%**	**50%**	**75%**		**DV**	**0%**	**25%**	**50%**	**75%**
Poplars	**0%**	—	0.989	0.747	0.644	**Andros-coggin**	**0%**	—			
	25%	0.261	—	0.905	0.832		**25%**	0.154	—		
	50%	**0.000**	**0.013**	—	0.998		**50%**	**0.029**	0.930	—	
	75%	**0.000**	**0.000**	0.100	—		**75%**	**0.004**	0.457	0.769	—
		Max 3						**Muhle Larsen**			
	DV	**0%**	**25%**	**50%**	**75%**		**DV**	**0%**	**25%**	**50%**	**75%**
Max 1	**0%**	—	0.805	0.133	**0.004**	**Max 4**	**0%**	—	0.960	0.084	**0.004**
	25%	0.979	—	0.586	0.060		**25%**	0.941	—	0.262	**0.026**
	50%	0.241	0.451	—	0.571		**50%**	0.125	0.352	—	0.761
	75%	**0.028**	0.077	0.780	—		**75%**	**0.011**	**0.050**	0.773	—

Table 4. Height differences between 75% defoliated and undefoliated plants (Δ 75-0) prior to the start of the growing season in 2009 after a one-time, two-time, or three-time defoliation treatment (* statistically significant difference according to t or Welch test).

	Study Site (Number of Defoliation Treatments)											
	Großthiemig (1 Defoliation Treatment)		Großthiemig (2 Defoliation Treatments)		Großthiemig (3 Defoliation Treatments)		Obercarsdorf (3 Defoliation Treatments)		Schönheide (3/2 Defoliation Treatments)		Total	
Genus/Variety	Δ 75-0 [cm]	Δ 75-0 [%]	Δ 75-0 [cm]	Δ 75-0 [%]	Δ 75-0 [cm]	Δ 75-0 [%]	Δ 75-0 [cm]	Δ 75-0 [%]	Δ 75-0 [cm]	Δ 75-0 [%]	Δ 75-0 [cm]	Δ 75-0 [%]
Poplars	−11.4	−6.3	−26.9 *	−14.2	−30.1 *	−16.6	−21.5 *	−24.0	−37.3 *	−28.3	−26.7 *	−18.3
Androscoggin	−7.3	−3.7	−21.8	−10.3	−36.1	−19.0	−38.3 *	−42.0	−46.0 *	−34.2	−33.3 *	−20.9
Max 1	−2.2	−1.3	−3.3	−2.0	−13.2	−8.4	−28.6	−31.0	−30.0 *	−22.6	−15.2	−11.2
Max 3	−9.9	−5.5	−38.3	−20.4	−31.0	−16.7	−3.0	−3.3	−41.1 *	−28.7	−29.5 *	−19.4
Max 4	−9.4	−5.3	−22.8	−13.0	−15.0	−9.0	−20.1	−23.1	−35.2 *	−25.2	−24.1 *	−16.7
Muhle Larsen	−33.8	−18.3	−56.5 *	−26.4	−54.7 *	−26.5	−26.3	−31.2	−34.5 *	−32.1	−34.9 *	−24.9
Willows	−13.1	−4.9	−24.2	−8.6	−29.6	−11.5	−15.3	−10.2	−7.2	−6.7	−16.6	−8.9
Sven	−20.2	−7.5	−44.3	−16.3	−15.6	−6.3	5.2	4.1	−4.4	−4.2	−10.2	−5.6
Tora	12.0	5.2	−58.3 *	−19.5	−62.0 *	−21.9	−39.5	−25.7	−18.5	−16.4	−33.0 *	−17.6
Tordis	−28.8	−9.7	39.0	14.4	−13.6	−5.6	−10.4	−6.5	1.6	1.5	−6.7	−3.5

3.3. *Plant Growth after the End of the Defoliation Treatments*

3.3.1. Harvest and Growth in Großthiemig

After the last data recording in February 2009, the short rotation coppice in Großthiemig was completely harvested manually to determine the weight of the aboveground biomass. However, it was not possible to determine the fresh weight of each individual plant. Instead, it was determined as the total weight of plants per variety, defoliation variant, and number of defoliation treatments. The options for statistical analyses are therefore limited and the sample size is very low. Not taking into account the number of defoliation treatments provides a sample size of n = 3, at which no statistically significant differences among defoliation variants are visible on the variety level (AN: $p = 0.079$–0.996), whereas on the tree genus level, undefoliated poplars had significantly greater fresh weights than 75%

defoliated poplars ($p = 0.045$) (Figure 4). With a p value of 0.052, the comparison between 25% and 75% is very close to a significant difference. Looking merely at the statistical comparison of undefoliated and 75% defoliated plants instead of analyzing the differences among all four defoliation variants, significantly greater weights for undefoliated plants were computed for Muhle Larsen (TT: $p = 0.014$) as well as for the total of all poplar varieties ($p = 0.011$).

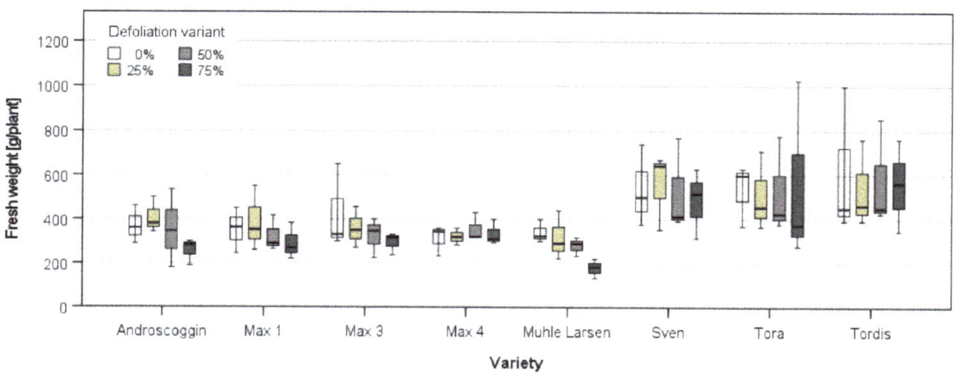

Figure 4. Fresh weight of aboveground biomass of plants on the study site in Großthiemig in February 2009 (n per variety and defoliation variant = 3 by not taking into account the number of defoliations).

Despite a relatively low number of statistically significant differences, looking at absolute numbers and not considering the number of defoliation treatments, all plants that experienced leaf loss had lower fresh weights than undefoliated plants (Table 5). Poplars that had 75% of their foliage removed reached a 25% lower fresh weight than undefoliated plants, and poplars that had 50 or 25% of their foliage showed a fresh weight reduction of 9% or 1% in comparison to undefoliated plants. The corresponding values for willows are 5%, 3%, and 5%, meaning defoliation-induced fresh weight reductions were lower than for poplars. Moreover, willows in total had a 65% greater fresh weight than poplars. An effect of the number of defoliation treatments on the fresh weight of poplars (AN: $p = 0.001$) and willows ($p = 0.000$) is also visible when looking at the total of all plants. For both tree genera, all pairwise comparisons result in significant differences, with the one exception of poplars defoliated one time compared to those defoliated three times.

Table 5. Mean fresh weight [g] of aboveground biomass of plants on the study site in Großthiemig with regard to the number of defoliation treatments in February 2009 (DV = defoliation variant, different letters indicate statistically significant differences).

	Poplars				Willows			
	Number of Defoliation Treatments				Number of Defoliation Treatments			
DV	1	2	3	Total	1	2	3	Total
0%	344.5	438.7	298.7	**360.6 (a)**	790.0	516.7	379.5	**562.1 (a)**
25%	323.0	424.7	328.1	**358.6 (ab)**	714.5	520.3	370.9	**535.3 (a)**
50%	319.1	380.0	283.0	**327.4 (ab)**	800.8	414.0	414.1	**543.0 (a)**
75%	265.3	313.3	235.5	**271.4 (b)**	805.3	483.9	312.8	**534.0 (a)**
Total	**313.0 (a)**	**389.2 (b)**	**286.3 (a)**	329.5	**777.7 (a)**	**483.7 (b)**	**369.3 (c)**	543.6

A detailed description of the dry weight data is omitted, since in hindsight we cannot completely rule out an error during data recording and analyses. One plausible result of these data is a statistically significant difference of the dry weight between 75% defoliated and undefoliated plants of Muhle Larsen (TT: $p = 0.010$), which was also computed for its fresh weight.

In June 2009, that is, three months after the harvest of the short rotation coppice in Großthiemig, the height of resprouting shoots and number of shoots per stool were recorded. Statistical analysis on the variety level only resulted in significant differences among defoliation variants for two groups (AN: Muhle Larsen/two defoliation treatments: $p = 0.024$, Max 3/total: $p = 0.025$) (Table 6). In contrast, on the tree genus level, significant differences in the number of shoots among the four defoliation variants were detected for poplars that had been defoliated twice ($p = 0.007$) and three times ($p = 0.023$), as well as in total ($p = 0.002$). In all three cases, undefoliated plants had a significantly greater number of shoots than 50% defoliated ($p = 0.009$–0.047) and 75% defoliated plants ($p = 0.004$–0.035). No effect of defoliations on the number of resprouting shoots after a harvest were recorded for willows.

Table 6. Mean number of resprouting shoots per stool on the study site in Großthiemig with regard to the number of defoliation treatments in June 2009 after the harvest in March 2009 with regard to the defoliation variant in 2007 and 2008.

	Number of Defoliation Treatments															
	1				2				3				Total			
	Defoliation Variant [%]															
Genus/Variety	**0**	**25**	**50**	**75**	**0**	**25**	**50**	**75**	**0**	**25**	**50**	**75**	**0**	**25**	**50**	**75**
Poplars	**8**	**8**	**8**	**7**	**11**	**7**	**6**	**7**	**7**	**6**	**5**	**5**	**8**	**7**	**7**	**6**
Androscoggin	7	8	7	7	16	7	7	6	7	7	6	4	9	7	6	6
Max 1	8	6	8	7	9	7	6	5	5	6	4	5	8	6	6	6
Max 3	11	10	10	7	13	9	7	9	9	8	6	7	11	9	8	7
Max 4	6	8	7	8	6	6	5	7	7	6	6	5	7	7	6	7
Muhle Larsen	6	7	6	6	8	7	4	4	5	5	5	5	6	6	5	5
Willows	**19**	**16**	**15**	**16**	**17**	**17**	**17**	**18**	**16**	**16**	**17**	**15**	**17**	**17**	**16**	**17**
Sven	22	18	16	17	18	18	20	18	17	18	20	19	19	18	19	18
Tora	16	15	12	19	15	16	14	19	17	15	15	12	16	15	14	17
Tordis	18	16	16	13	16	18	15	18	16	16	15	14	17	16	16	15

Even though poplars on average had an 80 cm lower height than willows in June 2009, they reached a 10 cm greater height in September 2009. A statistical analysis of height data resulted in significant differences for the two-time defoliated plants of Max 1 (AN: $p = 0.040$) and the three-time defoliated plants of Max 3 ($p = 0.002$) and Tora ($p = 0.026$) in June, while in September, significant height differences were computed for the three-time defoliated plants of Androscoggin ($p = 0.027$) and Tora ($p = 0.047$) (Figures A3 and A4 in Appendix A). When looking at pairwise comparisons, an effect of the number of defoliation treatments is visible again (Table 7). The more often plants were defoliated, the more often significant effects on plant heights were recorded.

Table 7. *p* values of pairwise comparisons of plant heights among former defoliation variants (DVs) for both tree genera (green) as well as for all varieties with at least one significant value (orange) on the study site in Großthiemig in June and in September 2009 after a harvest in March 2009 with regard to the number of defoliation treatments in 2007 and 2008 via Tukey HSD test (bold values indicate statistically significant differences).

	June 2009						September 2009				
Großthiemig (1 defoliation treatment)											
		Willows						**Willows**			
	DV	0%	25%	50%	75%		**DV**	0%	25%	50%	75%
Poplars	0%	—	0.931	0.992	0.991	**Poplars**	0%	—	0.988	1.000	0.996
	25%	1.000	—	0.989	0.810		25%	0.670	—	0.972	0.945
	50%	0.787	0.794	—	0.940		50%	0.983	0.459	—	0.999
	75%	1.000	1.000	0.784	—		75%	0.948	0.937	0.808	—
Großthiemig (2 defoliation treatments)											
		Willows						**Willows**			
	DV	0%	25%	50%	75%		**DV**	0%	25%	50%	75%
Poplars	0%	—	0.979	0.973	0.960	**Poplars**	0%	—	0.854	0.995	0.960
	25%	0.992	—	1.000	1.000		25%	0.953	—	0.942	0.990
	50%	0.890	0.972	—	1.000		50%	0.998	0.985	—	0.994
	75%	0.865	0.960	1.000	—		75%	0.959	0.717	0.894	—
	DV	0%	25%	50%	75%			no case of pairwise comparisons among defoliation variants with at least one significant value			
Max 1	0%	—									
	25%	0.100	—								
	50%	0.904	**0.030**	—							
	75%	0.843	0.338	0.457	—						
Großthiemig (3 defoliation treatments)											
		Willows						**Willows**			
	DV	0%	25%	50%	75%		**DV**	0%	25%	50%	75%
Poplars	0%	—	0.518	0.956	0.147	**Poplars**	0%	—	0.746	0.794	**0.014**
	25%	0.998	—	0.832	0.878		25%	0.946	—	1.000	0.179
	50%	0.092	0.149	—	0.385		50%	**0.040**	0.156	—	0.156
	75%	**0.018**	**0.034**	0.930	—		75%	**0.013**	0.064	0.983	—
		Tora						Tora			
	DV	0%	25%	50%	75%		**DV**	0%	25%	50%	75%
Max 3	0%	—	**0.023**	0.798	0.179	**Andros-coggin**	0%	—	0.057	0.315	0.083
	25%	0.507	—	0.194	0.777		25%	0.351	—	0.831	0.997
	50%	0.223	**0.014**	—	0.683		50%	0.099	0.955	—	0.913
	75%	0.077	**0.003**	0.984	—		75%	**0.019**	0.642	0.882	—

Merely comparing undefoliated plants with plants with the greatest leaf loss, statistically significant differences only exist for plants that had been defoliated three times and when looking at the total of plants (Tables 8 and 9). In general, height differences increase with an increasing number of defoliation treatments on the tree genus level, whereas this trend is hardly visible on the variety level. It is noticeable that the impact of defoliations on poplars is slightly greater than on willows, and that most of the significant height differences that were recorded in June still persisted in September. Furthermore, looking at the absolute figures of three-time defoliated plants, all varieties in June and all varieties with the exception of Max 4 in September showed reduced heights for 75% defoliated compared to undefoliated plants. Absolute height differences of undefoliated and 75% defoliated plants increased greatly between June and September.

Table 8. Height differences of 75% defoliated and undefoliated plants (Δ 75-0) on the study site in Großthiemig in June 2009 after a harvest in March 2009 with regard to the number of defoliation treatments in 2007 and 2008 (* statistically significant difference according to the *t* test).

	Number of Defoliation Treatments							
	1		2		3		Total	
Genus/Variety	Δ 75-0 [cm]	Δ 75-0 [%]	Δ 75-0 [cm]	Δ 75-0 [%]	Δ 75-0 [cm]	Δ 75-0 [%]	Δ 75-0 [cm]	Δ 75-0 [%]
Poplars	**−0.1**	**−0.1**	**−4.2**	**−3.6**	**−12.5 ***	**−10.8**	**−4.9 ***	**−4.3**
Androscoggin	−10.5	−8.8	−8.9	−7.5	−13.5	−11.9	−11.4 *	−9.7
Max 1	−2.5	−2.2	6.7	5.8	−20.1 *	−17.9	−6.3	−5.6
Max 3	5.6	4.8	−3.8	−3.2	−17.7 *	−14.4	−4.5	−3.8
Max 4	3.3	2.9	4.1	3.3	−1.1	−1.0	1.9	1.6
Muhle Larsen	6.0	6.1	−16.2	−14.1	−8.3	−7.4	−3.4	−3.1
Willows	**2.7**	**1.5**	**3.3**	**1.7**	**−16.9**	**−8.5**	**−5.4**	**−2.8**
Sven	−2.2	−1.3	−8.6	−4.6	−23.8	−12.7	−13.0	−7.2
Tora	33.1	18.5	6.3	3.1	−23.5*	−10.4	2.1	1.0
Tordis	−14.4	−7.1	10.8	5.6	−3.5	−1.9	−3.7	−1.9

Table 9. Height differences of 75% defoliated and undefoliated plants (Δ 75-0) on the study site in Großthiemig in September 2009 after a harvest in March 2009 with regard to the number of defoliation treatments (* statistically significant difference according to the *t* test).

	Number of Defoliation Treatments							
	1		2		3		Total	
Genus/Variety	Δ 75-0 [cm]	Δ 75-0 [%]	Δ 75-0 [cm]	Δ 75-0 [%]	Δ 75-0 [cm]	Δ 75-0 [%]	Δ 75-0 [cm]	Δ 75-0 [%]
Poplars	**−5.3**	**−1.8**	**−6.9**	**−2.4**	**−33.6 ***	**−11.9**	**−15.3 ***	**−5.3**
Androscoggin	−7.1	−2.6	−37.8	−13.2	−45.6*	−16.8	−26.1 *	−9.4
Max 1	−24.1	−7.6	28.8	9.8	−57.0 *	−19.0	−25.8	−8.4
Max 3	−24.5	−7.6	27.9	11.3	−28.7	−9.5	−21.1	−6.9
Max 4	15.6	5.2	−7.5	−2.3	5.1	1.8	8.4	2.8
Muhle Larsen	34.6	17.0	−16.2	−6.2	−35.6	−13.9	−2.8	−1.2
Willows	**3.2**	**1.2**	**5.0**	**1.8**	**−28.9 ***	**−10.2**	**−9.8**	**−3.5**
Sven	14.6	6.2	0.0	0.0	−26.5	−10.5	−6.4	−2.6
Tora	34.1	13.4	14.4	5.2	−30.0 *	−9.6	2.1	0.7
Tordis	−30.0	−9.8	0.0	0.0	−30.2	−10.7	−23.0 *	−7.8

3.3.2. Growth in Obercarsdorf and Schönheide

After the last defoliation treatment in August 2008 and the final recording of height growth in April 2009, plants in the short rotation coppices in Obercarsdorf and Schönheide were measured again in April 2010 to study a potential long-term effect of defoliations. By this time a mechanical ground vegetation removal had led to plant losses and considerably lower plant numbers in Obercarsdorf than in Schönheide. Statistical analysis only resulted in significant height differences among defoliation variants for Max 1 (AN: $p = 0.039$), Max 3 ($p = 0.007$), and Muhle Larsen (0.023) in Schönheide (Figure 5). No such differences were computed for the plants in Obercarsdorf. Pairwise comparisons generally still show a decrease of p values with increasing difference in leaf loss (Table 10). Despite the statistically significant result for Muhle Larsen on the study site in Schönheide on the group level, no significant differences were detected with pairwise comparisons. However, comparison of plant heights between undefoliated and 25% defoliated plants with 75% defoliated plants are relatively close to a significant result ($p = 0.055/0.056$).

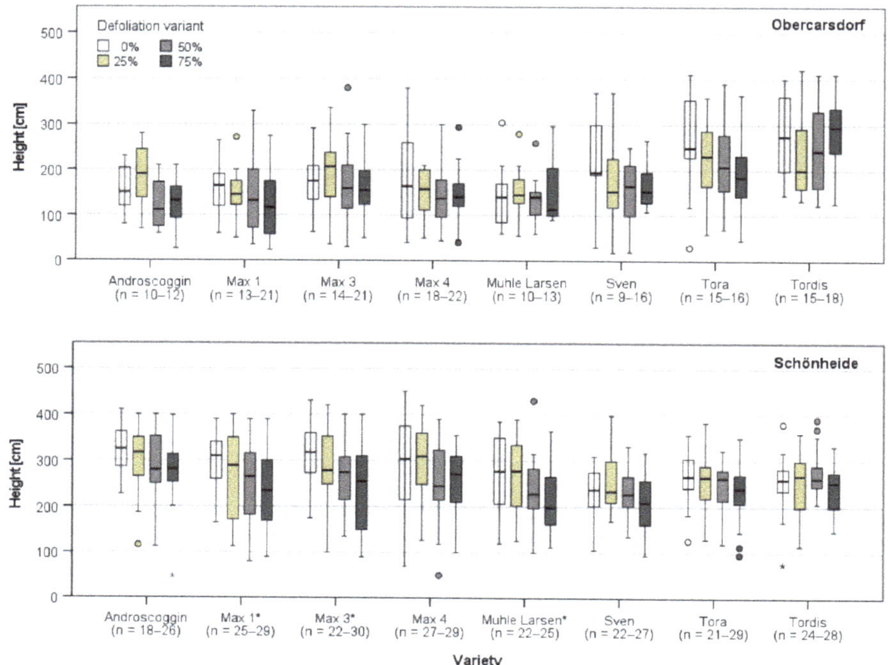

Figure 5. Height of poplar and willow varieties on the study sites in Obercarsdorf and Schönheide prior to the start of the growing season in April 2010 with regard to the defoliation variant in 2007 and 2008 (n = range of number of plants per defoliation variant, * statistically significant difference according to ANOVA).

Table 10. *p* values of pairwise comparisons of plant heights among former defoliation variants (DVs) for both tree genera (green) as well as for all varieties with at least one significant value (orange) on the study sites in Obercarsdorf and Schönheide prior to the start of the growing season in April 2010 via Tukey HSD test (bold values indicate statistically significant differences).

Genus Level							**Variety Level**				
				Obercarsdorf							
			Willows								
	DV	**0%**	**25%**	**50%**	**75%**						
Poplars	**0%**	—	0.124	**0.032**	0.165		no case of pairwise comparisons among defoliation variants with at least one significant value				
	25%	1.000	—	0.970	0.997						
	50%	0.335	0.282	—	0.909						
	75%	0.155	0.123	0.978	—						
				Schönheide							
			Willows						**Max 3**		
	DV	**0%**	**25%**	**50%**	**75%**		**DV**	**0%**	**25%**	**50%**	**75%**
Poplars	**0%**	—	0.988	0.999	0.088	**Max 1**	**0%**	—	0.725	0.114	**0.006**
	25%	0.632	—	0.967	**0.041**		**25%**	0.496	—	0.614	0.097
	50%	**0.000**	**0.026**	—	0.116		**50%**	0.153	0.900	—	0.683
	75%	**0.000**	**0.000**	0.376	—		**75%**	**0.028**	0.501	0.890	—

Comparing merely undefoliated and 75% defoliated plants results in considerably more significant height differences than analyzing differences among all four defoliation variants (Table 11). Significant differences were now also computed for the short rotation coppice in Obercarsdorf, where plants that had 75% of their foliage removed in 2007 and 2008 reached an almost one-third lower height in spring 2010 than undefoliated plants.

Looking at the average of both study sites, a significant effect of previous defoliation treatments on plant height was still visible after a year without such a treatment for the poplar varieties Max 3 and Max 4 as well as the willow varieties Sven and Tora. While the height reduction percentages of the five poplar varieties are relatively close to each other with 14 to 22%, the very low reduction of about 1% for Tordis in comparison to a reduction of about 17% for Sven and Tora is noticeable for the three willow varieties.

Table 11. Height differences between 75% defoliated and undefoliated plants (Δ 75-0) on the study sites in Obercarsdorf and Schönheide prior to the start of the growing season in April 2010 (* significant difference according to the *t* test or Welch test).

	Study Site					
	Obercarsdorf		Schönheide		Total	
Genus/Variety	Δ 75-0 [cm]	Δ 75-0 [%]	Δ 75-0 [cm]	Δ 75-0 [%]	Δ 75-0 [cm]	Δ 75-0 [%]
Poplars	**−23.0** *	**−13.9**	**−55.2** *	**−18.5**	**−41.8** *	**−17.0**
Androscoggin	−33.8	−21.1	−42.9 *	−13.4	−43.3	−16.5
Max 1	−33.4	−21.4	−63.7 *	−21.2	−33.2	−14.2
Max 3	−17.4	−9.9	−71.6 *	−23.0	−56.7 *	−21.9
Max 4	−40.0	−21.7	−36.2	−12.3	−43.3 *	−17.3
Muhle Larsen	7.3	5.2	−55.9 *	−20.5	−31.0	−13.6
Willows	**−41.2** *	**−15.7**	**−22.3** *	**−8.8**	**−29.3** *	**−11.5**
Sven	−65.4 *	−28.1	−28.4	−12.0	−40.7 *	−17.3
Tora	−81.3 *	−30.4	−27.6	−10.5	−46.2 *	−17.4
Tordis	4.9	1.7	−12.5	−4.9	−3.5	−1.3

4. Discussion

The results of this four-year study clearly show negative effects of defoliation on the height, fresh weight, and number of resprouting shoots of poplar and willow varieties, and that these effects explicitly depend on the site, tree species, and variety as well as the extent and number of defoliations. Several other studies on poplars and willows have also provided evidence for a plant growth reduction caused by defoliation and, in accordance with this study, for an increasing reduction of different growth parameters with an increasing extent of defoliation [25,41,44]. However, studies differ with regard to the minimum extent of leaf loss from which a significant effect on plant growth has to be expected. While some studies have already detected significant effects at 10–25% leaf loss [32,39,42], others have not recorded notable effects at defoliation levels of 40 and 50% but only starting from 75% [37,46]. In this study, significant effects were mainly detected at a defoliation level of 75% as well, but in several cases also at a level of 50%, in particular with plants that had been defoliated three times within two growing seasons. Only rarely did we record significant effects on plant height at a defoliation level of 25%. These exclusively occurred with the Tora variety, even though Bell et al. (2006) found the least effects of defoliation on this variety [24]. Anttonen et al. (2002) concluded that there is no consistently valid threshold value for negative effects of defoliation on plant growth but that instead it varies depending on the particular growth parameter [36].

When comparing the results of this study with literature, it has to be taken into account that other studies were often based on a different number of defoliation treatments per growing season and on different overall experiment durations. The maximum of three defoliation treatments within two growing seasons in this study lies below the number of treatments in many other studies that included two treatments within one growing season [24,42,44,47]. Moreover, Kendall et al. (1998) conclude that even two defoliation treatments within one growing season is not enough to simulate the natural defoliation by leaf beetles, which lasts for a longer period of time within the growing season [48]. This is why the defoliation treatment in some studies with poplars and willows was carried out four or five times within one growing season [25,32,41]. Therefore, we assume that a higher number of defoliation treatments in this study, for example two times instead of one time in the first growing season, would have led to a better reproduction of the natural defoliation on these sites and would have resulted in even greater effects on the plant growth parameters and a higher number of statistically significant differences between defoliation variants. This conclusion is also confirmed by comparing the growth parameters of the plants in Großthiemig defoliated once, twice, or three times among each another. However, we deliberately refrained from a second defoliation treatment in the first growing season to ensure the survival of plants despite the partly unfavorable site and climatic conditions, and guarantee the general feasibility of this study. An extension of the defoliation treatments to a third growing season was not possible, amongst others due to the great heights and leaf masses of plants.

Nevertheless, the data from this study in Saxony and Brandenburg are well in line with the results of similar studies. On the three study sites, the mean height reduction of poplars that had been defoliated three times to an extent of 75% was between 17 and 28%, and the maximum height reduction between 27 and 42%, in comparison to the plants that had not been defoliated, whereas the same defoliation treatment had a lower effect on willows and resulted in an average reduction of about 10%, with maximum values ranging from 22–26%. Correspondingly, Gao et al. (1985), Tucker et al. (2004), and Bassman et al. (1982) recorded height reductions between 20 and 31% for poplars with a defoliation level of 75% [37,42,46], and Kendall et al. (1998) determined a 15% height reduction for willows with a defoliation level of 70% [48]. With regard to the aboveground biomass, Reichenbacker et al. (1996) documented a reduction of 33% for poplars with a defoliation level of 75% in comparison to the zero variant [41], and Bell et al. (2006) and Kendall et al. (1998) a reduction of 31% and 36–72% for willows with a defoliation level of 70 and 75%, respectively [24,48]. The biomass reduction between 32 and 39% caused by severe defoliation by the leaf beetle *Phratora vulgatissima* on *Salix viminalis* lies in a similar range [49]. In comparison, the results of the fresh weight determination in this study on the plants at Großthiemig show considerably lower reduction values, at least partially. A reduction of 25% was recorded for all poplars, and 21% when merely considering the poplars that had been defoliated three times, whereas the corresponding values for willows are 5 and 18%. Reasons for these differences to other studies may be the lower age and heights of the plants in this study, which are associated with lower diameters so that height differences have a less pronounced effect on the biomass yield.

Our data further shows that the actual effects of defoliation on plants depend on numerous factors. The reaction of plants to herbivory varies according to the prevailing conditions, which can result in different growth losses at similar defoliation levels [49,50]. Since the total size of the photosynthetically active leaf area determines the yield production of plants [51], it is generally assumed that the reduction of leaf area by phyllophagous insects or leaf-infecting fungi reduces plant growth due to a reduction of the photosynthetic capacity [52–54]. However, under certain circumstances, leaf losses can be adjusted by compensatory growth, but the ability for it depends on several abiotic and biotic factors. Regarding abiotic factors, site conditions, such as the availability of soil water, the soil nutrient, and heavy metal content, play an important role [47,49,50]. Each deviation from the site optimum causes stress [55], which negatively affects the compensatory growth of plants [47]. The effect of site conditions is also well reflected in the results of this study. In particular with regard to the poplars, which have somewhat higher nutrient requirements compared to willows [56], a greater height reduction was recorded on the two study sites with rather unfavorable conditions, Obercarsdorf and Schönheide, in comparison to the site at Großthiemig, where conditions were more favorable for plant growth. Correspondingly, only the fertilized three-year-old birch plants in a study were able to fully compensate a 25% defoliation, whereas this defoliation resulted in a significant biomass yield reduction of unfertilized plants [36]. In another study, no effect of a medium-level defoliation of poplars by *Clostera inclusa* (Lepidopotera: Notodontidae) was only detected on the one study site with excellent conditions [57]. Besides the general growth conditions, the time of defoliation also influences its effects on plant growth. The earlier a defoliation takes place, the better plants are able to recover, that is, an early defoliation promotes the chances for compensatory growth [39,50,58]. The greatest impact on plant growth was recorded when poplars were defoliated during the most productive growth period between the beginning and the middle of summer [25]. The time of defoliation was no target parameter of this study, but the first and last defoliation treatment in July 2007 and August 2008 lay within the period mentioned by Larsson (1983) [25]. Only the defoliation treatment carried out between the beginning and middle of June 2008 was prior to this period and may have been balanced out more easily by compensatory growth.

Regarding biotic factors that have an influence on the compensatory growth of plants and therefore the effects of defoliation, tree species and variety play an especially important role. In a study with 11 willow varieties, significant differences in the reduction of plant height and biomass production caused by defoliation were documented [44]. On a few varieties, growth was not reduced but instead increased in comparison to the undefoliated plants, similar to the findings of another study on willows [38]. In contrast, the study by Bell et al. (2006) showed a negative reaction to simulated defoliation for all five willow varieties included, yet to a different extent [24]. In this study too, noticeable differences in the reaction to defoliation were recorded among varieties. While trends were relatively similar on all three study sites after the first defoliation treatment, there was no longer a consistent reaction of the individual varieties among sites after the third and last treatment. We noticed, however, that overall defoliation had a greater impact on the growth of poplar than on willow varieties. Comparing the heights of 75% defoliated plants to that of undefoliated ones, the promotion of plant growth by defoliation described by the two studies mentioned above [38,44] almost exclusively occurred with willow varieties, and on all three willow varieties included in this study. In particular with regard to the data recorded after the first defoliation treatment, an influence of the site conditions is visible as well. While 75% defoliated plants of all three willow varieties had a greater height as the undefoliated plants on the site with the most favorable conditions, those at the site with the least favorable conditions showed a reduced height growth. One reason for the generally better ability of willows for compensatory growth in comparison to poplars may be their superior regeneration capacity [5,59], which does not only apply after harvests but apparently also after defoliations. In contrast, only a single case of an increased growth of 75% defoliated plants compared to undefoliated plants occurred with poplars, namely with the Max 3 variety at the site in Obercarsdorf after the first defoliation treatment. Overall, the statement that faster growing poplar varieties suffer from greater height reductions by defoliations than slower growing varieties [46] was not confirmed by the data of this study. For example, plants of the Muhle Larsen variety often had significantly lower heights than other varieties but nevertheless showed rather great defoliation-induced height reductions.

The data of this study also suggests a lasting effect of defoliation events on the height growth of poplars and willows, and even on the number and height of resprouting shoots of plants that were harvested after those events. Accordingly, yield losses caused by artificial defoliation of willows during the first three-year growth period still persisted after the second three-year growth period without defoliation treatments [24]. Defoliation-induced reductions of root growth and drought tolerance are assumed to be some of the reasons for these long-term effects [41,60].

5. Conclusions

The results of this study confirm the literature stating that even minor leaf loss can have an impact on plant growth in short rotation coppices, which may also last. However, the actual effects of defoliation on a plant depend on numerous external and internal factors, which determine the ability of the plant for compensatory growth. In some cases, leaf loss can be fully compensated or, in single cases, even overcompensated. According to the results from Saxony and Brandenburg, this particularly applies to willow varieties.

Due to the illustrated complexity of the reaction of plants to leaf loss, an exact quantification of potential growth losses with regard to the extent of leaf loss as well as a specification of threshold values that, for example, indicate when control measures against phyllophagous pest insects in short rotation coppices are advisable, are hardly possible. In general, our data indicates that poplars are more susceptible to defoliation than willows, meaning control measures need to be applied earlier, and that defoliation levels above 50% often lead to significant growth reductions. Willows seem to be more resilient, which makes the need for interventions in these coppices less probable. However, the fact that the susceptibility to defoliation increases with decreasing site quality and with an increasing number of defoliation events applies to both tree genera. Ultimately, only the regular survey of plant growth and damage on site and their comparison to coppices on similar sites can help to assess the effects of existing leaf loss on the growth of the plants and the ability of the coppice for compensatory growth. In some cases, it can be profitable to already initiate countermeasures at moderate leaf loss. This particularly applies to situations when the survival of plants in newly established plantations is at risk due to leaf loss. If available, we also recommend considering the predictions of prognosis models on the weather-dependent population growth of the main insect pests during the decision process for or against control measures.

Author Contributions: Conceptualization, M.G.M., C.E.H. and D.L.; methodology, C.E.H.; software, C.E.H.; validation, M.G.M., C.E.H. and D.L.; formal analysis, C.E.H.; investigation, C.E.H.; resources, M.G.M. and C.E.H.; data curation, C.E.H.; writing—original draft preparation, C.E.H.; writing—review and editing, C.E.H. and D.L.; visualization, C.E.H. and D.L.; supervision, M.G.M.; project administration, M.G.M. and C.E.H.; funding acquisition, M.G.M. All authors have read and agreed to the published version of the manuscript.

Funding: This research was carried out within the AGROWOOD project funded by the German Federal Ministry of Education and Research (BMBF) (grant number 0330710 A).

Acknowledgments: We sincerely thank the Research Institute for Post-Mining Landscapes in Finsterwalde and the P&P Tree Nursery in Großthiemig for providing poplar cuttings at no charge, and the LTS GmbH in Groß Luja, the organic farm Biohof Böhme in Obercarsdorf and the P&P Tree Nursery in Großthiemig for providing the sites and the soil preparation for our sample coppices at no charge.

Conflicts of Interest: The authors declare no conflict of interest. The funders had no role in the design of the study; in the collection, analyses, or interpretation of data; in the writing of the manuscript, or in the decision to publish the results.

Appendix A

Table A1. Mean percentage of surviving plants on the three study sites in July/August 2007 four months after the establishment of the short rotation coppices (bold values indicate mean values of all varieties of a tree genus, different capital letters indicate statistically significant differences between tree genera, different small letters indicate statistically significant differences between varieties within tree genera).

Genus/Variety	Großthiemig		Obercarsdorf		Schönheide	
	Study Site					
Poplars	**93.3**	**A**	**82.0**	**A**	**81.4**	**A**
Androscoggin	89.8	a	65.6	a	70.3	a
Max 1	96.1	ab	87.5	bc	85.9	bc
Max 3	94.5	ab	93.7	c	89.1	c
Max 4	98.4	b	87.5	bc	88.3	c
Muhle Larsen	87.5	a	75.8	ab	73.4	ab
Willows	**96.4**	**B**	**84.1**	**A**	**84.1**	**A**
Sven	94.5	a	82.0	a	83.6	a
Tora	96.1	a	76.6	a	82.8	a
Tordis	98.4	a	93.7	b	85.9	a

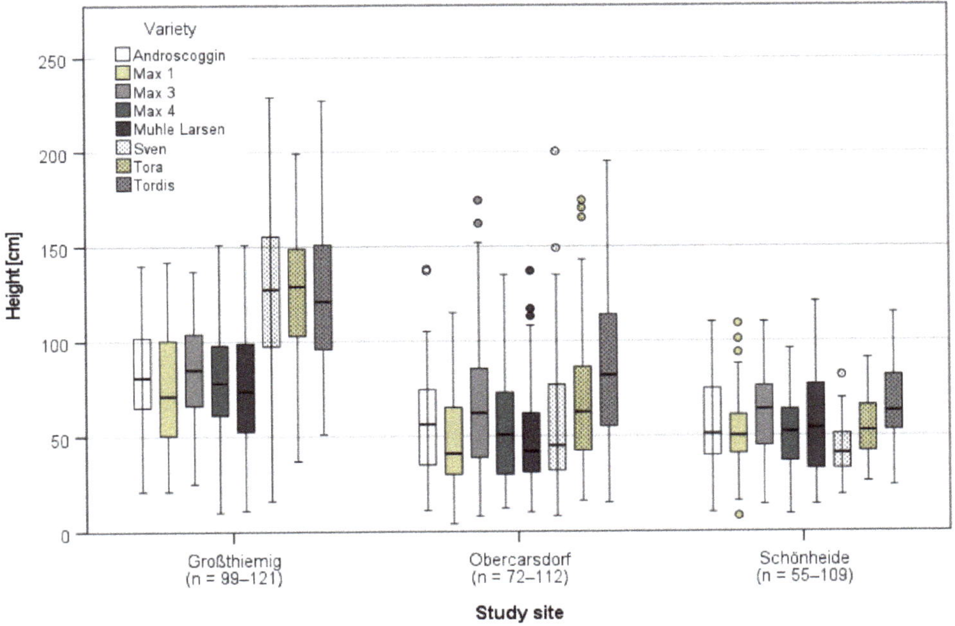

Figure A1. Height of poplar and willow varieties on the three study sites in summer 2007 prior to the start of the defoliation treatments (n = range of number of plants per variety).

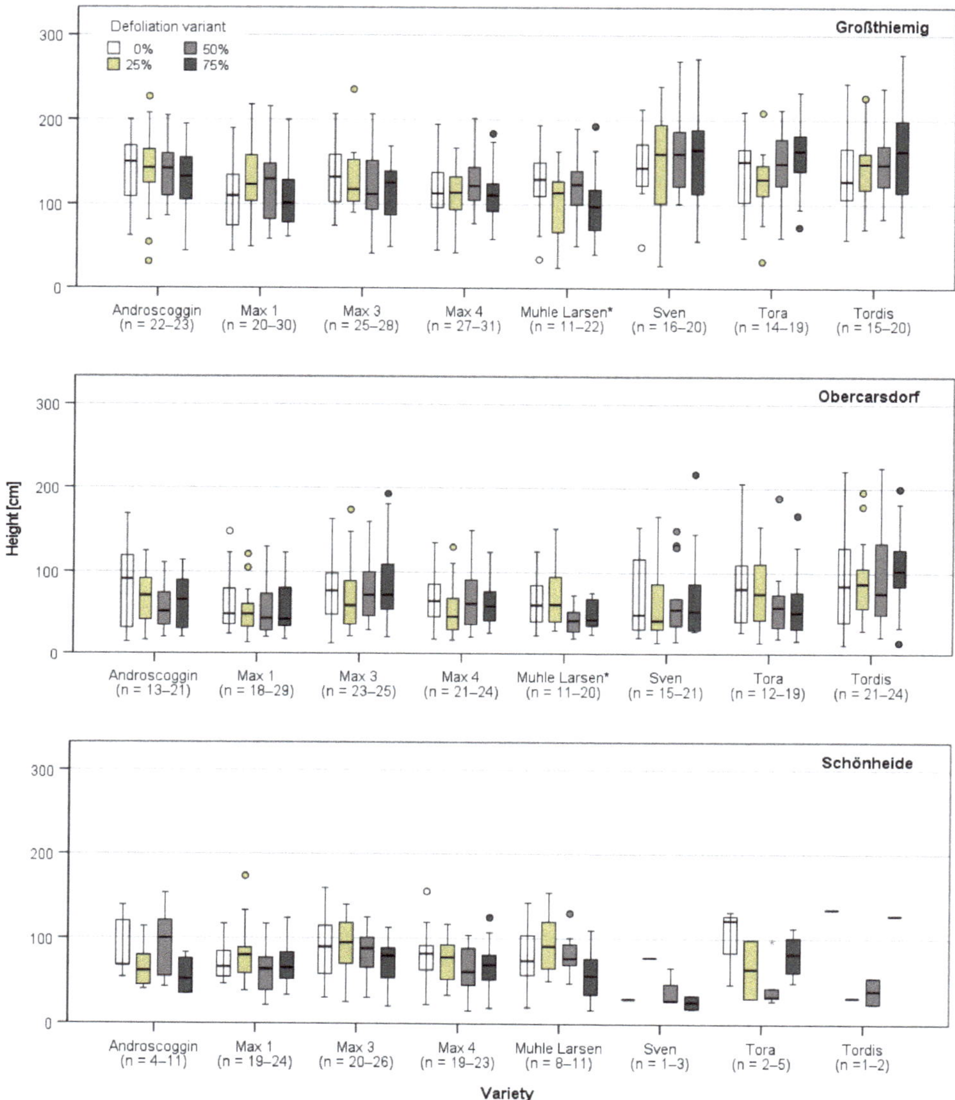

Figure A2. Height of poplar and willow varieties on the three study sites in December 2007 after the first defoliation treatment in July/August 2007, with the exception of the willow varieties in Schönheide (n = range of number of plants per defoliation variant, * statistically significant difference according to ANOVA).

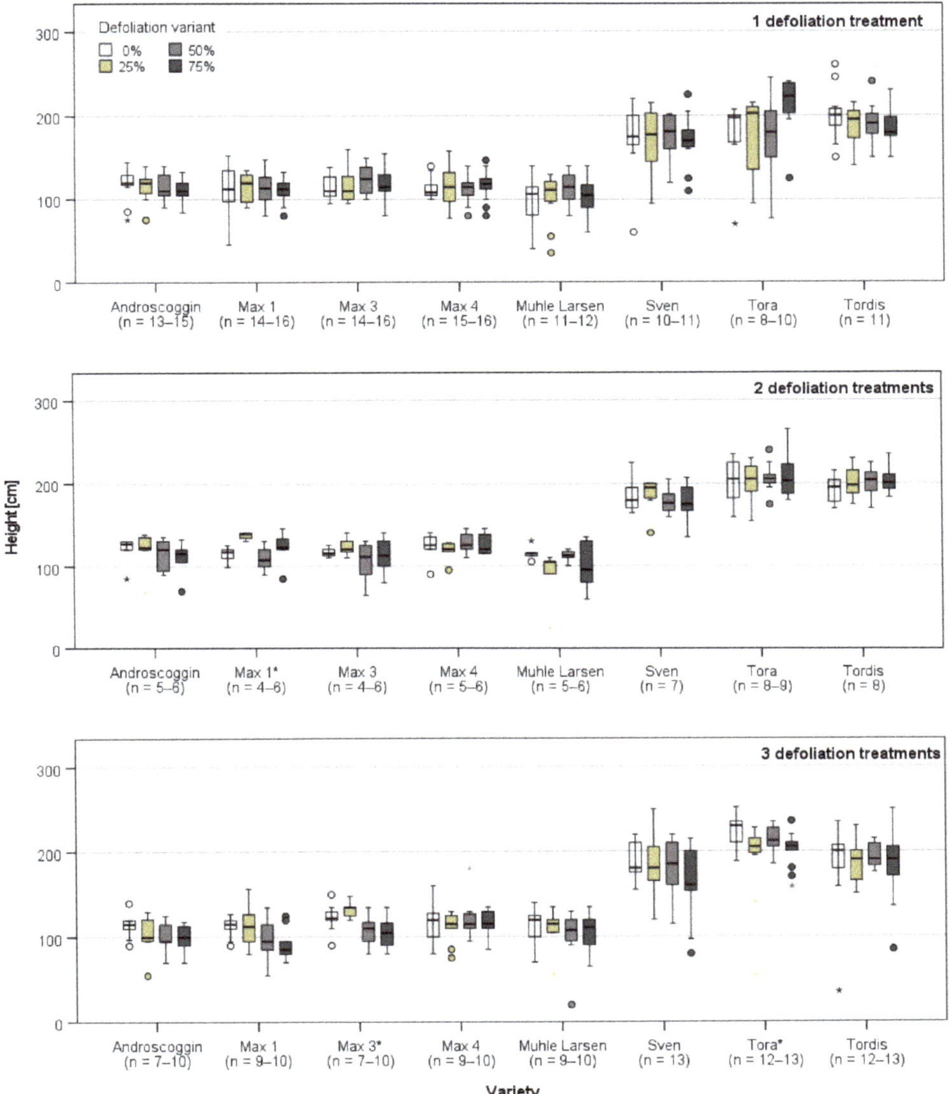

Figure A3. Height of poplar and willow varieties on the study site in Großthiemig in June 2009 after a harvest of the short rotation coppice in March 2009 with regard to the number of defoliation treatments and the defoliation treatment variant in 2007 and 2008 (n = range of number of plants per defoliation variant, * statistically significant difference according to ANOVA).

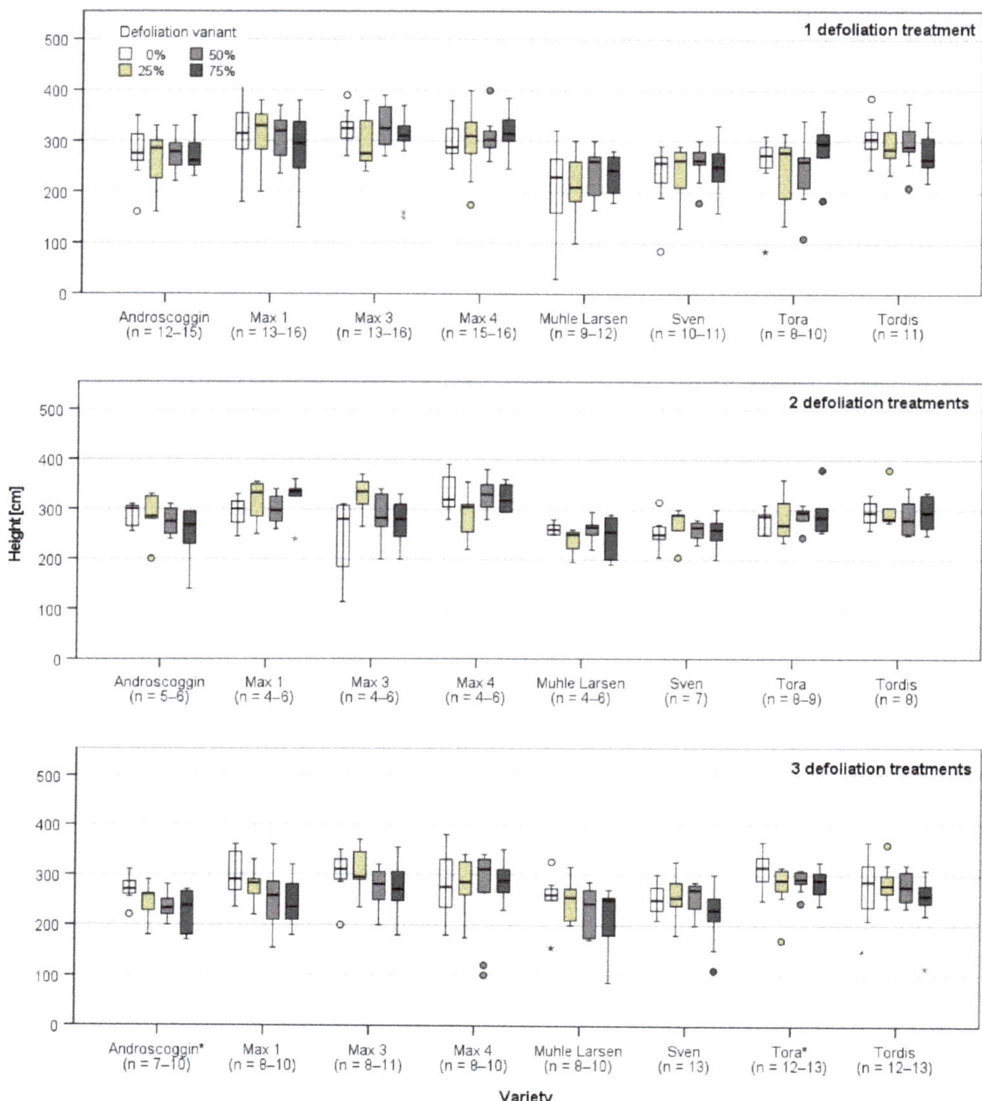

Figure A4. Height of poplar and willow varieties on the study site in Großthiemig in September 2009 after a harvest of the short rotation coppice in March 2009 with regard to the number of defoliation treatments and the defoliation treatment variant in 2007 and 2008 (n = range of number of plants per defoliation variant, * statistically significant difference according to ANOVA).

References

1. Landgraf, D.; Carl, C.; Neupert, M. Biomass yield of 37 different SRC poplar varieties grown on a typical site in north eastern Germany. *Forests* **2020**, *11*, 1048. [CrossRef]
2. Bielefeldt, J.; Bolte, A.; Busch, G.; Dohrenbusch, A.; Kroiher, F.; Lamersdorf, N.; Schulz, U.; Stoll, B. *Energieholzproduktion in der Landwirtschaft—Chancen und Risiken aus Sicht des Natur—und Umweltschutzes*; Naturschutzbund Deutschland, e.V.: Berlin, Germany, 2008.

3. Ball, J.P.; Carle, J.; del Lungo, A. Contribution of poplars and willows to sustainable forestry and rural development. *Unasylva Engl.* **2005**, *56*, 3–9.
4. Dimitriou, I.; Aronsson, P. Willows for energy and phytoremediation in Sweden. *Unasylva Engl.* **2005**, *56*, 47–50.
5. Hofmann, M. *Energieholzproduktion in der Landwirtschaft*; Fachagentur Nachwachsende Rohstoffe e.V.: Gülzow, Germany, 2010.
6. Isebrands, J.G.; Richardson, J. (Eds.) *Poplars and Willows: Trees for Society and the Environment*; CABI: Wallingford, UK, 2014. Available online: http://www.fao.org/3/i2670e/i2670e.pdf (accessed on 23 August 2021).
7. Wolf, H.; Schildbach, M.; Hartmann, K.-U. Plantagenbaumarten und deren Züchtung. In *AGROWOOD: Kurzumtriebsplantagen in Deutschland und Europäische Perspektiven*; Bemmann, A., Knust, C., Eds.; Weißensee-Verlag: Berlin, Germany, 2010; pp. 30–43.
8. Schwerdtfeger, F. Pappelkrankheiten und Pappelschutz. In *Das Pappelbuch*; Hesmer, H., Ed.; Verlag des Deutschen Pappelvereins: Bonn, Germany, 1951; pp. 155–186.
9. Prien, S.; Sann, H. Probleme des Forstschutzes in forstlichen Plantagen. In *Probleme und Methodische Fragen der Plantagenwirtschaft mit Forstlichen Baumarten*; Walter, F., Ed.; Agrarwissenschaftliche Gesellschaft der Deutschen Demokratischen Republik: Dresden, Germany, 1981; pp. 51–57.
10. Andow, D.A. Vegetational Diversity and Arthropod Population Response. *Annu. Rev. Entomol.* **1991**, *36*, 561–586. [CrossRef]
11. Führer, E.; Bacher, H. Biotische Schadrisiken in Energieholzplantagen. *Anz. Schädlingskunde* **1991**, *64*, 1–8. [CrossRef]
12. Coyle, D.R.; Nebeker, T.E.; Hart, E.R.; Mattson, W.J. Biology and management of insect pests in North American intensively managed hardwood forest systems. *Annu. Rev. Entomol.* **2005**, *50*, 1–29. [CrossRef]
13. Splechtna, B.; Glatzel, G. Optionen der Bereitstellung von Biomasse aus Wäldern und Energieholzplantagen für die Energetische Nutzung: Szenarien, ökologische Auswirkungen, Forschungbedarf. In *Materialien der Interdisziplinären Arbeitsgruppe Zukunftsorientierte Nutzung Länderlicher Räume—LandInnovation Nr. 1*; Berlin-Brandenburgische Akademie der Wissenschaften: Berlin, Germany, 2005.
14. Brändle, M.; Brandl, R. Species richness of insects and mites on trees: Expanding Southwood. *J. Anim. Ecol.* **2001**, *70*, 491–504. [CrossRef]
15. Bußler, H. Schmetterlinge und Käfer an der Schwarzpappel. *LWF Wissen* **2006**, *52*, 35–38.
16. Hacker, H. Die Insektenwelt der Weiden. *LWF Wissen* **2000**, *24*, 26–29.
17. Kennedy, C.E.J.; Southwood, T.R.E. The number of species of insects associated with British trees: A reanalysis. *J. Anim. Ecol.* **1984**, *53*, 455–478. [CrossRef]
18. Delplanque, A. *Les Insectes Associés aux Peupliers*; Editions Memor: Bruxelles, Belgium, 1998.
19. Richter, D. *Insektenschäden in Weidenkulturen und Maßnahmen zu Ihrer Verhütung*; Merkblatt 30; Institut für Forstwissenschaften Eberswalde, Abteilung Forstschutz Gegen Tierische Schädlinge: Eberswalde, Germany, 1959.
20. Helbig, C.; Müller, M.; Landgraf, D. Insect pests in short rotation coppice in Germany—An overview of their effects, risk potential and control measures. In *Short Rotation Forestry and Agroforestry: An Exchange of Experience between CDM Countries and Europe*; Benwood Project Consortium: Barolo, Italy, 2011; pp. 46–56.
21. Jaskiewicz, B.; Gorska Drabik, E.; Golan, K. Szkodniki zagrazajace uprawom wierzby wiciowej [*Salix viminalis*]. *Ochr. Rosl.* **2004**, *48*, 24–25.
22. Georgi, R.; Müller, M. Biotic risk factors in short rotation coppice in Germany: Current situation, new findings and future prospects. In *Bioenergy from Dendromass for the Sustainable Development of Rural Areas*; Butler Manning, D., Bemmann, A., Bredemier, M., Lamersdorf, N., Ammer, C., Eds.; Wiley-VCH: Weinheim, Germany, 2015; pp. 199–216.
23. Helbig, C.; Müller, M. Naturale Risiken und Grundzüge des Schadensmanagements in Kurzumtriebsplantagen. In *AGROWOOD: Kurzumtriebsplantagen in Deutschland und Europäische Perspektiven*; Bemmann, A., Knust, C., Eds.; Weißensee-Verlag: Berlin, Germany, 2010; pp. 74–87.
24. Bell, A.C.; Clawson, S.; Watson, S. The long-term effect of partial defoliation on the yield of short-rotation coppice willow. *Ann. Appl. Biol.* **2006**, *148*, 97–104. [CrossRef]
25. Larsson, S. Effects of artificial defoliation on stem growth in *Salix smithiana* grown under intensive culture. *Oecologica Appl.* **1983**, *4*, 343–349.
26. Crawley, M.J. Reduction of oak fecundity by low-density herbivore populations. *Nature* **1985**, *314*, 163–164. [CrossRef]
27. Kosola, K.R.; Dickmann, D.I.; Paul, E.A.; Parry, D. Repeated insect defoliation effects on growth, nitrogen acquisition, carbohydrates, and root demography of poplars. *Oecologia* **2001**, *129*, 65–74. [CrossRef] [PubMed]
28. Gieger, T.; Thomas, F.M. Effects of defoliation and drought stress on biomass partitioning and water relations of *Quercus robur* and *Quercus petraea*. *Basic Appl. Ecol.* **2002**, *3*, 171–181. [CrossRef]
29. Chen, Z.; Kolb, T.E.; Clancy, K.M. Effects of artificial and western spruce budworm (Lepidoptera: Tortricidae) defoliation on growth and biomass allocation of Douglas-fir seedlings. *J. Econ. Entomol.* **2002**, *95*, 587–594. [CrossRef]
30. Baldwin, I.T. Herbivory simulations in ecological research. *Trends Ecol. Evol.* **1990**, *5*, 91–93. [CrossRef]
31. Wright, S.L.; Hall, R.W.; Peacock, J.W. Effect of simulated insect damage on growth and survival of northern red oak (*Quercus rubra* L.) seedlings. *Environ. Entomol.* **1989**, *18*, 235–239. [CrossRef]
32. Powers, S.J.; Peacock, L.; Yap, M.L.; Brain, P. Simulated beetle defoliation on willow genotypes in mixture and monotype plantations. *Ann. Appl. Biol.* **2006**, *148*, 27–38. [CrossRef]
33. Paré, P.W.; Tumlinson, J.H. Plant volatiles as a defense against insect herbivores. *Plant Physiol.* **1999**, *121*, 325–331. [CrossRef] [PubMed]

34. Picard, S.; Chenault, J.; Augustin, S. Short time induction defense of a leuce poplar clone (*P. tremula* x *P. tremuloides*) against *Chrysomela tremulae*-propagation and evolution. *Acta Hortic.* **1994**, *381*, 540–543. [CrossRef]
35. Coyle, D.R.; McMillin, J.D.; Hall, R.B.; Hart, E.R. Cottonwood leaf beetle (Coleoptera: Chrysomelidae) defoliation impact on *Populus* growth and above-ground volume in a short-rotation woody crop plantation. *Agric. For. Entomol.* **2002**, *4*, 293–300. [CrossRef]
36. Anttonen, S.; Piispanen, R.; Ovaska, J.; Mutikainen, P.; Saranpaa, P.; Vapaavuori, E. Effects of defoliation on growth, biomass allocation, and wood properties of *Betula pendula* clones grown at different nutrient levels. *Can. J. For. Res.* **2002**, *32*, 498–508. [CrossRef]
37. Gao, R.T.; Qin, X.X.; Li, J.Z.; Hao, W.Q.; Wang, X.Z. A preliminary study on the relationship between artificial defoliation of poplar trees and their growth. *Sci. Silvae Sin.* **1985**, *21*, 199–205.
38. Houle, G.; Simard, G. Additive effects of genotype, nutrient availability and type of tissue damage on the compensatory response of *Salix planifolia* ssp. *planifolia* to simulated herbivory. *Oecologia* **1996**, *107*, 373–378. [CrossRef]
39. Kendall, D.A. *Phenology and Population Dynamics of Willow Beetles (Coleoptera: Chrysomelidae) in Short-Rotation Coppiced Willows at Long Ashton*; ETSU B/M4/00487/14/REP; Energy Technology Support Unit: Harwell, UK, 1996.
40. Osier, T.L.; Lindroth, R.L. Long-term effects of defoliation on quaking aspen in relation to genotype and nutrient availability: Plant growth, phytochemistry and insect performance. *Oecologia* **2004**, *139*, 55–65. [CrossRef]
41. Reichenbacker, R.R.; Schultz, R.C.; Hart, E.R. Artificial defoliation effect on *Populus* growth, biomass production, and total nonstructural carbohydrate concentration. *Environ. Entomol.* **1996**, *25*, 632–642. [CrossRef]
42. Tucker, S.A.; Nebeker, T.E.; Warriner, M.D.; Jones, W.D.; Beatty, T.K. Effects of artificial defoliation on the growth of cottonwood: Simulation of cottonwood leaf beetle defoliation. In *Proceedings of the 12th Biennial Southern Silvicultural Research Conference*; Biloxi, MS, USA, 24–28 February 2003, General Technical Report Southern Research Station, USDA Forest Service, SRS-71; USDA Forest Service: Asheville, NC, USA, 2004; pp. 169–171.
43. Deutscher Wetterdienst. Klimadaten zum Direkten Download. Vieljährige Mittelwerte für den Zeitraum 1981–2021. Available online: https://www.dwd.de/DE/leistungen/cdc/cdc_ueberblick-klimadaten.html?nn=17626&lsbId=344084 (accessed on 23 August 2021).
44. Peacock, L.; Herrick, S.; Harris, J. Interactions between the willow beetle *Phratora vulgatissima* and different genotypes of *Salix viminalis*. *Agric. For. Entomol.* **2002**, *4*, 71–79. [CrossRef]
45. *IBM SPSS Statistics for Windows*; Version 26.0.; IBM SPSS: Armonk, NY, USA, 2019.
46. Bassman, J.; Myers, W.; Dickmann, D.; Wilson, L. Effects of simulated insect damage on early growth of nursery-grown hybrid poplars in northern Wisconsin. *Can. J. For. Res.* **1982**, *12*, 1–9. [CrossRef]
47. Zvereva, E.L.; Kozlov, M.V. Effects of pollution-induced habitat disturbance on the response of willows to simulated herbivory. *J. Ecol.* **2001**, *89*, 21–30. [CrossRef]
48. Kendall, D.A.; Wiltshire, C.W.; Royle, D.J.; Paul, L.; Holdenrieder, O. Life-cycles and ecology of willow beetles on *Salix viminalis* in England. *Eur. J. For. Pathol.* **1998**, *28*, 281–288. [CrossRef]
49. Björkman, C.; Hoglund, S.; Eklund, K.; Larsson, S. Effects of leaf beetle damage on stem wood production in coppicing willow. *Agric. For. Entomol.* **2000**, *2*, 131–139. [CrossRef]
50. Maschinski, J.; Whitham, T.G. The continuum of plant responses to herbivory: The influence of plant association, nutrient availability, and timing. *Am. Nat.* **1989**, *134*, 1–19. [CrossRef]
51. Boelcke, B. Schnellwachsende Baumarten auf Landwirtschaftlichen Flächen; Ministeriums für Ernährung, Landwirtschaft, Forsten und Fischerei. 2006. Available online: https://docplayer.org/52963448-Schnellwachsende-baumarten-auf-landwirtschaftlichen-flaechen-leitfaden-zur-erzeugung-von-energieholz.html (accessed on 23 August 2021).
52. Begon, M.; Mortimer, M.; Thompson, D.J.; Müller, J. *Populationsökologie*; Spektrum-Lehrbuch; Spektrum Akademischer Verlag: Heidelberg, Germany, 1997.
53. Christersson, L.; Ramstedt, M.; Forsberg, J. Pests, diseases and injuries in intensive short-rotation forestry. In *Ecophysiology of Short Rotation Forest Crops*; Mitchell, C.P., Ed.; Elsevier Science Publisher: London, UK, 1992; pp. 185–215.
54. Jug, A. Ernährungs-und standortskundliche Untersuchungen. In *Modellvorhaben "Schnellwachsende Baumarten": Zusammenfassender Abschlussbericht*; Schriftenreihe Nachwachsende Rohstoffe; Hofmann, M., Ed.; Landwirtschaftsverlag: Münster, Germany, 1999; Volume 13, pp. 369–396.
55. Albrecht, J.; Böden, E. Zur Krankheitsanfälligkeit von zugelassenen Pappelklonen der Sektionen Aigeiros und Tacamahaca. *Holzzucht* **1988**, *42*, 25–29.
56. Schildbach, M.; Grünewald, H.; Wolf, H.; Schneider, B.-U. Begründung von Kurzumtriebsplantagen: Baumartenwahl und Anlageverfahren. In *Anbau und Nutzung von Bäumen auf Landwirtschaftlichen Flächen*; Reeg, T., Bemmann, A., Konold, W., Murach, D., Spieker, H., Eds.; Wiley-VCH: Weinheim, Germany, 2009; pp. 57–71.
57. Solomon, J.D.; Oliveria, F.L. *Evaluation of Poplar Tentmaker Defoliation on Growth and Survival of Older Cottonwood Plantations*; Research NoteSO-271; U.S. Department of Agriculture, Forest Service, Southern Forest Experiment Station: Asheville, NC, USA, 1993.
58. Bogacheva, I.A. Compensation of foliage losses in willow in the forest-tundra during an outbreak of leaf-eating insects. *Lesovedenie* **1994**, *6*, 62–69.

59. Boelcke, B. Ertragspotenzial und Ertragsaufbau von Weiden spec. in Kurzumtriebsplantagen. In *Anbau und Nutzung von Bäumen auf Landwirtschaftlichen Flächen: 1. Fachtagung*; Technische Universität Dresden: Tharandt, Germany, 2006; pp. 51–56.
60. Bach, C.E. Effects of herbivory and genotype on growth and survivorship of sand-dune willow (*Salix cordata*). *Ecol. Entomol.* **1994**, *19*, 303–309. [CrossRef]

 *forests* 

Article

Establishment of a Laboratory Scale Set-Up with Controlled Temperature and High Humidity to Investigate Dry Matter Losses of Wood Chips from Poplar during Storage

Albert Hernandez-Estrada [1], Ralf Pecenka [1,*], Sabrina Dumfort [2], Judith Ascher-Jenull [3], Hannes Lenz [1], Christine Idler [1] and Thomas Hoffmann [1]

1 Leibniz Institute for Agricultural Engineering and Bioeconomy (ATB), Max-Eyth-Allee 100, 14469 Potsdam, Germany; ahernandez@atb-potsdam.de (A.H.-E.); hlenz@atb-potsdam.de (H.L.); cidler@web.de (C.I.); thoffmann@atb-potsdam.de (T.H.)

2 Department for Environmental, Process and Energy Engineering, Management Center Innsbruck, Maximilianstrasse 2, 6020 Innsbruck, Austria; sabrina.dumfort@mci.edu

3 Department of Microbiology, University of Innsbruck, Technikerstrasse 25d, 6020 Innsbruck, Austria; judith.ascher@uibk.ac.at

* Correspondence: rpecenka@atb-potsdam.de; Tel.: +49-(0)331-5699-312

Abstract: The aim of this work was to improve the understanding of dry matter losses (DML) that occur in wood chips during the initial phase of storage in outdoor piles. For this purpose, a laboratory scale storage chamber was developed and investigated regarding its ability to recreate the conditions that chips undergo during the initial phase of outdoor storage. Three trials with poplar Max-4 (*Populus maximowiczii* Henry × *Populus nigra* L.) chips were performed for 6–10 weeks in the storage chamber under controlled temperature and assisted humidity. Two different set-ups were investigated to maintain a high relative humidity (RH) inside the storage chamber; one using water containers, and one assisted with a humidifier. Moisture content (MC) and DML of the chips were measured at different storage times to evaluate their storage behaviour in the chamber. Additionally, microbiological analyses of the culturable fraction of saproxylic microbiota were performed, with a focus on mesophilic fungi, but discriminating also xerophilic fungi, and mesophilic bacteria, with focus on actinobacteria, in two trials, to gain a view on the poplar wood chip-inhabiting microorganisms as a function of storage conditions (moisture, temperature) and time. Results show that DML up to 8.8–13.7% occurred in the chips within 6–10 storage weeks. The maximum DML were reached in the trial using the humidifier, which seemed a suitable technique to keep a high RH in the testing chamber, and thus, to analyse the wood chips in conditions comparable to those in outdoor piles during the initial storage phase.

Keywords: dry matter losses; short rotation coppices; poplar wood chips; laboratory scale; cultivable saproxylic microbiota

Citation: Hernandez-Estrada, A.; Pecenka, R.; Dumfort, S.; Ascher-Jenull, J.; Lenz, H.; Idler, C.; Hoffmann, T. Establishment of a Laboratory Scale Set-Up with Controlled Temperature and High Humidity to Investigate Dry Matter Losses of Wood Chips from Poplar during Storage. *Forests* **2022**, *13*, 459. https://doi.org/10.3390/f13030459

Academic Editor: Blas Mola-Yudego

Received: 26 January 2022
Accepted: 12 March 2022
Published: 15 March 2022

1. Introduction

Woody biomass from forestry and agriculture plays an important role in bioenergy technology to fulfill the targets for the global reduction of greenhouse gas emissions [1–3]. As such, to meet the growing demand for woody biomass, an increase of the production of fast-growing trees such as poplar (*Populus* spp. L.), willow (*Salix* spp. L.) and black locust (*Robinia pseudoacacia* L.) in plantations of short-rotation coppices (SRC) or agroforestry systems (AFS) is an important option [4]. A prerequisite for increasing wood production in agriculture, however, is the availability of suitable technologies [5–8] and competitiveness compared to other energy-supplying plants and traditional field crops [9,10]. Key problems interfering with the expansion of SRC cultivation are the high dry matter losses (DML) and quality losses during the storage of wood chips [11–13]. Wood chips from SRC are harvested in winter with moisture contents (MC) of 50–60% (wet-basis MC; poplar chips). Generally, the wood chips have to be stored and dried for six to nine months until the next heating season starts [14,15]. In order to guarantee low costs and easy handling, wood chips are typically stored to dry outdoors in large piles, and consequently, the wet chips are subjected to microbial attack [16–19]. Depending on the material properties of the wood chips, the design and the location of the storage piles, DML of up to 47% may occur [17,20–25].

1.1. DML and Temperature in Wood Chip Storage Piles

Due to the harvesting of SRC in winter, in some regions like central or northern Europe, the ambient temperature is below 0 °C for some weeks, and therefore the temperature in the storage pile stays, as well, below 0 °C [17,21,26]. This first storage phase is called 'frost phase' in the three-phase model by Idler et al. [19] (Figure 1; phase 0), and it lacks the microbial activity that contributes to DML of the wood chips [17,21,26].

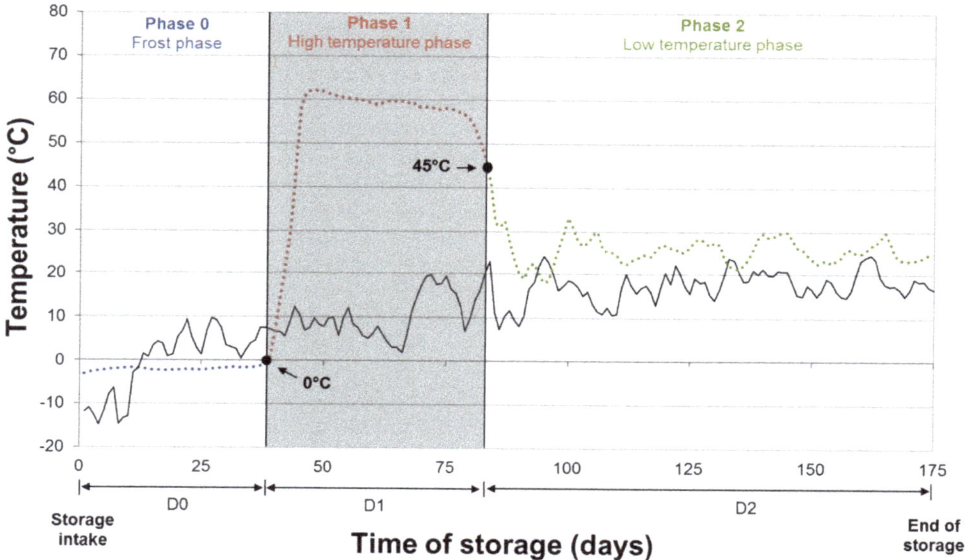

Figure 1. Typical temperature phases of a storage pile of poplar chips harvested in winter, set outdoors in Central Europe. (Modified from Idler et al. [19]).

With the rise of ambient temperature above 0 °C, the temperature inside the wood chips storage pile also increases as a consequence of the occurring microbial activity [16,17,27,28]. This second storage phase (Figure 1; phase 1) is called 'high temperature phase' in Idler et al. [19], and it is characterized by a fast rise of the pile temperature up to 50–60 °C [13,16–18]. This rise of temperature inside the storage pile is the result of both chemical and microbial exothermic reactions that occur in the wood chips when exposed outdoors. However, the temperature progression may differ depending on, e.g., tree variety, chip size and pile set-up [13,16,17,29]. The third storage phase proposed by Idler et al. [19] starts when the pile temperature decreased below 45 °C (Figure 1; phase 2). This 'low temperature phase' was characterized by a progressive decrease of the pile temperature, that gradually approached the ambient temperature, in line with the observations by, e.g., Barontini et al. [13], Manzone et al. [12], and Whittaker et al. [18].

For example, in storage piles of wood chips from eucalyptus with small size, i.e., P16, the high temperature phase can last for five months [29]. On the other hand, storage piles made with larger chips from poplar, i.e., P63 and P45, showed lower increases above ambient temperature in the phase 1 (Figure 1) that lasted for less than two months [17,30]. Similarly, when piles were set with wood chips produced from different parts of the trees (e.g., stem and crown), the pile temperature differed between the storage piles according to the different tree fractions [13]. However, even if this high temperature phase (Figure 1; phase 1) did not last very long, it implied DML of 7–10% for poplar chips stored outdoors [17,30,31]. For example, in the outdoor storage pile of P31 poplar chips done by Pecenka et al. [17], the high temperature phase lasted approximately for nine weeks. In this period, sampling was conducted after five and nine weeks from the storage intake, in order to assess DML at the respective storage times, revealing 5.6% and 9.4% DML, respectively [17].

DML of wood chips stored outdoors in piles can account for up to 47% [17,21–25,29], although there is a large variability, as reviewed, e.g., in Whittaker et al. [22] and Therasme et al. [25]. This variability may depend, among many other parameters, on the size, location and the set-up of the piles [27,29,32,33], and the chip size and the wood properties determined by, e.g., variety and tree age [13,17,18,24,29]. Furthermore, different methods used to determine DML [13,24,31,34] may show different DML values. In this regard, Whittaker et al. [22] provided a comprehensive review of the different parameters used in storage of wood chips and corresponding DML. Likewise, Dimitriou and Rutz [14] and Veste and Böhm [15] offered comprehensive information about SRC and AFS in practice.

1.2. Microorganisms and Degradation of Wood Chips

Wood-inhabiting microorganisms involved in the complex and dynamic process of wood decomposition are bacteria, archaea [35–39], and fungi [36,40,41]. These saproxylic microbiota possess a variety of hydrolytic and oxidative enzymes to degrade the main wood components cellulose, hemicellulose and lignin. The primary degradation of wood is carried out by different fungi, due to their diverse set of enzymes, as well as their ability to colonize wood with the use of hyphae [42]. The primary wood decomposers can be classified into white-(degradation of mainly lignin), soft-(degradation of mainly cellulose) and brown-rot (degradation of mainly cellulose) fungi. Even though bacteria are able to decompose wood, their contribution to wood degradation is negligible with respect to the fungal one [36,41,42], as reflected by so far much less scientific attention. Despite this, there is growing interest on and recent evidence about the role of bacteria within the wood-decomposition and complex bacteria–fungi interactions governing the decomposition of dead-wood [43–45]. Bacteria thrive on nutrients available in the wood cells [35,41] and are able to colonize wood in conditions with low availability of oxygen, or even in the absence of oxygen (anaerobiosis) [46], which are inhibitory to most fungi [35,47]. Saproxylic bacteria can be classified due to their functional role in wood decomposition into (i) bacteria that increase the water permeability of wood without affecting its structural integrity; (ii) bacteria with (albeit limited) decomposition ability; (iii) bacteria that stimulate fungal decomposition; and (iv) bacteria that inhibit fungal decomposition [35,43].

Among the most abundant and dominant bacterial phyla generally involved in the complex and dynamic decomposition of dead-wood are the Acidobacteria, Alphaproteobacteria, Gammaproteobacteria, Actinobacteria, and Bacteroidetes [43,44]. The abundance of such microorganisms generally increases with each progressing stage of decomposition. The dynamic microbial succession is due to changing physico-chemical properties during dead-wood decay, with nitrogen content and pH among the master-variables shaping the microbial community in terms of composition and abundance [38,45,48,49]. Actinobacteria (Actinomycetes), commonly found in soil and in wood in contact with soil [41], form mycelia-like filaments [41,42]. This explains—also from a morphologic point of view—their role as wood decomposers, similar to fungi, already in the early stages of decomposition [42]. On the other hand, most of deadwood-decaying bacteria, including Archaea, thrive on fungal metabolites, thus occurring mainly at later stages of wood decomposition [38]. As already mentioned, among the main driving factors governing the composition and activity of microbial communities in a wood chip storage pile is the chip size [19,21,30]. For example, the work done by Idler et al. [19] in storage piles of 500–1000 m^3 of P31 and P45 poplar chips revealed that mesophilic fungi in small chips (P31) developed in higher concentration when the pile temperature was relatively low, i.e., between 20 °C and 30 °C, and the chips' MC was below 47%. On the other hand, in larger poplar chips (P45), mesophilic fungi developed in its maximum at 19–23 °C and 28–33% MC [19]. Moreover, the maximum content of mesophilic fungi was clearly lower in the chips with the larger size P45 than in the small P31 chips (5.8 $_{lg}$CFU g^{-1} and 6.9 $_{lg}$CFU g^{-1}, respectively; Idler et al. [19]). Furthermore, thermophilic fungi are also present in storage piles of wood chips, as it was reported in, e.g., Idler et al. [19] and Pecenka et al. [30]. In fact, considering the elevated temperatures occurring in outdoor piles during storage, thermotolerant and thermophilic wood-degrading microorganisms play a key role [48,50].

1.3. Aim of This Work

As mentioned before, DML depend on multiple factors such as weather conditions, pile size and wood variety. At ambient temperatures above 0 °C, the fast rise in temperature occurring in wood chip piles and lasting up to 60–90 days is characteristic for wood chips from SRC stored and naturally dried in outdoor piles (high temperature phase; Figure 1). This high temperature phase is of particular interest, because approximately 50% of the total storage losses occur during this relatively short storage time. To what extent this temperature rise caused by microbiological and chemical processes leads to DML is not very well known. Detailed measurements based on frequent sampling at a practice scale are very labour intensive and potentially lead to disturbances of the microbiological and chemical processes in the pile. Furthermore, due to the annually changing weather conditions, it is hardly possible to repeat a storage test under the very same conditions. Therefore, this work aimed at investigating the possibilities and limitations of setting up laboratory-scale trials, in order to provide a better understanding of the DML of poplar chips in outdoor storages.

This work focused on the high temperature storage phase (Figure 1; phase 1), and the trials were set in a storage chamber with controlled temperature and assisted air humidity. The storage conditions in the chamber were chosen based on previous measurements in large-scale outdoor storage trials of poplar chips in central Europe (Figure 1). The aim was to recreate the conditions that wood chips in the inner layers of the pile experience in the high temperature phase when stored in such outdoor piles. In addition to the measurement of DML and chips' MC, microbiological analyses were performed. These analyses aimed to assess the culturable fraction of the autochthonous saproxylic microorganisms present in the poplar chips at storage intake and during the storage, as a function of storage conditions (moisture, temperature) and time. The focus was on xerophilic and mesophilic fungi, and (actino)bacteria. Xerophilic fungi were of particular interest due to the drying of the chips during storage.

2. Materials and Methods

2.1. Raw Material and Chip Preparation

The trees used in this work were poplar Max-4 (*Populus maximowiczii* Henry × *Populus nigra* L.) from a SRC plantation established in March 2016 at the Leibniz Institute for Agricultural Engineering and Bioeconomy (ATB) in Potsdam, Germany (52.49° N, 13.01° E). Trees with a stem diameter at the cutting height of 7–10 cm were selected for each trial and were harvested manually at the times corresponding to each of the trials T1–T3. Table 1 shows the harvesting date and storage times of each trial, as well as the corresponding samplings and sampling frequencies.

Table 1. Storage trials T1–T3 and the corresponding samplings.

Trial No.	Harvest and Storage Intake (Date)	Storage (Weeks)	Samplings		
			MC and DML	Microbiological and pH Analyses	Ash and Elemental Analysis
T1	14 June 2018	6	Storage intake (n = 6; only MC) and after 1, 2, 3, 4, 5, and 6 storage weeks (n = 3; sample bag method) Chips	Storage intake (n = 2) and after 6 storage weeks (n = 3) Chips and wood and bark fractions	Storage intake (n = 2) Chips
T2	20 August 2018	10	Storage intake (n = 6; only MC) and after 2, 4, 6, 8, and 10 storage weeks (n = 3; sample bag method) Chips	Storage intake (n = 2) and after 2, 4, 6, 8 and 10 * storage weeks (n = 3) Chips and wood and bark fractions	Storage intake (n = 2) Chips
T3	16 July 2019	9	Storage intake (n = 6; only MC) and after 3, 6, and 9 storage weeks (n = 3; sample bag method) Chips	Not performed	Storage intake (n = 2) and after 3, 6, and 9 storage weeks (n = 3) Chips

* Microbiological analysis not performed at storage week 10, only pH analysis; MC: moisture content (wet-basis); DML: dry matter loss.

Before chipping the trees, the leaves were removed to avoid any potential effect on DML of the wood chips during the storage. The chips were produced with a stationary woodchipper HE100 500 STA (JENZ GmbH, Petershagen, Germany) equipped with an automatic feeding chain conveyor, which provided a constant feeding rate into the chipping drum. Pecenka et al. [43] offered a detailed overview of the woodchipper; however, the screen used in the outtake by Pecenka et al. [51] was not used in producing the chips of this work. The produced chips were analysed regarding particle size distribution using the oscillating screen method (standard EN ISO 17827-1:2016 [52]) and classified according to the standard EN ISO 17225-1:2014 [53].

The initial MC of the fresh wood chips (MC_{in}) was determined as the mean value of six samples of 250–450 g using a Sartorius TE3102S balance (Göttingen, Germany) with the oven dry method at 105 °C based on EN ISO 18134-2:2017 [54]. The chips undergoing storage used the method based on sample bags, which is often used to calculate DML of wood chips in large outdoor storage trials (e.g., [26,27,30]). Sample bags were prepared right after the chipping process, each one containing about 2 kg of wood chips, and the mass of each sample bag was measured individually (m_{in}; excluding the mass of the net bag) using a Sartorius BP 12000-S balance (Göttingen, Germany).

2.2. Testing Set-Up and Sampling for the DML Calculations

The storage trials were done in a programmable incubation chamber Binder KB-400 (Binder, Tuttlingen, Germany) providing a storage volume of 400 L. The set-ups used are shown schematically in Figure 2. When using such a closed chamber for storage investigations at lab-scale, the conditions that prevail inside a storage pile in the first few weeks after harvest in winter at practice scale should be reproduced as close as possible. This can avoid some of the important known problems of outdoor storage trials with wood chips at a practice scale, such as the large heterogeneity of samples received from a large storage pile or the very labour-intensive sampling. Accordingly, before starting the storage trials, the chamber was pre-conditioned at 5 °C, to start the trials at similar conditions to outdoor storage trials when harvesting in winter. At these conditions, the prepared sample bags were placed in the storage chamber, distributing the three sample bags corresponding to each sampling time in different shelves, and left overnight (about 15–18 h) to condition the poplar chips at 5 °C.

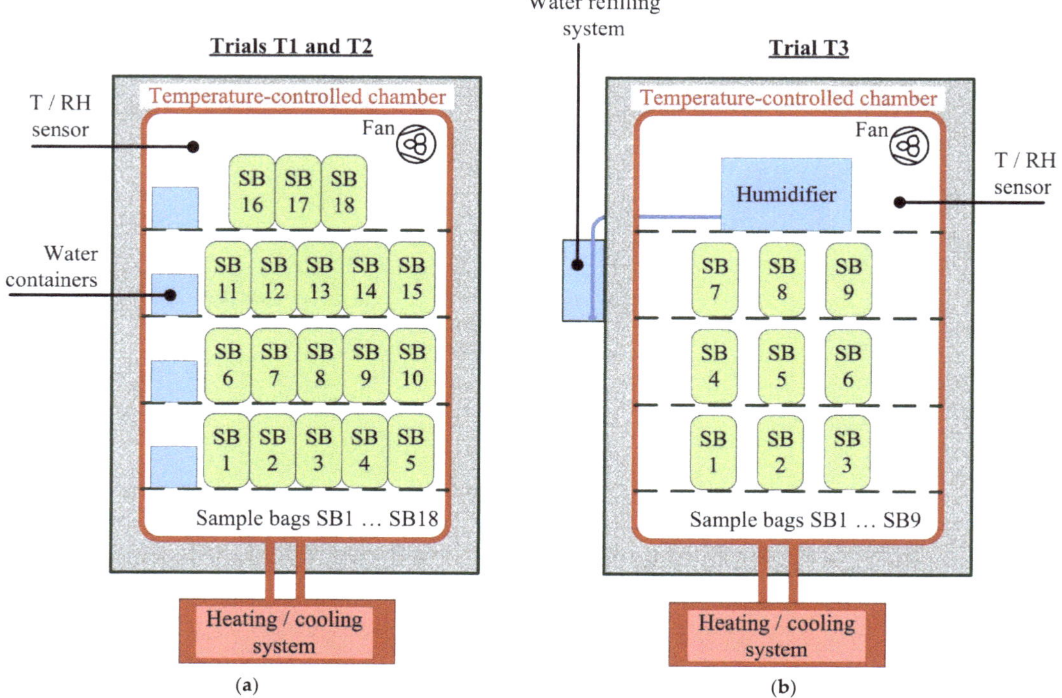

Figure 2. Storage of sample bags in a chamber under controlled temperature and assisted humidity. (**a**) Set-up used in trials T1 and T2 with water containers placed in the chamber, and (**b**) set-up using a humidifier in trial T3. T: temperature; RH: relative humidity; SB: sample bag.

The programmed storage plan started by progressively rising the chamber temperature from 5 °C to 55 °C during the first 48 h. Afterwards, the chamber temperature was kept constant at 55 °C for the rest of the storage experiment, i.e., six weeks (Trial T1 in Table 1), nine weeks (Trial T2 in Table 1) or ten weeks (Trial T3 in Table 1). The internal fan of the incubation chamber was set at the minimum speed of the chamber control, which was 620 rpm (measured with a hand-held tachometer, model 6611, from Veeder-Root GmbH, Neuhausen, Germany). This configuration allowed minimized air circulation inside the chamber and, on the other hand, contributed to having comparable storage conditions for all sample bags in the chamber.

Two storage set-ups were evaluated in order to assist in having a high humidity atmosphere inside the chamber, comparable to the moisture conditions in outdoor piles. In the trials T1 and T2 (Table 1), four containers with distilled water placed inside the chamber were used (Figure 2a). In the alternative set-up, trial T3, we used an in-house developed humidifier (Figure 2b). This humidifier was actively transferring moisture to the chamber atmosphere, and was equipped with a water refilling system to assure a sufficient water supply (Figure 3). An Almemo ZA 9020-FS Thermo R2E4 sensor was used in the trials to measure the temperature inside the storage chamber, and the relative humidity (RH) was measured with Almemo FHAD36RAS (trials T1 and T2) and Almemo FH0D46C sensors (trial T3), together with an Almemo 3290-8 datalogger. The sensors and the datalogger were manufactured by Ahlborn Mess-und Regelungstechnik (Holzkirchen, Germany).

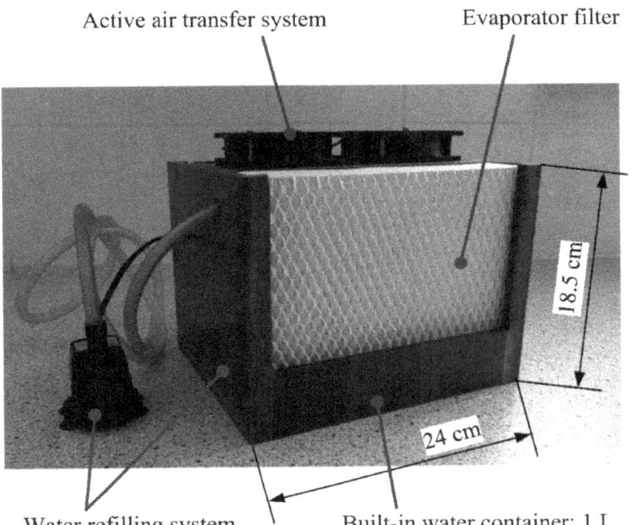

Figure 3. Humidifier used in the set-up of trial T3 and details of the components.

The method to calculate DML was done according to the Equation (1), with the MC and the masses of the sample bags at the storage intake and at the sampling time (after storage). The mass of the net bag was subtracted from the mass measurements, and the calculations were done only considering the mass of the chips contained in each sample bag. At each sampling time (see Table 1), the corresponding sample bags were unloaded from the chamber and the mass of each sample bag was measured ($m_{out,i}$; excluding the mass of the net bag). Right after opening the sample bags and before any further manipulation, the sampling for the microbiological and pH analyses was done (see Section 2.3). With the remaining material, the MC of the chips ($MC_{out,i}$ in the Equation (1)) was determined for each sample bag and expressed as the mean value (n = 4), using the oven dry method at 105 °C.

$$DML_i = \left[1 - \frac{m_{out,\,i}\,(100 - MC_{out,i})}{m_{in}(100 - MC_{in})}\right] \times 100 \quad (1)$$

where
DML_i, dry matter loss of the chips in the sample bag at the storage time i [%];
m_{in}, wet mass of the chips at the storage intake (excluding the net bag) [kg];
$m_{out,i}$, wet mass of the chips after storage at the time i (excluding the net bag) [kg];
MC_{in}, MC (wet-basis) of the chips at the storage intake [%]; and
$MC_{out,i}$, MC (wet-basis) of the chips in the sample bag after storage, at the time i [%].

2.3. Microbiological and pH Analyses

The sampling for the microbiological and pH analyses was done in the trials T1 and T2. For these analyses, not only the wood chips, but also wood and bark fractions of the same sample bags were investigated, due to the different nature and chemical composition of both materials. Therefore, herein, the terms 'wood chips' or 'chips' refer to the entire material as obtained after the chipping process (Section 2.1), i.e., containing wood and bark. On the other hand, the term 'wood' refers to the wood fraction of the chips, i.e., resulting after removing the bark from the chips; likewise, the term 'bark' refers to the bark fraction prepared from the chips after removing the woody parts, and so containing only bark.

The sampling at the storage intake was done preparing two sets (n = 2) of the analysed samples (i.e., chips, wood, and bark; Table 1) and the mean is reported herein. The sampling of the bags undergoing storage was done right after opening the sample bags and before any further manipulation of the chips; from each sample bag, the sampling of the chips, wood and bark was prepared (Table 1) separately. Each sampling time included three sample bags (n = 3; Table 1), in order to report the means for the chips, and the fractions of wood and bark, separately (Table 1). The chips, wood and bark pieces were cut manually into smaller pieces (maximum size of 1 cm × 1 cm × 1 cm), avoiding any contamination.

Microbiological analysis consisted of dilution plating to assess the culturable fraction of the wood chip-inhabiting microbiota (saproxylic microbiota) as a function of storage conditions (temperature, moisture) and time. We used 20 g each of the cut chips, wood and bark, which were analysed separately. The samples were transferred to Erlenmeyer flasks containing 180 mL of sterile Ringer's solution (Merck, Darmstadt, Germany) and shaken at room temperature for 30 min in an orbital shaker TR-125 (Infors AG, Bottmingen, Switzerland) at 180 rpm. Afterwards, a 1:10 dilution series was prepared in Ringer's solution and, from the dilution series, aliquots of 100 µL were plated on the respective nutrient agars.

The wood chip samples were screened for their saproxylic microbiota discriminating fungi (mesophilic; xerophilic) and bacteria. Mesophilic fungi were isolated using selective malt extract agar (MEA; Merck, Darmstadt, Germany) containing 0.01% chloramphenicol (Fluka, Ulm, Germany), and incubated aerobically for seven days, at 25 °C in dark conditions. Dicloran glycerol agar (DG18; Oxoid, Hampshire, UK), a low water activity medium [55], was used for isolation and enumeration (CFUs) of xerophilic fungi under the conditions described before. The agar used to investigate mesophilic bacteria was plate count agar (PCA; Merck, Darmstadt, Germany), and the plates were incubated aerobically at 25 °C in dark conditions for four days. Under the same conditions, the presence of actinobacteria was investigated using casein-soja-pepton-agar (CASO; Merck, Darmstadt, Germany).

After the incubation period, all colonies were counted as colony forming units (CFU) per gram of sample (CFU g^{-1}; wet basis) and expressed as logarithmic values to base 10 ($_{lg}$CFU g^{-1}). Plates prepared for mesophilic and xerophilic fungi with fewer than 10 colonies or exceeding 150 colonies per plate were excluded from the calculations. Likewise, for mesophilic bacteria and actinobacteria, the plates containing less than 10 or more than 300 CFUs were not considered for the quantitative assessment. The detection limit was 100 CFU g^{-1}, and in case the plate of the first dilution did not present any colony of the target microorganism, the value of 99 CFU g^{-1} (i.e., 2 $_{lg}$CFU g^{-1}) was assigned. This is the minimum content of microorganisms that is possible to detect in a sample according to our methods applied for the sample preparation [19].

The pH analysis used 10 g of the cut materials mentioned above, i.e., chips, wood and bark, and the analysis was done separately for each prepared sample (see Table 1) according to DIN 38404-5:2009 [56]. The material was placed in an Erlenmeyer flask containing 100 mL of distilled water, and the mixture (1:10 wt/v) was shaken at 180 rpm for 15 min at room temperature in the orbital shaker mentioned before. Afterwards, the pH was measured with a WTW 3210 pH meter (Xylem Analytics, Weilheim, Germany).

2.4. Elemental and Ash Analyses

Ash and elemental analyses were performed to have a basic information about the material properties. Chips used previously to calculate the MC, at the storage intake and at the different storage times, were used, afterwards, in the elemental and ash analyses (see Table 1). For these analyses, the chips were ground to the maximum size of 1 mm using a Pulverisette 15 cutting mill (Fritsch, Oberstein, Germany).

The elemental analyser Vario EL III (Elementar Analysensysteme GmbH, Langenselbold, Germany) was used to perform the elemental analysis according to VDLUFA:1997 [57]. For each sample (see Table 1), two tin boats of 6 mm × 6 mm × 12 mm (article S22 137 419, Elementar Analysensysteme GmbH, Langenselbold, Germany) were prepared, each one containing 10 mg of the ground chips; therefore, the percentage of C, H, N and S is herein reported as a mean value (n = 2).

The ash content analysis was done with 2.5–5 g of the ground chips according to EN ISO 14775:2009 [58]. One measure was done for each prepared sample of the ground chips at the storage intake and the different storage times (Table 1).

2.5. Statistical Analysis

It was investigated for MC, DML, microbiological parameters and pH value whether the mean values of the measurement series differ significantly over storage time ($p < 0.05$). The mean value comparisons between weeks and trials, as well as between weeks and materials (chips, wood, bark), were carried out using the single values. The interactions week–trial and week–material were considered. For the evaluation, the procedure GLIMMIX of the statistical software SAS (Version 9.4, SAS Institute, Cary, NC, USA) was used in order to carry out a two factorial ANOVA.

3. Results and Discussion

3.1. Particle Size Distribution

Figure 4 shows the particle size distribution of the poplar chips used in each of the storage trials T1, T2 and T3 (Table 1), which complied with the wood chip class P31 according to EN ISO 17225-1:2014 [53]. The analysis of the particle size distribution showed no significant differences for the wood chips used for the trials T1–T3.

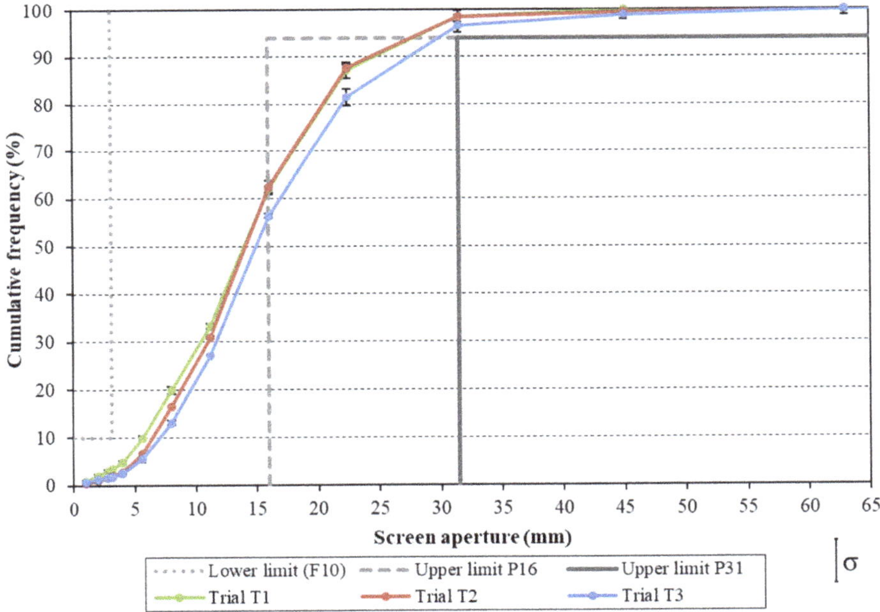

Figure 4. Particle size distribution of wood chips used in the storage trials T1–T3.

3.2. MC and DML of Poplar Chips, and RH Values, in the Storage Trials

Table 2 shows MC and DML values of the three storage trials T1, T2 and T3, as well as the RH measured during the storage trials. As supplementary information, detailed MC and DML obtained in the trials T1–T3 are reported in Appendix A (Tables A1–A3). The sample bags yielding a negative DML were considered as an experimental error and, consequently, excluded from the calculations. This was the case for one sample bag at the storage week 2 in the trial T2, that resulted in −2.1% DML (Table A2; Appendix A); therefore, the MC and DML of this storage week 2 in trial T2 is the mean of two sample bags (n = 2).

Table 2. MC, DML and RH of poplar chips of the storage trials T1–T3. MC and DML expressed as mean (n = 6 and n = 3 at the storage intake and storage times, respectively) ± standard deviation (SD).

	Trial T1			Trial T2			Trial T3		
Storage Time (Weeks)	**MC (%)**	**DML (%)**	**RH (%)**	**MC (%)**	**DML (%)**	**RH (%)**	**MC (%)**	**DML (%)**	**RH (%)**
Storage intake, 0	56.8 ± 0.7 ^{ab}	0.0 ^{A}	92.7%	52.3 ± 0.7 ^{ab}	0.0 ^{a}	88.0%	58.3 ± 0.7 ^{a}	0.0 ^{A}	
1	52.7 ± 1.0	4.2 ± 0.4	95.6%	-	-	95.0%	-	-	
2	47.8 ± 1.7 ^{abc}	6.4 ± 0.9 ^{BC}	93.9%	44.4 ± 0.5 ^{bcd}	0.6 ± 0.5 ^{A}	93.3%	-	-	
3	39.5 ± 1.3	7.5 ± 0.6	93.9%			91.6%	44.3 ± 1.3	11.0 ± 1.3	
4	32.4 ± 3.9 ^{de}	9.8 ± 0.9 ^{CD}	94.2%	37.8 ± 6.2 ^{dc}	5.0 ± 0.8 ^{B}	91.2%	-	-	98.5 to
5	25.1 ± 2.5	8.3 ± 0.8	93.6%	-	-	90.4%	-	-	100%
6	21.3 ± 1.6 ^{efg}	8.8 ± 0.8 ^{CD}	92.7% *	23.5 ± 3.2 ^{ef}	8.1 ± 1.4 ^{BC}	90.3%	31.9 ± 8.6 ^{ed}	11.6 ± 2.1 ^{DE}	
7	-	-	-	-	-	89.0%	-	-	
8	-	-	-	17.0 ± 0.8 ^{fg}	9.4 ± 0.6 ^{CD}	85.6%	-	-	
9	-	-	-	-	-	70.4%	24.5 ± 4.7 ^{ef}	13.7 ± 0.3 ^{E}	
10	-	-	-	9.9 ± 0.3 ^{g}	9.3 ± 1.2 ^{CD}	64.2%	-	-	

* Due to sensor failure, this RH value corresponds to two days before the end of the storage (week 6); MC: moisture content (wet-basis); DML: dry matter loss; RH: relative humidity; significant differences between trials and over storage are marked by different small letters for MC and different capital letters for DML (GLIMMIX, SAS 9.4).

The initial MC_{in} of the poplar chips were 56.8%, 52.3% and 58.3% (Table 2), respectively, for the trials T1, T2 and T3 (Table 1), which are in line with the wet-basis MC of poplar chips at harvest reported by, e.g., Barontini et al. [13], Pecenka et al. [30] and Pari et al. [24]. At the end of the respective six, ten and nine storage weeks, the wood chips showed a significantly reduced MC of 21.3%, 9.9% and 24.5% ($p < 0.05$; trials T1, T2 and T3 in Table 2), which represented a decrease of 5.9%, 4.2% and 3.8% MC per week on average (MC_{PWAV}), respectively. The lowest MC_{PWAV} obtained in the trial T3 (i.e., 3.8% MC_{PWAV}) suggested the impact of assisting the storage trial with a humidifier (Figure 2b).

Focusing on the six initial storage weeks allowed a better comparison of the results. Table 2 shows that chips' MC decreased by 5.9% and 4.8% MC_{PWAV} in the initial six storage weeks (trials T1 and T2 in Table 2). The lowest value, however, was obtained in trial T3, i.e., 4.4% MC_{PWAV} (from storage intake to storage week 6; Table 2), which indicated the impact of using the humidifier on lowering the rate at which chips lose MC. However, no significant differences could be detected for MC among all trials T1–T3 at storage week 6 ($p < 0.05$). Moreover, all these MC_{PWAV} of 4.4% to 5.9% in the initial 6–10 weeks showed a faster reduction in MC compared to practice storage trials of poplar chips in outdoor piles. Lenz et al. [26] measured 1.7% MC_{PWAV} in the initial six storage weeks of fine P31 poplar chips, and 2.2% MC_{PWAV} was reported by Pecenka et al. [17] in the initial nine storage weeks at practice conditions. This shows the slower drying rate of wood chips in outdoor storage piles at practice scale compared to chips investigated in the storage chamber trials T1–T3 at laboratory scale, and it highlights a limitation of the set-up used in the trials T1–T3 (Table 1).

After six storage weeks, the chips' wet-based MC in trials T1 and T2 were 21.3% and 23.5%, respectively (Table 2). These values correspond to 27.1% and 30.7% dry-basis MC (MC_{DB}), which are close to the fibre saturation point (FSP) of poplar wood, i.e., 32–35% MC_{DB} [59], and in agreement to the range of FSP reported in literature for wood, i.e., 21–35% MC_{DB} [59–61]. FSP in wood is the MC point at which the wood cell lumen is empty of water while the wood cell wall is water saturated [59,60], both named respectively 'free water' and 'bound water' [59,60]. The total value of FSP varies depending on wood species [59,61]. Moreover, FSP values reported in literature refer to 'only wood' specimens, whereas the MC values reported in Table 2 refer to the 'whole chip', i.e., containing wood and bark. However, the chips' MC in trials T1 and T2 after six weeks of storage, were in the range of the FSP data reported in literature. This might indicate that poplar chips in the trials T1 and T2 had very little free water in the cell lumens at this storage time.

On the other hand, at the storage week 6, chips in trial T3 showed a MC of 31.9% (Table 2), which equals to 46.8% MC_{DB}. This value is clearly higher than the FSP of 32–35% MC_{DB} for poplar wood [59], and it might indicate that chips in trial T3, at the storage week 6, still contained free water in the cell lumen. As long as the chips' MC is above the FSP, it can be assumed that a reduced availability of water is not a limiting factor for microbiological degradation processes in the chips [19,41]. Furthermore, these results suggest the suitability of the set-up assisted with the humidifier, in order to resemble the conditions that wood chips undergo during the initial storage weeks in outdoor piles.

DML of Poplar Chips

Focusing on DML, the three trials T1–T3, despite having some similitudes, did not follow a similar progression. For example, the DML of both trials T1 and T2 (Table 2) stabilized between 8.1% and 9.8%, regardless of the storage time. The stabilisation of DML in both trials T1 and T2 occurred approximately at the same time as the MC decreased to values near the FSP (Table 2). On the other hand, chips in the trial T3 showed 11.0% DML (Table 2) already at the storage week 3, which was higher than the DML measured in the storage trials T1 and T2 (see Table 2), and increased up to 13.7% at the end of the nine-week storage (Trial T3; Table 2). Results presented in Table 2 suggest that the high RH provided by the humidifier used in the trial T3 (Figure 2b) might contribute to higher DML in the wood chips stored in the trial T3 when compared to the trials T1 and T2. The statistical analysis of the results has shown that DML increased for all trials with storage time. While significant differences were found for the DML between trial T2 and T3 at the storage week 6, these were not visible comparing trial T1 and T3, as well as T1 and T2 (Table 2). However, the number of repetitions ($n = 3$) in every trial was limited due to the limited volume of the storage chamber, and a higher number of samples would have provided statistically more robust results.

The DML results obtained in the trial T3 (Table 2) were closer to the DML reported for outdoor pile storages, e.g., approximately 10% for the initial storage phase of 9–12 weeks in 500 m^3 storage piles of P31 poplar chips [17,30]. This would suggest that the use of the humidifier in the storage set-up of trial T3 (Figure 2b) is more suitable for storage trials at a laboratory scale.

3.3. Microbiological and pH Analysis

The results of the microbiological analyses done in the storage trials T1 and T2 (Table 1) are shown in Figure 5a–d, corresponding to mesophilic fungi, xerophilic fungi, mesophilic bacteria and actinobacteria, respectively. Detailed numerical values are presented in Appendix B (Tables A4 and A5).

Our isolation/cultivation experiments revealed that xerophilic and mesophilic fungi, and mesophilic bacteria and actinobacteria, were present in the fresh poplar chips at the storage intake, for both trials T1 and T2 (Figure 5a–d). The statistical analysis showed no significant differences between the trials ($p < 0.05$). However, the number of mesophilic bacteria and actinobacteria (Figure 5c,d, respectively) present in the chips at the storage intake were higher than the number of mesophilic fungi and xerophilic fungi (Figure 5a,b, respectively). Whilst the number of mesophilic fungi and xerophilic fungi were in the range of 3–5 $_{lg}$CFU g^{-1}, mesophilic bacteria and actinobacteria showed numbers between 4 $_{lg}$CFU g^{-1} and 7 $_{lg}$CFU g^{-1} (Figure 5).

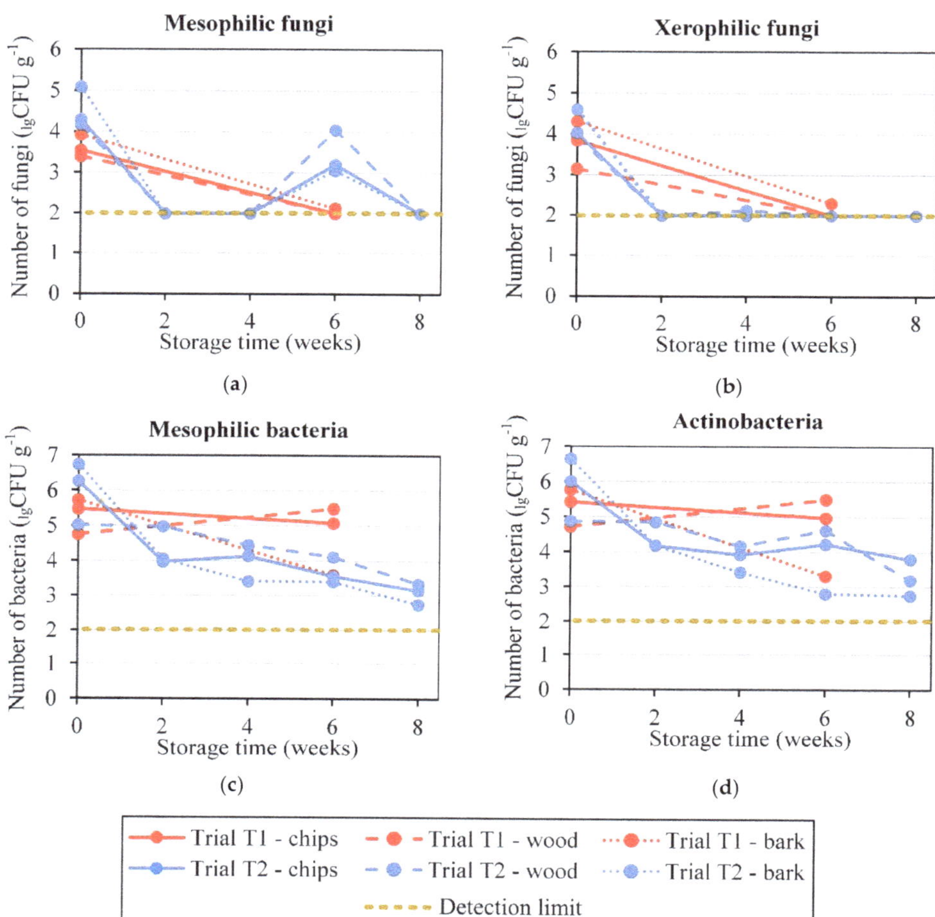

Figure 5. Results of the microbiological analyses of poplar chips, wood and bark of trials T1 and T2.

As a function of progressing wood decomposition during storage, bacteria and fungi trends became evident (Figure 5). Mesophilic bacteria and actinobacteria, despite showing a decrease over the course of storage, remained detectable until the end of storage, in both trials T1 and T2 (Figure 5c,d, respectively). This decrease over storage time proved to be significant for trial T1, but was not significant if all results from T1 and T2 were analysed together. On the other hand, xerophilic and mesophilic fungi (Figure 5a,b, respectively) dropped to the detection limit (i.e., 2 $_{lg}$CFU g^{-1}) during storage. One exception was at the storage week 6 in trial T2, where chips, wood and bark samples showed a presence of mesophilic fungi (Figure 5a) between 3 $_{lg}$CFU g^{-1} and 4 $_{lg}$CFU g^{-1}. This trend proved to be significant for both trials and all investigated fungi, if the outlier for mesophilic fungi in trial T2 at storage week 6 was excluded from the statistical analysis. However, this peak could be attributed to spores present in the sample at the sampling time, that started to grow during the lab analyses when the optimal conditions were provided. Furthermore, in line with our findings, the results of Idler et al. [19] for trials with P45 poplar chips stored outdoors demonstrated the presence of mesophilic fungi of around 3 $_{lg}$CFU g^{-1} during the initial 50 storage days at pile temperatures of 45–55 °C. This might explain the peak of mesophilic fungi observed at the storage week 6 in trial T2 (Figure 5a).

The decrease in the observed number of bacteria and fungi is a consequence of the test temperature being 55 °C. This temperature is too elevated for mesophilic microorganisms growing at temperatures between 20 °C and 50 °C and it is selective for thermotolerant and thermophilic microorganisms. The thermophilic phase is characterized by temperatures up to 50 °C or even 75 °C [62]. In fact, the temperature is a major abiotic factor selective for the majority of environmental microorganisms. Zöhrer et al. [48] investigated the fungal and bacterial community of poplar wood chips stored in 2019 in Austria by high throughput next-generation sequencing over a storage period of 120 days. Their research revealed a dynamic shift of the bacterial and fungal community over time being mainly a result of the storage temperature, pH and nutrient availability. After storage intake, and long-term storage relative abundancy of the predominant species decreased. However, it must be taken into account that molecular analyses based on the extraction of microbial DNA are culture independent. Furthermore, these techniques reveal all microorganisms, independent of their physiological status, not discriminating between living, dormant (e.g., spores) and dead microbial cells. In contrast, culture dependent methods, such as agar plate cultivation techniques which have been used here, are restricted to the cultivable fraction of the microbiota, embracing only a small fraction (1–10%) of the overall (soil) microbiota [63,64].

Comparing the results of the microorganisms isolated from the wood and bark fractions (Figure 5) at the storage intake, the presence of mesophilic fungi, xerophilic fungi, mesophilic bacteria and actinobacteria was generally significantly higher in bark than in wood. However, at the end of the storage trials, the wood fraction showed a higher presence of mesophilic bacteria and actinobacteria when compared to bark (Figure 5c,d, respectively), for both trials T1 and T2. On the other hand, mesophilic and xerophilic fungi (Figure 5a,b, respectively) decreased to the detection limit. However, at the end of the storage trial T1 (i.e., six weeks), xerophilic fungi showed a presence of 2.3 $_{lg}$CFU g^{-1} in the bark fraction (Table A4; Appendix B), while they decreased to the detection limit in the respective wood fraction (Figure 5b). The varying presence of microorganisms in wood and bark fractions can be attributed to differences in environmental conditions of the different wood fractions being already apparent at storage intake. In particular, the tree characteristics such as wood density, wood composition (regarding the different carbohydrates and lignin), as well as pH, shape the microbial communities and in turn determine the decomposition rates of different wood fractions [65–67]. Due to the complex structure of bark with a higher content of sugars and nutrients compared to heart- and sapwood, it hosts a diverse microbiological community with an especially high fungal species richness and higher abundancies [68]. However, in order to further interpret our complex and changing storage-condition dependent findings revealed by cultivation techniques, molecular analysis is needed.

pH Results

The results of the pH analysis showed that, at the storage intake, the pH of the chips in trials T1 and T2 were similar (i.e., 5.9 and 6.0, respectively; Figure 6). Moreover, wood fractions in both trials T1 and T2 showed slightly higher pH than bark fractions (6.0–6.2 and 5.7–5.8, respectively; Figure 6). However, the statistical analysis has demonstrated that differences between the fractions were not significant at storage intake (Table A6, Appendix B). Whereas, for trial T1, the pH values were measured at the storage intake and at the end of the trial only, the results presented for the trial T2 (Figure 6) allowed a more detailed analysis, due to the two-week sampling frequency used in trial T2 (Table 1).

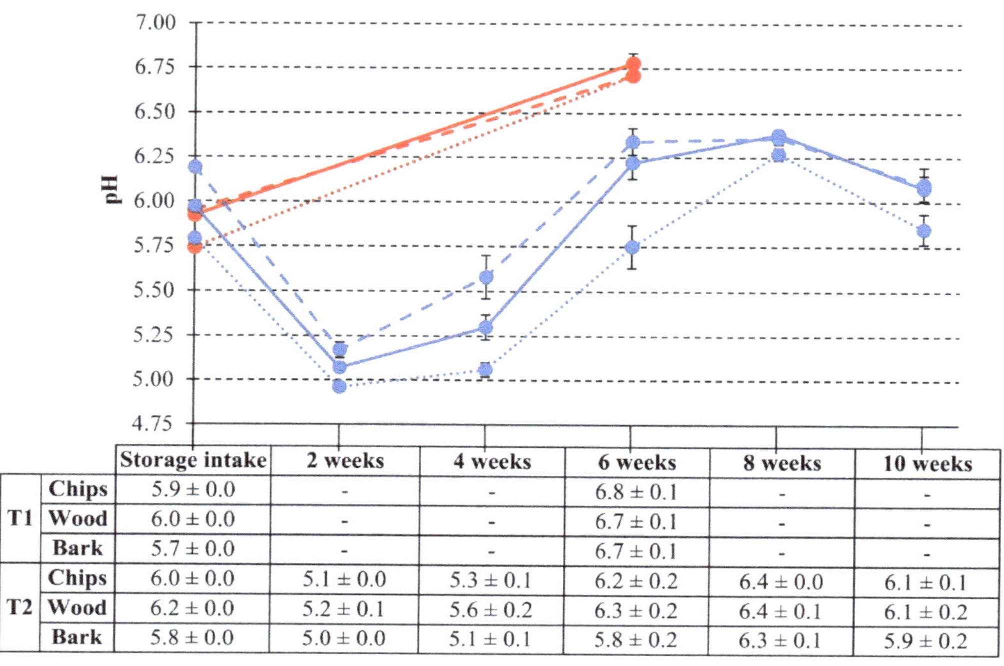

		Storage intake	2 weeks	4 weeks	6 weeks	8 weeks	10 weeks
T1	Chips	5.9 ± 0.0	-	-	6.8 ± 0.1	-	-
	Wood	6.0 ± 0.0	-	-	6.7 ± 0.1	-	-
	Bark	5.7 ± 0.0	-	-	6.7 ± 0.1	-	-
T2	Chips	6.0 ± 0.0	5.1 ± 0.0	5.3 ± 0.1	6.2 ± 0.2	6.4 ± 0.0	6.1 ± 0.1
	Wood	6.2 ± 0.0	5.2 ± 0.1	5.6 ± 0.2	6.3 ± 0.2	6.4 ± 0.1	6.1 ± 0.2
	Bark	5.8 ± 0.0	5.0 ± 0.0	5.1 ± 0.1	5.8 ± 0.2	6.3 ± 0.1	5.9 ± 0.2

Figure 6. The pH data of wood chips as a function of storage trials (T1 and T2) and time (0–10 weeks); presented as mean (n = 2 and n = 3 for the storage intake and storage times, respectively) ± standard deviation.

The pH values of the wood chips in the trial T2 increased between storage intake and storage week 8 (from 6.0 to 6.4; Figure 6), similarly to trial T1, where a pH of 6.8 was reached after six weeks of storage. However, the two-week sampling frequency used in the trial T2 allowed the observation that, in the initial two storage weeks, the pH decreased significantly to 5.0–5.2 for all the analysed samples, i.e., chips, wood and bark (Figure 6 and Table A6, Appendix B). This can be attributed to acidification by microorganism, especially brown-rot fungi, producing organic acids [69]. In their work, Humar et al. [69] compared pH values of wood with the rate of decay by different fungi, showing a significant pH drop before mass loss was exhibited. The pH reduction was caused by the production of oxalic acid by brown-rot fungi, playing an important role in the non-enzymatic stage of decay. After the storage week 2, the pH of chips, wood and bark in trial T2 increased up to values of 6.3–6.4 (Figure 6); however, although wood and bark followed a similar progression, they slightly differed in pH. The decrease of pH in the initial two storage weeks in trial T2 resembled the results reported in Pecenka et al. [17]. This study observed a similar decrease of the pH from 6.37 to 5.06 in the initial five storage weeks of P31 poplar chips stored in outdoor piles, followed by an increase to 7.48 in the following two months. Additionally, Humar et al. [69] showed that some wood-degrading fungal species, such as *Schizophyllum commune*, caused an increase in wood pH, since this fungi prefer pH values in the range of pH 5 and 7 [69]. Based on the fact that pH is among the most prominent factors shaping microbial communities, the observed changes in pH during the storage could have induced the observed shifts in the microbial community during the storage trial (Figure 6). During the initial stage with high dry matter losses, brown-rot fungi in particular produce an acidic environment for degrading wood cellulose; whereas, in later stages of lower degradation rates, fungi preferring higher pH values take over.

3.4. Ash and Elemental Analyses

Table 3 shows the results of the ash content and elemental analyses of the poplar chips of storage trials T1–T3, and the complete dataset of the elemental and ash analyses is shown in Appendix C (Table A7).

Table 3. Results of the elemental and ash analyses of poplar chips. (Mean ± standard deviation).

Trial	Storage Time (Weeks)	n	Ash (%) *	C (%) *	H (%) *	N (%) *	S (%) *	C:N Ratio **
T1	Storage intake, 0	2	2.34 ± 0.05	52.07 ± 0.03	4.79 ± 0.01	0.305 ± 0.001	0.119 ± 0.004	171:1 ± 0
T2	Storage intake, 0	2	2.22 ± 0.05	52.32 ± 0.18	4.85 ± 0.05	0.294 ± 0.043	0.113 ± 0.000	182:1 ± 27:1
T3	Storage intake, 0	2	3.12 ± 0.16	51.36 ± 0.08	4.65 ± 0.04	0.274 ± 0.008	0.107 ± 0.004	187:1 ± 5:1
	3	3	2.71 ± 0.14	51.91 ± 0.06	4.76 ± 0.06	0.328 ± 0.047	0.111 ± 0.006	161:1 ± 21:1
	6	3	3.19 ± 0.21	51.68 ± 0.15	4.65 ± 0.03	0.272 ± 0.035	0.105 ± 0.007	194:1 ± 28:1
	9	3	2.97 ± 0.47	51.50 ± 0.12	4.71 ± 0.04	0.326 ± 0.022	0.107 ± 0.004	159:1 ± 12:1

* Expressed as percentage of the dry matter fraction; ** C:N ratios calculated as the mean value of the C:N ratios of the single measures.

The ash contents at the storage intake of the trials T1–T3 (Table 3) are within the range of literature data for poplar chips, i.e., 2–3.02% [13,17,24,31]. Moreover, the analysis at different storage times in the storage trial T3 showed that the progression of the ash content oscillated between 2.71% and 3.19% (Table 3). These results differ from the trend reported in literature for chips stored in piles outdoors, that showed that ash content increased with increasing storage time [13,24,31]. For example, Pari et al. [24] investigated storage of poplar chips produced from the stem showing an initial ash content of 3.02%, whereas, after 18 months of storage, this value increased to 3.36%. Similarly, Lenz et al. [34] analysed the ash content of P31 poplar chips stored outdoors for nine months and found that the ash content increased from 2.07% at the storage intake to 2.81% at the end of the storage. However, our reported ash contents (Table 3) are still in the typical range for poplar wood from SRC.

Results from the elemental analysis of the trials T1–T3 (Table 3) showed that C:N ratios at the storage intake were between 171:1 and 187:1, being in the typical range for C:N ratios reported in literature for poplar chips, which range from 113:1 to 224:1 [17,24,26,70]. Moreover, C:N ratios at different storage times in trial T3 (Table 3) did not show a clear trend and oscillated in the same range of the above reported literature.

3.5. Limitations of the Storage Chamber Trials

The experiments have shown that the storage conditions which occur inside of a large wood chips pile under practice conditions after harvesting in winter in central Europe can be reproduced well with the investigated storage chamber set-up, particularly when using the humidifier (Figure 3). As shown, DML, pH and the development of microorganisms are within the ranges that are known from other outdoor storage experiments with poplar wood chips under comparable temperature–humidity conditions [13,17–19,26]. However, as mentioned before, the MC in the chamber set-ups decreased slightly faster than in outdoor storage trials. In addition, not all influencing variables that are effective in practice could be recreated accordingly. These are, in particular, the influence of changing weather conditions, such as heat transfer due to solar radiation, rewetting from rain, changing air flow conditions due to pile compaction as a result of large pile sizes, or changing gas concentrations in the pile (e.g., CO_2 and O_2) due to microbiological–chemical degradation processes [22,26]. However, the influence of changing weather conditions in particular is limited to the outer pile layers and has only little influence on the processes inside large wood chip pile which have been the focus of the experiments presented here [18,22,34]. For reliable statistical evaluation, a larger volume of storage chamber should be used, in order to be able to increase the number of samples per each sampling date. Furthermore, even if no negative effects of opening the door during sampling could be determined, additional measures to reduce the air exchange during sampling and to monitor the testing atmosphere inside the chamber appear to be advisable as well.

4. Conclusions

The three storage trials presented in this work were conducted to investigate DML occurring in poplar chips in the early (high temperature) storage phase, using a laboratory set-up in a chamber under controlled temperature. This experimental set-up enabled regular sampling at short intervals during wood chip storage, and thus contributed to a better understanding of processes occurring during storage of poplar chips. The use of an incubation chamber seemed a good starting method for this purpose. In addition, the assistance of a humidifier proved to be a good system to keep a high RH, i.e., resembling the conditions that wood chips in outdoor storage piles undergo during the early storage phase (phase 1). Results showed that DML of 8.8–13.7% occurred in the poplar chips stored in the chamber for 6 to 10 weeks, which is a substantial portion of the total DML reported for outdoor storages of wood chips, that may account for up to 47% DML in storage piles set for 6–12 months. Microbiological results showed good correlation with results from a practice scale for the aspects considered in this work. Thus, overall, the storage chamber set-up has the potential to gain new and much more detailed information about the microbiological and chemical processes during storage. However, further research coupling isolation-cultivation techniques and cultivation independent molecular techniques (high throughput next generation sequencing) is needed, in order to unravel the complex and dynamic microbial processes (microbial succession) during storage. In conclusion, this work highlighted the possibilities of setting up such laboratory scale storage trials, in order to investigate the DML occurring in wood chips when drying in outdoor storage piles. Compared to large practice scale experiments, the proposed setting enables a better monitoring of the overall system under controlled conditions, and a more accurate sampling at shorter time intervals with a significantly reduced workload. The potential of the proposed laboratory scale set-up could be further improved by implementing additional storage parameters into the monitoring. For example, monitoring and control of CO_2 and O_2 concentrations would be suitable in order to simulate storage conditions even more accurately.

Author Contributions: Conceptualization, A.H.-E., R.P., S.D., H.L., C.I. and T.H.; methodology, investigation and validation, A.H.-E., R.P., S.D., J.A.-J., H.L., C.I. and T.H.; writing—original draft preparation, A.H.-E. and R.P.; writing—review and editing, A.H.-E., R.P., S.D., J.A.-J., H.L., C.I. and T.H.; supervision, R.P. and C.I.; funding acquisition, S.D. All authors have read and agreed to the published version of the manuscript.

Funding: This research was funded by the Austrian Research Promotion Agency (FFG), Energieforschung program, project number 858837. The publication of this article was funded by the Open Access Fund of the Leibniz Association, Germany.

Institutional Review Board Statement: Not applicable.

Acknowledgments: Teodor Teodorov, Peter Kaulfuß and Helmuth Carl (ATB) are thanked for the technical assistance in performing the storage trials. Katrin Busse (ATB) is thanked for performing the microbiological analyses. Miriam Felgentreu, Mandy Jäkel and Giovanna Rehde (ATB) are thanked for conducting the ash and elemental analyses.

Conflicts of Interest: The authors declare no conflict of interest. The funders had no role in the design of the study; in the collection, analyses, or interpretation of data; in the writing of the manuscript, or in the decision to publish the results.

Appendix A

Table A1. Moisture content and dry matter losses of the sample bags in trial T1.

MC_{in} = 56.8% ± 0.7%			
Storage Time (Weeks)	**Sample Bag**	**MC_{out} (%)**	**DML_i (%)**
1	1	53.8	4.1
	2	51.3	3.9
	3	53.1	4.7
	Mean ± SD	52.7 ± 1.0	4.2 ± 0.4
2	1	50.1	7.4
	2	46.6	5.3
	3	46.6	6.6
	Mean ± SD	47.8 ± 1.7	6.4 ± 0.9
3	1	40.6	8.0
	2	40.2	7.6
	3	37.7	6.7
	Mean ± SD	39.5 ± 1.3	7.5 ± 0.6
4	1	35.7	10.7
	2	34.7	10.0
	3	27.0	8.6
	Mean ± SD	32.4 ± 3.9	9.8 ± 0.9
5	1	21.6	7.8
	2	27.4	9.4
	3	26.4	7.7
	Mean ± SD	25.1 ± 2.5	8.3 ± 0.8
6	1	21.8	8.5
	2	19.1	8.1
	3	22.9	9.9
	Mean ± SD	21.3 ± 1.6	8.8 ± 0.8

MC_{in}: moisture content (wet-basis) of poplar wood chips at the storage intake; MC_{out}: moisture content (wet-basis) of poplar wood chips at the sampling time; DML: dry matter loss; SD: standard deviation.

Table A2. Moisture content and dry matter losses of the sample bags in trial T2.

MC_{in} = 52.3% ± 0.7%			
Storage Time (Weeks)	**Sample Bag**	**MC_{out} (%)**	**DML_i (%)**
2	1	45.0	1.1
	2	43.9	0.1
	3	39.9 *	−2.1 *
	Mean ± SD	44.4 ± 0.5	0.6 ± 0.5
4	1	43.1	3.8
	2	41.3	5.8
	3	29.1	5.2
	Mean ± SD	37.8 ± 6.2	5.0 ± 0.8
6	1	25.0	10.0
	2	26.4	7.6
	3	19.1	6.7
	Mean ± SD	23.5 ± 3.2	8.1 ± 1.4

Table A2. *Cont.*

MC_{in} = 52.3% ± 0.7%			
Storage Time (Weeks)	**Sample Bag**	MC_{out} (%)	DML_i (%)
8	1	17.6	9.2
	2	15.9	8.9
	3	17.6	10.2
	Mean ± SD	17.0 ± 0.8	9.4 ± 0.6
10	1	10.3	9.7
	2	9.7	7.7
	3	9.8	10.6
	Mean ± SD	9.9 ± 0.3	9.3 ± 1.2

* Experimental error; results excluded from the analysis; MC_{in}: moisture content (wet-basis) of poplar wood chips at the storage intake; MC_{out}: moisture content (wet-basis) of poplar wood chips at the sampling time; DML: dry matter loss; SD: standard deviation.

Table A3. Moisture content and dry matter losses of the sample bags in trial T3.

MC_{in} = 58.3% ± 0.7%			
Storage Time (Weeks)	**Sample Bag**	MC_{out} (%)	DML_i (%)
3	1	44.5	12.0
	2	45.8	11.8
	3	42.5	9.3
	Mean ± SD	44.3 ± 1.3	11.0 ± 1.3
6	1	28.0	13.6
	2	43.9	12.4
	3	23.9	8.7
	Mean ± SD	31.9 ± 8.6	11.6 ± 2.1
9	1	21.1	13.9
	2	31.1	13.2
	3	21.1	13.9
	Mean ± SD	24.5 ± 4.7	13.7 ± 0.3

MC_{in}: moisture content (wet-basis) of poplar wood chips at the storage intake; MC_{out}: moisture content (wet-basis) of poplar wood chips at the sampling time; DML: dry matter loss; SD: standard deviation.

Appendix B

Table A4. Results of the microbiological analysis of chips, wood and bark in trial T1; presented as mean ± standard deviation, both in colony forming units per gram of sample (CFU g^{-1}). In brackets, the mean is expressed as logarithmic to base 10 ($_{lg}$CFU g^{-1}). Different small letters indicate significant differences between storage intake and end of storage ($p < 0.05$). Different capital letters indicate significant differences between wood fractions at storage intake respectively at week 6 ($p < 0.05$).

Storage Time (Weeks)	Sample	Mesophilic Fungi (CFU g^{-1} ± CFU g^{-1}) [$_{lg}$CFU g^{-1}]	Xerophilic Fungi (CFU g^{-1} ± CFU g^{-1}) [$_{lg}$CFU g^{-1}]	Mesophilic Bacteria (CFU g^{-1} ± CFU g^{-1}) [$_{lg}$CFU g^{-1}]	Actinobacteria (CFU g^{-1} ± CFU g^{-1}) [$_{lg}$CFU g^{-1}]
Storage intake, 0	Chips	$3.5 \times 10^3 \pm 5.0 \times 10^2$ [3.5] $^{a\,A}$	$7.0 \times 10^3 \pm 2.0 \times 10^3$ [3.8] $^{a\,A}$	$3.0 \times 10^5 \pm 9.0 \times 10^4$ [5.5] $^{a\,A}$	$2.7 \times 10^5 \pm 7.5 \times 10^4$ [5.4] $^{a\,A}$
	Wood	$2.5 \times 10^3 \pm 5.0 \times 10^2$ [3.4] $^{a\,A}$	$1.4 \times 10^3 \pm 4.0 \times 10^2$ [3.1] $^{a\,A}$	$5.7 \times 10^4 \pm 1.1 \times 10^4$ [4.8] $^{a\,A}$	$5.3 \times 10^4 \pm 1.2 \times 10^4$ [4.7] $^{a\,A}$
	Bark	$8.5 \times 10^3 \pm 4.5 \times 10^3$ [3.9] $^{a\,A}$	$2.1 \times 10^4 \pm 4.5 \times 10^3$ [4.3] $^{a\,A}$	$5.4 \times 10^5 \pm 1.5 \times 10^5$ [5.7] $^{a\,A}$	$5.9 \times 10^5 \pm 1.7 \times 10^5$ [5.8] $^{a\,A}$

Table A4. *Cont.*

Storage Time (Weeks)	Sample	Mesophilic Fungi (CFU g^{-1} ± CFU g^{-1}) [$_{lg}$CFU g^{-1}]	Xerophilic Fungi (CFU g^{-1} ± CFU g^{-1}) [$_{lg}$CFU g^{-1}]	Mesophilic Bacteria (CFU g^{-1} ± CFU g^{-1}) [$_{lg}$CFU g^{-1}]	Actinobacteria (CFU g^{-1} ± CFU g^{-1}) [$_{lg}$CFU g^{-1}]
6	Chips	ND $^{b\,A}$	ND $^{a\,A}$	$1.2 \times 10^5 \pm 3.2 \times 10^4$ [5.1] $^{a\,A}$	$9.5 \times 10^4 \pm 6.3 \times 10^4$ [5.0] $^{a\,AB}$
	Wood	ND $^{b\,A}$	ND $^{a\,A}$	$3.1 \times 10^5 \pm 8.1 \times 10^4$ [5.5] $^{a\,B}$	$3.2 \times 10^5 \pm 1.1 \times 10^5$ [5.5] $^{a\,B}$
	Bark	$1.3 \times 10^2 \pm 4.8 \times 10^1$ [2.1] $^{a\,A}$	$2.0 \times 10^2 \pm 8.2 \times 10^1$ [2.3] $^{a\,A}$	$3.9 \times 10^3 \pm 3.3 \times 10^2$ [3.6] $^{a\,C}$	$2.1 \times 10^3 \pm 2.4 \times 10^2$ [3.3] $^{a\,AC}$

ND: Not detected; detection limit 2.0 $_{lg}$CFU g^{-1}.

Table A5. Results of the microbiological analysis of the chips, wood and bark in trial T2; presented as mean ± standard deviation, both in colony forming units per gram of sample (CFU g^{-1}). In brackets, the mean is expressed as logarithmic to base 10 ($_{lg}$CFU g^{-1}). Different small letters indicate significant differences between storage intake and end of storage ($p < 0.05$). Different capital letters indicate significant differences between wood fractions at storage intake respectively at week 6 or 8 ($p < 0.05$).

Storage Time (Weeks)	Sample	Mesophilic Fungi (CFU g^{-1} ± CFU g^{-1}) [$_{lg}$CFU g^{-1}]	Xerophilic Fungi (CFU g^{-1} ± CFU g^{-1}) [$_{lg}$CFU g^{-1}]	Mesophilic Bacteria (CFU g^{-1} ± CFU g^{-1}) [$_{lg}$CFU g^{-1}]	Actinobacteria (CFU g^{-1} ± CFU g^{-1}) [$_{lg}$CFU g^{-1}]
Storage intake, 0	Chips	$2.0 \times 10^4 \pm 1.0 \times 10^4$ [4.3] $^{a\,A}$	$1.1 \times 10^4 \pm 2.3 \times 10^3$ [4.0] $^{a\,A}$	$1.9 \times 10^6 \pm 1.2 \times 10^6$ [6.3] $^{a\,A}$	$1.0 \times 10^6 \pm 1.8 \times 10^5$ [6.0] $^{a\,A}$
	Wood	$1.5 \times 10^4 \pm 5.0 \times 10^3$ [4.2] $^{a\,A}$	$8.5 \times 10^3 \pm 5.0 \times 10^2$ [3.9] $^{a\,A}$	$1.1 \times 10^5 \pm 2.5 \times 10^4$ [5.0] $^{a\,A}$	$7.5 \times 10^4 \pm 2.5 \times 10^4$ [4.9] $^{a\,AB}$
	Bark	$1.3 \times 10^5 \pm 7.5 \times 10^4$ [5.1] $^{a\,A}$	$4.0 \times 10^4 \pm 1.0 \times 10^4$ [4.6] $^{a\,A}$	$5.6 \times 10^6 \pm 1.1 \times 10^6$ [6.7] $^{a\,A}$	$4.5 \times 10^6 \pm 5.5 \times 10^5$ [6.6] $^{a\,C}$
2	Chips	ND	ND	$9.5 \times 10^3 \pm 4.1 \times 10^3$ [4.0]	$1.5 \times 10^4 \pm 5.0 \times 10^3$ [4.2]
	Wood	ND	ND	$9.6 \times 10^4 \pm 4.9 \times 10^4$ [5.0]	$7.2 \times 10^4 \pm 5.0 \times 10^4$ [4.9]
	Bark	ND	ND	$1.1 \times 10^4 \pm 5.8 \times 10^3$ [4.1]	$1.5 \times 10^4 \pm 5.9 \times 10^3$ [4.2]
4	Chips	ND	ND	$1.4 \times 10^4 \pm 9.6 \times 10^3$ [4.1]	$8.4 \times 10^3 \pm 5.5 \times 10^3$ [3.9]
	Wood	ND	$1.3 \times 10^2 \pm 4.8 \times 10^1$ [2.1]	$2.8 \times 10^4 \pm 3.5 \times 10^4$ [4.5]	$1.5 \times 10^4 \pm 1.0 \times 10^4$ [4.2]
	Bark	ND	ND	$2.6 \times 10^3 \pm 2.3 \times 10^3$ [3.4]	$2.6 \times 10^3 \pm 1.0 \times 10^3$ [3.4]
6	Chips	$1.6 \times 10^3 \pm 5.1 \times 10^2$ [3.2] $^{a\,A}$	ND $^{a\,A}$	$3.6 \times 10^3 \pm 3.1 \times 10^3$ [3.6] $^{a\,A}$	$1.7 \times 10^4 \pm 1.3 \times 10^4$ [4.2] $^{a\,A}$
	Wood	$1.1 \times 10^4 \pm 1.5 \times 10^4$ [4.1] $^{a\,A}$	ND $^{b\,A}$	$1.3 \times 10^4 \pm 1.5 \times 10^4$ [4.1] $^{a\,A}$	$4.2 \times 10^4 \pm 5.5 \times 10^4$ [4.6] $^{a\,A}$
	Bark	$1.1 \times 10^3 \pm 1.1 \times 10^3$ [3.1] $^{a\,A}$	ND $^{a\,A}$	$2.5 \times 10^3 \pm 3.2 \times 10^3$ [3.4] $^{a\,A}$	$6.3 \times 10^2 \pm 2.6 \times 10^2$ [2.8] $^{a\,A}$
8	Chips	ND $^{a\,A}$	ND $^{a\,A}$	$1.4 \times 10^3 \pm 9.4 \times 10^2$ [3.2] $^{a\,A}$	$6.3 \times 10^3 \pm 6.3 \times 10^3$ [3.8] $^{a\,A}$
	Wood	ND $^{a\,A}$	ND $^{b\,A}$	$2.2 \times 10^3 \pm 2.8 \times 10^3$ [3.3] $^{a\,A}$	$1.6 \times 10^3 \pm 1.5 \times 10^3$ [3.2] $^{a\,A}$
	Bark	ND $^{a\,A}$	ND $^{a\,A}$	$5.7 \times 10^2 \pm 4.7 \times 10^1$ [2.8] $^{a\,A}$	$5.7 \times 10^2 \pm 1.2 \times 10^2$ [2.8] $^{a\,A}$

ND: Not detected; detection limit 2.0 $_{lg}$CFU g^{-1}.

Table A6. Results of the pH analysis of the chips, wood and bark in trial T1 and T2; presented as mean ± standard deviation. Different letters indicate significant differences within one trial over storage time ($p < 0.05$).

Storage Time (Weeks)	Sample	Trial T1	Trial T2
Storage intake, 0	Chips	5.9 ± 0.0 [a]	6.0 ± 0.0 [abcde]
	Wood	6.0 ± 0.0 [a]	6.2 ± 0.0 [abcd]
	Bark	5.7 ± 0.0 [a]	5.8 ± 0.0 [cdef]
2	Chips	-	5.1 ± 0.0 [gh]
	Wood	-	5.2 ± 0.2 [gh]
	Bark	-	5.0 ± 0.0 [h]
4	Chips	-	5.3 ± 0.1 [gfh]
	Wood	-	5.6 ± 0.2 [egf]
	Bark	-	5.1 ± 0.1 [h]
6	Chips	6.8 ± 0.1 [b]	6.2 ± 0.2 [abcd]
	Wood	6.7 ± 0.1 [b]	6.3 ± 0.2 [abc]
	Bark	6.7 ± 0.1 [b]	5.8 ± 0.2 [def]
8	Chips	-	6.4 ± 0.0 [a]
	Wood	-	6.4 ± 0.1 [ab]
	Bark	-	6.3 ± 0.1 [abc]
10	Chips	-	6.1 ± 0.1 [abcde]
	Wood	-	6.1 ± 0.2 [abcd]
	Bark	-	5.9 ± 0.2 [bcde]

Appendix C

Table A7. Results of the elemental analysis and ash analysis.

Trial	Storage Time (Weeks)	Sample	Ash (%) *	C (%) *	H (%) *	N (%) *	S (%) *	C:N Ratio **
T1	Storage intake, 0	1	2.39	52.04	4.80	0.306	0.123	170
		2	2.29	52.10	4.77	0.304	0.116	171
		Mean ± SD	2.34 ± 0.05	52.07 ± 0.03	4.79 ± 0.01	0.305 ± 0.001	0.119 ± 0.004	171:1 ± 0
T2	Storage intake, 0	1	2.27	52.14	4.89	0.337	0.113	155
		2	2.16	52.50	4.80	0.251	0.113	210
		Mean ± SD	2.22 ± 0.05	52.32 ± 0.18	4.85 ± 0.05	0.294 ± 0.043	0.113 ± 0.000	182:1 ± 27:1
T3	Storage intake, 0	1	3.28	51.44	4.62	0.282	0.111	183
		2	2.96	51.28	4.69	0.267	0.102	192
		Mean ± SD	3.12 ± 0.16	51.36 ± 0.08	4.65 ± 0.04	0.274 ± 0.008	0.107 ± 0.004	187:1 ± 5:1
	3	1	2.89	51.85	4.68	0.395	0.119	131
		2	2.55	51.88	4.78	0.289	0.110	180
		3	2.70	52.00	4.82	0.302	0.104	172
		Mean ± SD	2.71 ± 0.14	51.91 ± 0.06	4.76 ± 0.06	0.328 ± 0.047	0.111 ± 0.006	161:1 ± 21:1
	6	1	2.97	51.86	4.69	0.223	0.095	233
		2	3.46	51.68	4.61	0.297	0.111	174
		3	3.12	51.50	4.65	0.296	0.111	174
		Mean ± SD	3.19 ± 0.21	51.68 ± 0.15	4.65 ± 0.03	0.272 ± 0.035	0.105 ± 0.007	194:1 ± 28:1
	9	1	2.52	51.46	4.70	0.339	0.106	152
		2	3.61	51.39	4.66	0.344	0.111	149
		3	2.76	51.66	4.75	0.294	0.103	176
		Mean ± SD	2.97 ± 0.47	51.50 ± 0.12	4.71 ± 0.04	0.326 ± 0.022	0.107 ± 0.004	159:1 ± 12:1

* Expressed as percentage of the dry matter fraction; ** C:N ratios calculated as the mean value of the C:N ratios of the single measures; SD: standard deviation.

References

1. Kriegler, E.; Weyant, J.P.; Blanford, G.J.; Krey, V.; Clarke, L.; Edmonds, J.; Fawcett, A.; Luderer, G.; Riahi, K.; Richels, R.; et al. The role of technology for achieving climate policy objectives: Overview of the EMF 27 study on global technology and climate policy strategies. *Clim. Chang.* **2014**, *123*, 353–367. [CrossRef]
2. Rose, S.K.; Kriegler, E.; Bibas, R.; Calvin, K.; Popp, A.; van Vuuren, D.P.; Weyant, J. Bioenergy in energy transformation and climate management. *Clim. Chang.* **2014**, *123*, 477–493. [CrossRef]
3. Amanatidis, G. *European Policies on Climate and Energy towards 2020, 2030 and 2050*; European Parliament, Policy Department for Economic, Scientific and Quality of Life Policies: Brussels, Belgium, 2019.
4. European Commission. *Impact Assessment—A Policy Framework for Climate and Energy in the Period from 2020 up to 2030*; European Commission: Brussels, Belgium, 2014.
5. Berhongaray, G.; El Kasmioui, O.; Ceulemans, R. Comparative analysis of harvesting machines on an operational high-density short rotation woody crop (SRWC) culture: One-process versus two-process harvest operation. *Biomass Bioenergy* **2013**, *58*, 333–342. [CrossRef]
6. Landgraf, D.; Carl, C.; Neupert, M. Biomass yield of 37 different SRC poplar varieties grown on a typical site in North Eastern Germany. *Forests* **2020**, *11*, 1048. [CrossRef]
7. Pecenka, R.; Hoffmann, T. Harvest technology for short rotation coppices and costs of harvest, transport and storage. *Agron. Res.* **2015**, *13*, 361–371.
8. Stolarski, M.J.; Niksa, D.; Krzyzaniak, M.; Tworkowski, J.; Szczukowski, S. Willow productivity from small- and large-scale experimental plantations in Poland from 2000 to 2017. *Renew. Sustain. Energy Rev.* **2019**, *101*, 461–475. [CrossRef]
9. Manzone, M.; Bergante, S.; Facciotto, G. Energy and economic sustainability of woodchip production by black locust (*Robinia pseudoacacia* L.) plantations in Italy. *Fuel* **2015**, *140*, 555–560. [CrossRef]
10. Rosenqvist, H.; Berndes, G.; Börjesson, P. The prospects of cost reductions in willow production in Sweden. *Biomass Bioenergy* **2013**, *48*, 139–147. [CrossRef]
11. Noll, M.; Jirjis, R. Microbial communities in large-scale wood piles and their effects on wood quality and the environment. *Appl. Microbiol. Biotechnol.* **2012**, *95*, 551–563. [CrossRef] [PubMed]
12. Manzone, M.; Balsari, P.; Spinelli, R. Small-scale storage techniques for fuel chips from short rotation forestry. *Fuel* **2013**, *109*, 687–692. [CrossRef]
13. Barontini, M.; Scarfone, A.; Spinelli, R.; Gallucci, F.; Santangelo, E.; Acampora, A.; Jirjis, R.; Civitarese, V.; Pari, L. Storage dynamics and fuel quality of poplar chips. *Biomass Bioenergy* **2014**, *62*, 17–25. [CrossRef]
14. Dimitriou, I.; Rutz, D. *Sustainable Short Rotation Coppice—A Handbook*; SRC+: Munich, Germany, 2015.
15. Veste, M.; Böhm, C. *Agrarholz—Schnellwachsende Bäume in der Landwirtschaft*; Springer Spektrum: Berlin/Heidelberg, Germany, 2018.
16. Scholz, V. *Lagerung von Holzhackschnitzeln—Experten-Workshop*; Bornimer Agrartechnische Berichte-Heft 63, Leibniz-Institut für Agrartechnik Potsdam-Bornim e.V. (ATB): Potsdam, Germany, 2008.
17. Pecenka, R.; Lenz, H.; Idler, C. Influence of the chip format on the development of mass loss, moisture content and chemical composition of poplar chips during storage and drying in open-air piles. *Biomass Bioenergy* **2018**, *116*, 140–150. [CrossRef]
18. Whittaker, C.; Yates, N.E.; Powers, S.J.; Misselbrook, T.; Shield, I. Dry matter losses and quality changes during short rotation coppice willow storage in chip or rod form. *Biomass Bioenergy* **2018**, *112*, 29–36. [CrossRef]
19. Idler, C.; Pecenka, R.; Lenz, H. Influence of the particle size of poplar wood chips on the development of mesophilic and thermotolerant mould during storage and their potential impact on dry matter losses in piles in practice. *Biomass Bioenergy* **2019**, *127*, 105273. [CrossRef]
20. Gigler, J.K.; van Loon, W.K.P.; Vissers, M.M.; Bot, G.P.A. Forced convective drying of willow chips. *Biomass Bioenergy* **2000**, *19*, 259–270. [CrossRef]
21. Jirjis, R. Effects of particle size and pile height on storage and fuel quality of comminuted Salix viminalis. *Biomass Bioenergy* **2005**, *28*, 193–201. [CrossRef]
22. Whittaker, C.; Macalpine, W.; Yates, N.E.; Shield, I. Dry matter losses and methane emissions during wood chip storage: The impact on full life cycle greenhouse gas savings of short rotation coppice willow for heat. *Bioenergy Res.* **2016**, *9*, 820–835. [CrossRef] [PubMed]
23. Krzyzaniak, M.; Stolarski, M.J.; Niksa, D.; Tworkowski, J.; Szczukowski, S. Effect of storage methods on willow chips quality. *Biomass Bioenergy* **2016**, *92*, 61–69. [CrossRef]
24. Pari, L.; Scarfone, A.; Santangelo, E.; Gallucci, F.; Spinelli, R.; Jirjis, R.; Del Giudice, A.; Barontini, M. Long term storage of poplar chips in mediterranean environment. *Biomass Bioenergy* **2017**, *107*, 1–7. [CrossRef]
25. Therasme, O.; Volk, T.A.; Eisenbies, M.H.; San, H.; Usman, N. Hot water extracted and non-extracted willow biomass storage performance: Fuel quality changes and dry matter losses. *Front. Energy Res.* **2020**, *7*, 165. [CrossRef]
26. Lenz, H.; Idler, C.; Hartung, E.; Pecenka, R. Open-air storage of fine and coarse wood chips of poplar from short rotation coppice in covered piles. *Biomass Bioenergy* **2015**, *83*, 269–277. [CrossRef]
27. Hofmann, N.; Mendel, T.; Schulmeyer, F.; Kuptz, D.; Borchert, H.; Hartmann, H. Drying effects and dry matter losses during seasonal storage of spruce wood chips under practical conditions. *Biomass Bioenergy* **2018**, *111*, 196–205. [CrossRef]

28. Barontini, M.; Crognale, S.; Scarfone, A.; Gallo, P.; Gallucci, F.; Petruccioli, M.; Pesciaroli, L.; Pari, L. Airborne fungi in biofuel wood chip storage sites. *Int. Biodeterior. Biodegrad.* **2014**, *90*, 17–22. [CrossRef]
29. Pari, L.; Bergonzoli, S.; Cetera, P.; Mattei, P.; Alfano, V.; Rezaei, N.; Suardi, A.; Toscano, G.; Scarfone, A. Storage of fine woodchips from a medium rotation coppice eucalyptus plantation in central Italy. *Energies* **2020**, *13*, 2355. [CrossRef]
30. Pecenka, R.; Lenz, H.; Idler, C.; Daries, W.; Ehlert, D. Development of bio-physical properties during storage of poplar chips from 15 ha test fields. *Biomass Bioenergy* **2014**, *65*, 13–19. [CrossRef]
31. Lenz, H.; Pecenka, R.; Hartung, E.; Idler, C. Development and test of a simplified method to calculate dry matter loss during open-air storage of poplar wood chips by analysing ash contents. *Biomass Bioenergy* **2016**, *94*, 258–267. [CrossRef]
32. Anerud, E.; Bergstrom, D.; Routa, J.; Eliasson, L. Fuel quality and dry matter losses of stored wood chips-influence of cover material. *Biomass Bioenergy* **2021**, *150*, 106109. [CrossRef]
33. Manzone, M.; Balsari, P. Poplar woodchip storage in small and medium piles with different forms, densities and volumes. *Biomass Bioenergy* **2016**, *87*, 162–168. [CrossRef]
34. Lenz, H.; Pecenka, R.; Idler, C.; Dumfort, S.; Whittaker, C.; Ammon, C.; Hartung, E. Continuous weighing of a pile of poplar wood chips—A comparison of methods to determine the dry matter losses during storage. *Biomass Bioenergy* **2017**, *96*, 119–129. [CrossRef]
35. Greaves, H. The bacterial factor in wood decay. *Wood Sci. Technol.* **1971**, *5*, 6–16. [CrossRef]
36. Blanchette, R.A.; Nilsson, T.; Daniel, G.; Abad, A. Biological degradation of wood. In *Archaeological Wood*; American Chemical Society: Washington, DC, USA, 1990; Volume 225, pp. 141–174.
37. Singh, A.P.; Kim, Y.S.; Singh, T. Chapter 9—Bacterial degradation of wood. In *Secondary Xylem Biology*; Kim, Y.S., Funada, R., Singh, A.P., Eds.; Academic Press: Boston, MA, USA, 2016; pp. 169–190.
38. Pastorelli, R.; Paletto, A.; Agnelli, A.E.; Lagomarsino, A.; De Meo, I. Microbial communities associated with decomposing deadwood of downy birch in a natural forest in Khibiny Mountains (Kola Peninsula, Russian Federation). *For. Ecol. Manag.* **2020**, *455*, 117643. [CrossRef]
39. Moll, J.; Kellner, H.; Leonhardt, S.; Stengel, E.; Dahl, A.; Bassler, C.; Buscot, F.; Hofrichter, M.; Hoppe, B. Bacteria inhabiting deadwood of 13 tree species are heterogeneously distributed between sapwood and heartwood. *Environ. Microbiol.* **2018**, *20*, 3744–3756. [CrossRef] [PubMed]
40. Daniel, G. Chapter 8—Fungal degradation of wood cell walls. In *Secondary Xylem Biology*; Kim, Y.S., Funada, R., Singh, A.P., Eds.; Academic Press: Boston, MA, USA, 2016; pp. 131–167.
41. Schmidt, O. *Wood and Tree Fungi*; Springer: Berlin/Heidelberg, Germany, 2006.
42. Eriksson, K.-E.L.; Blanchette, R.; Ander, P. *Microbial and Enzymatic Degradation of Wood and Wood Components*; Springer: Berlin/Heidelberg, Germany, 1990.
43. Johnston, S.R.; Boddy, L.; Weightman, A.J. Bacteria in decomposing wood and their interactions with wood-decay fungi. *FEMS Microbiol. Ecol.* **2016**, *92*, fiw179. [CrossRef] [PubMed]
44. Tlaskal, V.; Baldrian, P. Deadwood-inhabiting bacteria show adaptations to changing carbon and nitrogen availability during decomposition. *Front. Microbiol.* **2021**, *12*, 685303. [CrossRef] [PubMed]
45. Gómez-Brandón, M.; Probst, M.; Siles, J.A.; Peintner, U.; Bardelli, T.; Egli, M.; Insam, H.; Ascher-Jenull, J. Fungal communities and their association with nitrogen-fixing bacteria affect early decomposition of Norway spruce deadwood. *Sci. Rep.* **2020**, *10*, 8025. [CrossRef] [PubMed]
46. Van der Lelie, D.; Taghavi, S.; McCorkle, S.M.; Li, L.L.; Malfatti, S.A.; Monteleone, D.; Donohoe, B.S.; Ding, S.Y.; Adney, W.S.; Himmel, M.E.; et al. The metagenome of an anaerobic microbial community decomposing poplar wood chips. *PLoS ONE* **2012**, *7*, e36740. [CrossRef]
47. Kretschmar, E.I.; Gelbrich, J.; Militz, H.; Lamersdorf, N. Studying bacterial wood decay under low oxygen conditions—Results of microcosm experiments. *Int. Biodeterior. Biodegrad.* **2008**, *61*, 69–84. [CrossRef]
48. Zöhrer, J.; Probst, M.; Dumfort, S.; Lenz, H.; Pecenka, R.; Insam, H.; Ascher-Jenull, J. Molecular monitoring of the poplar wood chip microbiome as a function of storage strategy. *Int. Biodeterior. Biodegrad.* **2021**, *156*, 105133. [CrossRef]
49. Probst, M.; Gómez-Brandón, M.; Bardelli, T.; Egli, M.; Insam, H.; Ascher-Jenull, J. Bacterial communities of decaying Norway spruce follow distinct slope exposure and time-dependent trajectories. *Environ. Microbiol.* **2018**, *20*, 3657–3670. [CrossRef] [PubMed]
50. Dumfort, S.; Pecenka, R.; Ascher-Jenull, J.; Peintner, U.; Insam, H.; Lenz, H. The potential of calcium hydroxide to reduce storage losses: A four months monitoring study of spruce wood chip piles at industrial scale. *Fuel* **2021**, *298*, 120738. [CrossRef]
51. Pecenka, R.; Lenz, H.; Jekayinfa, S.O.; Hoffmann, T. Influence of tree species, harvesting method and storage on energy demand and wood chip quality when chipping poplar, willow and black locust. *Agriculture* **2020**, *10*, 116. [CrossRef]
52. *EN ISO 17827-1*; Solid Biofuels—Determination of Particle Size Distribution for Uncompressed Fuels—Part 1: Oscillating Screen Method Using Sieves with Apertures of 3.15 mm and above. ISO: Geneva, Switzerland, 2016.
53. *EN ISO 17225-1*; Solid Biofuels—Fuel Specifications and Classes—Part 1: General Requirements. ISO: Geneva, Switzerland, 2014.
54. *EN ISO 18134-2*; Solid Biofuels—Determination of Moisture Content—Oven Dry Method—Part 2: Total Moisture—Simplified Method. ISO: Geneva, Switzerland, 2017.
55. Hocking, A.D.; Pitt, J.I. Dichloran-glycerol medium for enumeration of xerophilic fungi from low-moisture foods. *Appl. Environ. Microb.* **1980**, *39*, 488–492. [CrossRef] [PubMed]

56. *DIN 38404-5:2009;* Deutsche Einheitsverfahren zur Wasser-, Abwasser- und Schlammuntersuchung—Physikalische und Physikalisch-Chemische Kenngrößen (Gruppe c)—Teil 5: Bestimmung des ph-Werts (c 5). DIN: Berlin, Germany, 2009.
57. VDLUFA. *Methodenbuch Band iii "Die Chemische Untersuchung von Futtermitteln" (the Chemical Inspection of Feedstuffs)*, 3rd ed.; Verband Deutscher Landwirtschaftlicher Untersuchungs- und Forschungsanstalten: Speyer, Germany, 1997; Chapter 4.1.2.
58. *EN ISO 14775;* Solid Biofuels—Determination of Ash Content. ISO: Geneva, Switzerland, 2009.
59. Lohmann, U. *Holz Handbuch;* DRW-Verlag: Leinfelden-Echterdingen, Germany, 2001.
60. Hill, C.A.S. *Wood Modification;* Wiley: West Sussex, UK, 2006.
61. Panshin, A.J.; de Zeeuw, C. *Textbook of Wood Technology*, 4th ed.; McGraw-Hill: New York, NY, USA, 1980.
62. Biyada, S.; Merzouki, M.; Demcenko, T.; Vasiliauskiene, D.; Ivanec-Goranina, R.; Urbonavicius, J.; Marciulaitiene, E.; Vasarevicius, S.; Benlemlih, M. Microbial community dynamics in the mesophilic and thermophilic phases of textile waste composting identified through next-generation sequencing. *Sci. Rep.* **2021**, *11*, 23624. [CrossRef] [PubMed]
63. Nannipieri, P.; Ascher-Jenull, J.; Ceccherini, M.T.; Pietramellara, G.; Renella, G.; Schloter, M. Beyond microbial diversity for predicting soil functions: A mini review. *Pedosphere* **2020**, *30*, 5–17. [CrossRef]
64. Nannipieri, P.; Ascher, J.; Ceccherini, M.T.; Landi, L.; Pietramellara, G.; Renella, G. Microbial diversity and soil functions. *Eur. J. Soil Sci.* **2003**, *54*, 655–670. [CrossRef]
65. Hoppe, B.; Purahong, W.; Wubet, T.; Kahl, T.; Bauhus, J.; Arnstadt, T.; Hofrichter, M.; Buscot, F.; Kruger, D. Linking molecular deadwood-inhabiting fungal diversity and community dynamics to ecosystem functions and processes in Central European forests. *Fungal Divers.* **2016**, *77*, 367–379. [CrossRef]
66. Hoppe, B.; Kruger, D.; Kahl, T.; Arnstadt, T.; Buscat, F.; Bauhus, J.; Wubet, T. A pyrosequencing insight into sprawling bacterial diversity and community dynamics in decaying deadwood logs of Fagus sylvatica and Picea abies. *Sci. Rep.* **2015**, *5*, 9456. [CrossRef] [PubMed]
67. Kahl, T.; Arnstadt, T.; Baber, K.; Bassler, C.; Bauhus, J.; Borken, W.; Buscot, F.; Floren, A.; Heibl, C.; Hessenmoller, D.; et al. Wood decay rates of 13 temperate tree species in relation to wood properties, enzyme activities and organismic diversities. *For. Ecol. Manag.* **2017**, *391*, 86–95. [CrossRef]
68. Krah, F.S.; Seibold, S.; Brandl, R.; Baldrian, P.; Muller, J.; Bassler, C. Independent effects of host and environment on the diversity of wood-inhabiting fungi. *J. Ecol.* **2018**, *106*, 1428–1442. [CrossRef]
69. Humar, M.; Petric, M.; Pohleven, F. Changes of the pH value of impregnated wood during exposure to wood-rotting fungi. *Holz Roh Werkst.* **2001**, *59*, 288–293. [CrossRef]
70. Hartmann, H.; Reisinger, K.; Turowski, P.; Roßmann, P. *Handbuch Bioenergie Kleinanlagen;* FNR: Guelzow, Germany, 2013.

MDPI
St. Alban-Anlage 66
4052 Basel
Switzerland
Tel. +41 61 683 77 34
Fax +41 61 302 89 18
www.mdpi.com

Forests Editorial Office
E-mail: forests@mdpi.com
www.mdpi.com/journal/forests